THE POWER OF CALCULUS

WITHDRAWN

THE POWER OF
CALCULUS

FOURTH EDITION

Kenneth L. Whipkey
University of North Carolina at Wilmington

Mary Nell Whipkey
Youngstown State University

JOHN WILEY & SONS
New York
Chichester
Brisbane
Toronto
Singapore

Production supervised by Ellen C. Baron
Manuscript edited by Deborah Herbert
Cover and text designed by Ann Marie Renzi

Library of Congress Cataloging-in-Publication Data
Whipkey, Kenneth L
  The power of calculus.

  Includes index.
  1. Calculus. I. Whipkey, Mary Nell. II. Title.
QA303.W47 1986   515   85-13022
ISBN 0-471-06382-7

Printed in the United States of America

10 9 8 7 6 5 4 3

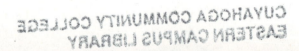

To our parents with gratitude

PREFACE

This edition, as did the earlier ones, evolved from our experiences in teaching calculus to students of business, economics, biology, and social studies. As a text it represents our belief that students using carefully worked examples, models pertinent to their discipline, and relevant problems will develop an appreciation of the power, elegance, and efficiency of the calculus.

We are grateful to the hundreds of thousands of students and to the instructors who used the first, second, and third editions. Their opinions have been reflected here. We have kept their most favored features, such as carefully worked out examples, while adding hundreds of real life problems in business, economics, biology, and the social studies and additional graphs and illustrations. In rewriting the text material it has often been possible to reduce the mathematical level and the amount of notation, thus affording the instructor who wishes to spend more time on the models and applied problems an opportunity to do so.

We hope this volume accomplishes our objectives. If errors are found, Mr. Whipkey still holds Mrs. Whipkey responsible and Mrs. Whipkey accordingly blames Mr. Whipkey.

Kenneth L. Whipkey
Mary Nell Whipkey

ACKNOWLEDGMENTS

We express our appreciation to the many professors and students who used the previous editions and related their experiences to us, and to Gary Ostedt and Fred Corey, former mathematics editors of Wiley, and to Carolyn Moore, present mathematics editor, for their help and support.

Although final decisions on content and style were made by us and are our responsibility, in this and previous editions we had the assistance or opinions of the following persons.

Reviewers

Yousef Alavi, Western Michigan University; Sue Boren, The University of Tennessee at Martin; Kenneth S. Brown, Cornell University; Robert S. Doran, Texas Christian University; Joseph A. Gallian, University of Minnesota; Richard A. Giesen, Ferrum College; Sylvia Goodrich, University of Texas at Austin; Thomas Hern, Bowling Green State University; John E. Hunter, Michigan State University; S.W. Kodis, Cabrillo College; Theodore Laetsch, University of Arizona; S.M. Lukawecki, Clemson University; Gus Mavrigian, Youngstown State University; Dan C. Messerschmidt, Westminster College; John Morton, Baylor University; Walter Neath, Chico State College; Joseph A. Nordstrom, University of Houston, Quantitative Management Science; Michael Sullivan, Chicago State University; Arthur Schwartz, University of Michigan at Ann Arbor; Wei Shen Hsia, University of Alabama; and Ray Wilson, Central Piedmont Community College.

Commentators

Jean J. Pederson, University of Santa Clara; Albert C. Reynolds, The University of Tulsa; Father George Brunish, George Conway, Barbara Faires, Warren Hickman, Thomas Nealeigh, Pat Johnson, and Jean Mitchell, Westminster College; John Buoni, John Cleary, Theodosius Demen, Raymond Hurd, Stephen Kozarich, Nicholas Mortellaro, and James Poggione, Youngstown State University.

Production

Kathy Anderson, Joe Leone, Alyce Marcotuli, Adele Marcotuli, and Theresa Presecan.

We welcome any suggestions and comments that users of this edition may have. We look forward to receiving them!

K.L.W.
M.N.W.

ORGANIZATION OF THE BOOK

Since some students are reviewing mathematics as they take this course, we have resisted the temptation to deal with certain topics simultaneously. Therefore, we have discussed one topic at a time. The spiral effect is used. Thus concepts build from chapter to chapter, culminating in Chapter 6 where previously discussed techniques are reviewed and reapplied.

Changes from the Third Edition

We have consulted with experts in the fields of business and economics to make certain our terminology, notation, and usage is consistent with current practices in their disciplines. Moreover, all of our problems relating to business and economics have been individually reviewed by a professor of economics and business. The exercises are abundant, properly graded, and contain more applied problems with applications to business, economics, and the social and biological sciences. More than 225 graphs and drawings are included for clarity and to illustrate the concepts related to the calculus. There is an abundance of worked-out examples with an extensive variety and many are applied problems. These examples have more steps and provide better preparation for working the exercises; they are designed to help the student learn the material. At the request of students, Appendix I contains a review of basic mathematical skills including factoring, operations on algebraic fractions, solutions to equations, and work in exponents, radicals, and logarithms with all answers given in the book. The definitions and statements of theorems are clearly stated and are accessible.

Worked-Out Examples

Extensive examples are given. The concepts are reexplained more frequently than in the usual calculus textbook. Since the steps in each example are complete, there are no gaps in the solutions. The purpose of each example is stated in a separate sentence. This crystallizes the main idea that the example conveys.

Problems

The exercises are thorough and occur throughout the chapters. Also, at the end of each chapter, there is a set of optional review problems. Most instructors will not have the class time to formally assign and go over each chapter review. However, students have requested that we include such a review so that they may, on their own, gain additional practice. (The appendix also contains the brief review of algebra, exponents, and logarithms that many students requested.) A knowledge of trigonometry is not required for the text itself; however, there is a self-contained trigonometry section in the appendix. Applied problems do not require an extensive background in any of the social science or business courses, since all terms from these disciplines are defined. In a one- or two-term course it is not feasible to teach complicated business or social science theory. We believe that applications requiring complicated theory are better handled in later courses.

K.L.W.
M.N.W.

NOTES TO THE INSTRUCTOR

Chapter 1 may be used at the discretion of the instructor. Although it may be assumed that students are prepared to begin their collegiate mathematics with a calculus-level course, we have found that this is not always true. Therefore, Chapter 1 is included for the students who need additional background and algebra review before undertaking calculus.

Although the text was designed for a one-term course of approximately 50 class sessions of 50 minutes each, it is adaptable to other time requirements. A short course of 37 sessions may be devised by limiting the study basically to Chapters 2 to 6. Also a two-term course is appropriate by including all optional material and the appendices, and by pacing the course at an optimal rate. By omitting most optional sections and proofs of lesser importance, and by leaving the worked examples for the students to study, the following is a realistic syllabus for approximately a 50-session course of about 50 minutes each.

Chapter 1	7 lectures	Preparation for the Calculus to include Slope, Straight Line, Functions, Inequalities
2	7 lectures	Limits, Differentiation of Functions of One Independent Variable
3	6 lectures	Rules of Differentiation
4	10 lectures	Extremization of Functions and Applications of Differential Calculus
5	12 lectures	The Indefinite Integral, the Definite Integral, the Fundamental Theorem of Integral Calculus, Applications of the Definite Integral
6	8 lectures	The Logarithmic and Exponential Functions, Differential Equations, Growth and Decay Problems

K.L.W.
M.N.W.

CONTENTS

THE POWER OF CALCULUS

ONE

Review and Preparation for the Study of Calculus

1.1 THE ARREST OF A SEVENTEENTH CENTURY NONCONFORMIST AND THE DEVELOPMENT OF ANALYTIC GEOMETRY

You, René Descartes, are hereby ordered to appear before the magistrates to answer charges brought against you, to wit, that you are an atheist, that you lead an unsettled, irresponsible, or disreputable life, moving from place to place without a fixed home, and that you are completely given up to dissipation and licentiousness.

VERSUS

But, there are other men who attain greatness because they embody the potentiality of their own day and magically reflect the future. They express the thoughts which will be everybody's two or three centuries after them. Such a one was Descartes.

Thomas Huxley

Such a one was Descartes—ex-soldier, philosopher, tutor to Queen Christina of Sweden, mathematician, and the person credited with the invention of analytic geometry. However, analytic geometry, which became the foundation for the calculus, was not entirely the brainchild of Descartes. It built on the works of previous seventeenth-century mathematicians who were concerned with two basic problems:

Basic Problem I. Given any curve, find the slope of the tangent line drawn to the curve at a point on the curve.

Example. Given curve C. Find the slope of the tangent line, t, to C at point P.

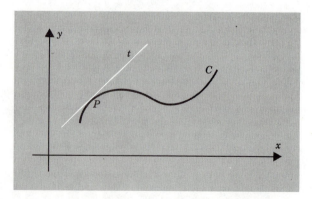

Basic Problem II. Given any curve C, find the area under that curve.

Example. Given curve C. Find area A.

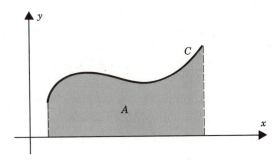

Both of these problems had to wait for their solution until the calculus was developed. Therefore, the analytic geometry of Descartes was a necessary prelude to the development of the calculus. Descartes' great contribution was that he gave us a way of dealing with curves in general. Thus, instead of handling circles separately, lines separately, and so forth, as we are forced to do in high school geometry, we can handle whole groups of figures simultaneously.

Incidentally, the previously mentioned charges against Descartes were dropped by the government of Holland. Nevertheless, he was overjoyed to leave Holland, at the invitation of the Queen of Sweden, to become a royal tutor. Little did he know that the Queen required her lessons to be given at 5 a.m. Descartes, always a late riser, had his life-long routine shattered and died of pneumonia four months later. (There may be a moral here concerning math classes given before 9 a.m.)

1.2   REBELLION AT THE UNIVERSITY AND THE FORMALIZATION OF THE CALCULUS

An honest courage in these matters will secure all, having law on our sides.

Isaac Newton

These were Newton's words about the situation at Cambridge University and his role as a leader of opposition to the monarchy. The government, headed by James II, was attempting to control thinking at the university and to dictate the choice of university personnel. Fortunately, this time, it was James II who left the country. Unfortunately, for mathematics, Newton's role in the university rebellion caused him to devote the rest of his life to governmental service. For it was his prominence in the rebellion that led to Newton's election as representative from Cambridge to the Convention Parliament. During the next 30 years, Newton served first as Warden and then as Master of the Mint. Most of his free time during this period was spent in studying theology and philosophy. This was in contrast to the prerebellion days when he devoted most of his time to the study of scientific and mathematical theory.

During the prerebellion days Newton had formalized the calculus. Concurrently, Leibnitz in Germany had also been able to solve the two Basic Problems: (I) finding the slope of the tangent line to a curve, and (II) finding the area under a curve. The "tools" that Newton and Leibnitz independently invented to solve these two basic problems are now called the *derivative* and the *integral*.

Basic Problem	*Tool Invented to Solve Basic Problem*
I. Find the slope of the tangent line to a curve.	Derivative
II. Find the area under a curve.	Integral

related

One of Newton's and Leibnitz's great claims to fame and honor is that they recognized the two basic problems are related. Therefore, the tools invented to solve these problems, the derivative and the integral, are also related.

Moreover, one of the great bonanzas of history is that the derivative and integral, which were invented to solve two particular problems, have applications to a great number of different problems in diverse academic fields.

1.3   THE POWER OF THE CALCULUS AND THE RATIONALE FOR ITS STUDY

The power of the calculus is derived from two sources. The first is that the derivative and the integral can be used to solve a multitude of problems in many different academic disciplines. The second source of power is found in the relevancy of the calculus to the problems facing mankind. Among the present-day applications of the calculus are the building of abstract models for the study of the ecology of populations, cybernetics and its social impact on man, management practices, economics, and medicine. Examples of basic problems that you will encounter and be able to solve are:

A manufacturing company has found that by ordering goods at precise times they can make a larger profit on their product. This is often called inventory control. This particular company has determined that the following function describes profit as a function of inventory ordering time: $P(t) = 12t - .25t^2$, where t is the time in days after the inventory drops below 2000 parts. Determine the time when parts should be reordered to maximize the profit.

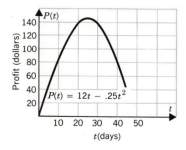

From the graph, it appears that parts should be reordered approximately 25 days after the inventory drops below 2000 parts. By using calculus, you can easily determine that by reordering parts after 24 days the profit will be maximized.

A large city is hit by a flu epidemic and people are becoming ill at a rate of $270 - 9t^2$ people per day. Approximately how many people will have the flu in the first two weeks of the epidemic?

If a country's population is increasing at a continuous rate of 4 percent each year, in how many years will the country's population double?

Indeed, it is the power of the calculus that demands we become familiar with its fundamental concepts. However, we will not study the subject as it originated but will take advantage of the improvements made in the calculus from Newton's time to the present.

Therefore, as students of the calculus, we are dealing with ideas that have evolved over hundreds of years and that were formalized by some of the greatest geniuses of all time. As we deal with the products of their thinking, we should not be dismayed that certain ideas and concepts may at first and even at subsequent readings seem hazy and confusing to us. The key to understanding these concepts is to repeatedly return to each idea until it becomes meaningful. That the task is not impossible is embodied by the basic premise for the existence of this book. For, in writing this book, we assume that it is realistic to present the great ideas of the calculus in a meaningful manner that will minimize symbol shock and function fatigue and also will meet the recommendations of the Social Science Research Council (SSRC) and the Committee on the Undergraduate Program in Mathematics (CUPM).

As students in management, business, education, and the social sciences, it is important to realize, as your professional groups have long recognized, the need for greater training in mathematics. The Social Science Research Council has for years made this recommendation. Also, the fields of biology and medicine are experiencing a need for more advanced mathematics. Moreover, most graduate schools in business are requiring sophisticated courses in mathematics, and often a graduate student may shorten his or her program by exhibiting adequate preparation in the calculus. Even for those students not contemplating graduate school, the calculus is the foundation for many upper-division courses, particularly probability and statistics, which are now required in many curricula.

Before one may reconstruct the greatness, orderliness, and power of the calculus, an awareness of certain fundamental techniques and definitions must be attained. Those of you who have made adequate and recent preparation for this excursion in calculus may find the rest of this chapter a review of previous work. However, if your mathematics is weak or rusty through nonuse, it will be an experience that will pay future dividends and, as such, is relevant to the goals of this book.

Also, Appendix I, Review of Basic Mathematical Skills, includes an extensive review of the basic algebraic skills needed throughout this book. Some of the topics discussed are factoring, algebraic simplification, the solution of algebraic equations, the use of exponents and radicals, and work with logarithms. Many students using this book have requested that these topics be included for easy reference. You may find it helpful to review these topics first; and as you study this text you will find it desirable, from time to time, to make use of this appendix.

1.4 THE SET OF REAL NUMBERS— A REVIEW OF BASIC IDEAS

The number system used in elementary calculus is called the set of *real numbers, R*. We will now briefly review the properties and construction of *R*. To do so, we begin with the set of natural numbers.

The set of natural numbers is the set composed of the numbers with which we learned to count. The set of *natural numbers* is represented by

$$N = \{1,2,3,4,5,6,7,8, \ldots\}$$

Here, the three dots, . . . , or ellipsis, mean that the numbers are continued in the usual manner.

When we have only the set of natural numbers with which to work, subtraction is not always possible. For example, $7 - 7$ and $2 - 5$ have no answers among the natural numbers. To do subtraction problems, we invent the set of integers. The set of *integers* will be designated by

$$I = \{. \ . \ . \ ,-3,-2,-1,0,+1,+2,+3, \ . \ . \ .\}$$

We will assume that the integer $+1$ behaves just like the natural number 1 and, therefore, $+1$ and 1 may be used interchangeably. The same behavior holds true for $+2$ and 2, $+3$ and 3, and so forth.

With the set of integers, division is not always possible. For example, $\frac{6}{2}$ does find a solution among the integers, but $\frac{6}{4}$ or $\frac{3}{2}$ does not. Therefore, we invent the set of rational numbers, *Q*. A *rational number* can be expressed as $\frac{a}{b}$ where a and b are both integers and $b \neq 0$.

Notice that the divisor b cannot equal zero. It can be proved that division by zero is impossible (see Exercise 1.4–1.5, Problem 2). There is no way that we can repair or modify our number system to divide by zero. Remember: Never divide by zero.

We emphasize, the rational numbers, *Q*, are numbers that can be written, in at least one way, as the quotient of two integers (excluding division by zero).

Notice that every integer is a rational number. Examples of rational numbers are:

Rational Number	Representation as Quotient of Two Integers
$+8$	$\dfrac{+16}{+2}$ or $\dfrac{-8}{-1}$ or . . .
$-.5$	$\dfrac{-1}{2}$ or $\dfrac{3}{-6}$ or . . .
$.333\overline{3}$	$\dfrac{+1}{+3}$ or . . .

(The bar above the 3 means that the 3 repeats forever.)

0	$\dfrac{0}{7}$ or . . .
$2.141414\overline{14}$	$\dfrac{+212}{+99}$ or . . .

(The 14 repeats forever.)

There are real numbers that are not rational numbers. The real numbers that cannot be expressed as the quotient of two integers will be called *irrational numbers*. Examples of irrational numbers are $\sqrt{3}$, $1 + \sqrt{2}$, π, ... Note that some nth roots of rational numbers are rational and others are irrational.

nth Root	Answer and Nature of Root
$\sqrt{4}$	2, a rational number
$\sqrt{3}$	1.732 . . . , an irrational number
$\sqrt[3]{8}$	2, a rational number
$\sqrt[3]{-8}$	-2, a rational number
$\sqrt[3]{7}$	1.912 . . . , an irrational number

The numbers used in this course are the set of real numbers. Symbolically, we have

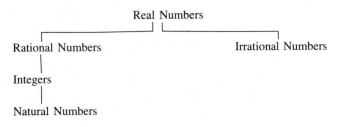

Number Line

To give a pictorial representation of a real number, it is necessary to establish a correspondence between the set of real numbers and the set of points on a number line. Such a correspondence, called a one-to-one correspondence, means that for each element of one set there corresponds one and only one element of another set. This idea of a one-to-one correspondence is of particular importance when we construct a *number line*. We begin this construction by taking a straight line and choose on it an origin, 0.

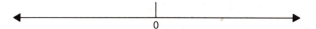

Select a convenient length for one unit and plot on the number line the points corresponding to the set of integers.

Proceed by imagining that we mark off points corresponding to the rational numbers. After doing this, there are some points that have not been used; these points correspond to the irrational numbers. However, once these unused points have been assigned to the irrational numbers, this will exhaust the available points of the number line. Therefore, there exists a fundamental one-to-one correspondence between the points of the number line and the set of real numbers, R.

The Fundamental Correspondence between the Set of Real Numbers and the Set of Points on the Number Line Is:

1. To every real number (rational or irrational) there corresponds one and only one point of the number line.

and

2. Every point on the number line represents one and only one real number.

For example, this means that there exists only one point on the number line that represents the real number 2.314159632 and, if we pick a specific point of a number line, there is one and only one real number associated with that point.

In summary, the numbers that we will use in this course are the set of real numbers. A real number may be either rational or irrational. If the real number is rational it may be expressed, in at least one way, as the quotient of two integers (denominator $\neq 0$). However, if the real number is irrational it cannot be represented by the quotient of two integers.

When each real number is represented by a point on a number line, this exhausts the points of the number line and establishes a one-to-one correspondence between the set of real numbers and the set of points of the number line.

Thus, to every real number, rational or irrational, there corresponds one and only one point of the number line and to every point of the number line there is associated one and only one real number.

Order Relationships

Given two real numbers, we would like to develop a method for determining the relationship between them. Therefore, to compare the magnitudes of two quantities, we establish an "order relationship" between any two real numbers and represent these order relationships on number lines.

Given any two real numbers, a and b, one and only one of the following relationships is true:

a is greater than b, written $a > b$, or

a is less than b, written $a < b$, or

a is equal to b, written $a = b$.

We will say a is greater than b iff (if and only if) a lies to the right of b on the number line. Similarly, a is less than b iff b lies to the right of a on the number line.

This topic will be discussed further in the section dealing with inequalities.

Example 1

Purpose To use the order relationship to determine intervals on the number line:

Problem A Represent the points on the number line where $-2 \leqslant x \leqslant 2$, read x is greater than or equal to -2 and less than or equal to 2.

Solution

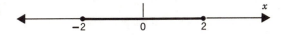

Problem B Represent the points on the number line where $-1 < x < 3$, read x is greater than -1 and less than 3.

Solution

Note The open "circles" at -1 and 3 mean -1 and 3 do not belong to the interval.

1.5 RECTANGULAR COORDINATE SYSTEM AND GRAPHING

The previously developed number lines will serve as building blocks for the development of a rectangular coordinate system. This system is needed to represent pictorially the relationship between two quantities, for instance, the relationship between wages paid and hours worked, the rate at which cancerous cells are produced and time in which they are produced, and the like.

Two number lines intersecting at a right angle at their origins divide two-dimensional space into four quadrants and form a two-dimensional Rectangular Cartesian Coordinate System. The horizontal number line is often called the x axis or, more correctly, the axis of abscissas. The vertical number line is often called the y axis or, more correctly, the axis of ordinates (see Figure 1.1).

We locate a point by giving an ordered pair of numbers, (x,y), where x represents the abscissa or distance of the point from the vertical axis. Also, y represents the ordinate or distance of the point from the horizontal axis.

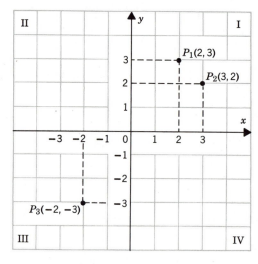

Figure 1.1

Thus, in quadrant I, P_1 has coordinates $(2,3)$ and P_2 has coordinates $(3,2)$ (*Note:* the order in which the coordinates are listed is important) and, in quadrant III, P_3 has coordinates $(-2,-3)$.

One principal use of the Rectangular Cartesian Coordinate System is the graphing of equations. The *graph of an equation* is the set of all points (x,y) whose coordinates are numbers satisfying the equation.

Example 1

Purpose To illustrate the graphing of an equation:

Problem Under a specified set of physical conditions a biologist finds that the rate at which cancerous cells are produced is given by: $R = t^2 + 2t + 1$ where t is measured in hours and R is measured in units of 1000 cells per hour. Find R at $t = 0$, 1, and 3; and graph the equation. Note that t is greater than or equal to zero.

Solution

$$\text{If } t = 0, R = 1$$
$$\text{If } t = 1, R = 1^2 + 2(1) + 1 = 4$$
$$\text{If } t = 3, R = 3^2 + 2(3) + 1 = 16$$

Therefore $(0,1)$, $(1,4)$, $(3,16)$ satisfy the equation. The graph of the equation includes these ordered pairs.

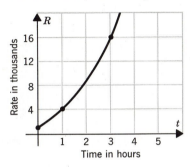

Rate at which cancer cells are produced.

Example 2

Purpose To use the graph of an equation to obtain desired information:

Problem In business-decision models a quadratic profit equation is often utilized. Let $y = 5x - .0125x^2$ be such an equation formulated by the management of the C & S Manufacturing Corporation. Here, x represents the number of skimobiles produced and y represents the profit in hundreds of dollars. Show graphically that y appears to reach a maximum value when $x = 200$.

Solution Find some ordered pairs that satisfy this equation.

$$y = 5x - .0125x^2$$
$$\text{Let } x = 0: \quad y = 5(0) - .0125(0)^2$$
$$y = 0$$
$$\text{Let } y = 0: \quad 0 = 5x - .0125x^2$$
$$0 = x(5 - .0125x)$$
$$x = 0 \qquad 5 - .0125x = 0$$
$$- .0125x = -5$$
$$x = \frac{-5}{-.0125}$$
$$x = 400$$
$$\text{Let } x = 100: \quad y = 5(100) - .0125(100)^2$$
$$y = 500 - 125$$
$$y = 375$$

Let $x = 200$: $\quad y = 5(200) - .0125(200)^2$

$$y = 1000 - 500$$

$$y = 500$$

Therefore, $(0,0)$, $(100,375)$, $(200,500)$, and $(400,0)$ are among the set of points that satisfy the equation.

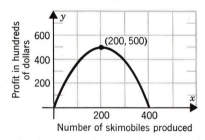

The profit y, in hundreds of dollars, appears to reach a maximum value of 500 when $x = 200$.

Notice from the above example that intercepts are often used in graphing. An *x intercept* is the x coordinate of a point at which the graph crosses or touches the x axis. The *x intercepts* are found by letting $y = 0$ and in the above example occur at $x = 0$ and $x = 400$. In a similar manner, the *y intercept* is the y coordinate of a point at which the graph crosses or touches the y axis. The *y intercept* is found by letting $x = 0$ and in the above examples occurs at $y = 0$. In Exercise 1.4–1.5, see Problem 12 for a review discussion of the equation of a parabola and its graph.

The Distance between Two Points

Having completed the development of the rectangular coordinate system and the plotting of points, we now examine the distance between two points. Of particular interest is the change in x (the difference between two x coordinate values) and the change in y. For this and other applications where distance is important we explore the distance between two points. For our purposes we deal only with an undirected or nonnegative distance.

First, we consider the distance between two points that lie on a horizontal line (a line parallel to the x axis). $P_1(x_1,y_1)$ and $P_2(x_2,y_1)$ are two distinct points with P_1 and P_2 lying on a horizontal line, and P_2 to the right of P_1 (that is, $x_2 > x_1$).

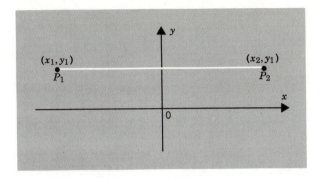

We define the horizontal distance between P_1 and P_2 to be: $x_2 - x_1$, or the x coordinate of the right-hand point minus the x coordinate of the left-hand point: $x_{\text{right}} - x_{\text{left}}$. In a similar manner, when $P_1(x_1, y_1)$ and $P_2(x_1, y_2)$ are on a vertical line (a line parallel to the y axis) with P_2 lying above P_1, we define the distance between P_1 and P_2 to be: $y_2 - y_1$. Or, the distance between two points on a vertical line is equal to the y coordinate of the top point minus the y coordinate of the bottom point: $y_{\text{top}} - y_{\text{bottom}}$.

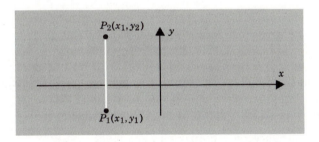

Now, if the two points, $P_1(x_1, y_1)$ and $P_2(x_2, y_2)$, are on a slanting line (a line not parallel to either the x axis or the y axis) we will derive a formula to obtain the undirected, nonnegative distance between P_1 and P_2.

Distance Formula

If $P_1(x_1, y_1)$ and $P_2(x_2, y_2)$ are points in a plane, then the undirected distance between them is given by

$$d = \sqrt{(x_2 - x_1)^2 + (y_2 - y_1)^2}.$$

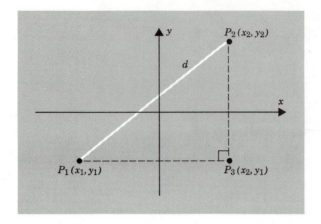

Draw P_1P_3 parallel to the axis of abscissas and P_3P_2 parallel to the axis of ordinates. Notice the coordinates of P_3. The desired distance, d, is the hypotenuse of the right triangle $P_1P_2P_3$. The Pythagorean theorem[1] can be used to find d, since we know the distances P_1P_3 and P_3P_2.

$$P_1P_3 = x_{\text{right}} - x_{\text{left}} = x_2 - x_1$$
$$P_3P_2 = y_{\text{top}} - y_{\text{bottom}} = y_2 - y_1$$
$$d^2 = (x_2 - x_1)^2 + (y_2 - y_1)^2$$
$$d = \sqrt{(x_2 - x_1)^2 + (y_2 - y_1)^2}$$

Example 3

Purpose   To find the distance between two points on horizontal, vertical, and slanting lines:

Problem A   Find the distance between $P_1(2,3)$ and $P_2(3,3)$.

Solution   This is distance on a horizontal line.

$$d = x_{\text{right}} - x_{\text{left}} = 3 - 2 = 1$$

Problem B   Find the distance between $P_1(2,3)$ and $P_2(2,-3)$.

Solution   This is distance on a vertical line.

$$d = y_{\text{top}} - y_{\text{bottom}} = 3 - (-3) = 6$$

Problem C   Find the distance between $P_1(2,3)$ and $P_2(-1,-3)$.

Solution   This is distance on a slanting line.

$$d = \sqrt{(x_2 - x_1)^2 + (y_2 - y_1)^2} = \sqrt{(-1 - 2)^2 + (-3 - 3)^2} = \sqrt{45}$$
$$= \sqrt{9 \cdot 5} = 3\sqrt{5}$$

---

[1] If a right triangle has sides a, b, and c where c is the hypotenuse (longest side), then $c^2 = a^2 + b^2$.

or

$$d = \sqrt{(x_1 - x_2)^2 + (y_1 - y_2)^2} = \sqrt{[2 - (-1)]^2 + [3 - (-3)]^2} = \sqrt{45}$$
$$= 3\sqrt{5}$$

Exercise 1.4–1.5

1. Identify the following numbers as: natural, integer, rational, or irrational.

   (a) 2   (e) $\dfrac{\sqrt{3}}{2}$ (i) $\dfrac{7}{0}$

   (b) $\sqrt{4}$ (f) $2\sqrt{3}$ (j) $\dfrac{0}{7}$

   (c) $\sqrt{3}$ (g) -3975 (k) $\sqrt{4} + \sqrt{9}$

   (d) 0   (h) $7 - 2\frac{3}{4}$ (l) $2.16\overline{66}$

2. Prove: division by zero is impossible.

   (Use an indirect proof and the definition for division: $\dfrac{a}{b} = c$ if and only if

   $a = bc$ and c is a unique number.)

3. Draw a two-dimensional rectangular coordinate system. Designate the following points:
   $(0,0)$; $(3,4)$; $(-2,0)$; $(\frac{3}{2},-3)$; $(3,0)$; $(-2,-2)$.

4. Draw a two-dimensional rectangular coordinate system. Designate the following points:
   $(-\frac{3}{2},-3)$; $(4,0)$; $(0,-4)$; $(\sqrt{2},3)$; $(-1,5)$; $(2,-\frac{1}{3})$.

5. Draw a two-dimensional rectangular coordinate system. Given a point (a,b) in quadrant II, designate the following points: (b,a); $(-a,b)$; $(b,0)$; $(-a,-b)$; $(-b,a)$.

6. Draw a two-dimensional rectangular coordinate system.
   (a) Plot the following points:
   $(-2,-5)$; $(-1,-3)$; $(0,-1)$; $(1,1)$; $(2,3)$; $(3,5)$.
   (b) Note, this is the *graph* of these ordered pairs.

7. Economists are often interested in how a certain resource (money, land, food) is distributed among the population. This type of data is often presented by a graph where the x coordinate is a fraction of the population and the y coordinate is the fraction of the resource accounted for by the corresponding fraction of the population. Notice that when the fraction of the population being considered is 0 ($x = 0$), none of the resource is accounted for ($y = 0$). Also, when the whole population is considered ($x = 1$), all of the resource is accounted for ($y = 1$). Thus, this type of graph will always start at $(0,0)$ and end at $(1,1)$. A graph like this is called a Lorenz curve.

The following data relate a fraction of the population to the fraction of income accounted for. Plot these data points and connect them with a curve.

Fraction of the Population (x)	Fraction of Income (y)
0.0	0.00
0.2	0.04
0.4	0.16
0.5	0.25
0.6	0.36
0.8	0.64
1.0	1.00

8. A promotion agent finds that the more she advertises on television, the more goods she sells. The relationship may be expressed by $y = \frac{3}{2}x + 150$, where y is the number of goods sold per week, and x indicates the number of television commercials sponsored during the week. Sketch the graph of this equation.

9. By keeping coal as dry as possible, an electric company can produce more energy from a pound of coal. The equation, $E = -10x + 13{,}000$ applies. E gives the energy in British thermal units (Btu) evolved from a pound of coal, and x represents the pounds of water in 1000 pounds of coal. Draw a graph of this equation.

10. A laboratory rat learns to press a bar in a Skinner box for a reward of food. The equation $y = \frac{1}{2}x$ relates the number of times the bar is pressed in one minute (y) to the amount of food given for the response (x). Graph this equation. When four units of food are given, how many times will the rat press the bar in one minute?

11. The correlation between the grade point average earned in high school and the predicted score on a college entrance test is described by the graph $y = 125x$ (y is projected test scores; x is known grade point average). Graph this correlation equation. A student received an average of 3.2 throughout high school. What entrance score will he be likely to receive?

12. A *parabola* is defined as the set of all points whose distance from a fixed point F (the focus) and a fixed line d (the directrix) are equal. Show that an equation for a parabola with focus $(0,p)$ and directrix $y = -p$ is $x^2 = 4py$.

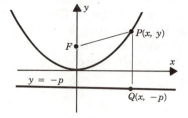

Using the definition of a parabola and the distance formula

$$\text{Distance } FP = \text{Distance } PQ$$
$$\sqrt{(x - 0)^2 + (y - p)^2} = y - (-p)$$

square both sides of the equation and simplify to $x^2 = 4py$.

A general quadratic equation of the form $y = ax^2 + bx + c$ is a parabola. For example, $y = x^2 + 2x + 3$ can be equivalently expressed as $y = x^2 + 2x + 1 + 2$ or $y = (x + 1)^2 + 2$.

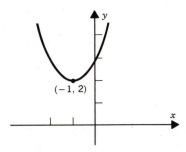

The important point $(-1,2)$ is called the *vertex* of the parabola. It is the lowest point on this graph. The vertex may also be the highest point. Consider $y = -x^2 + 2$.
Graphically:

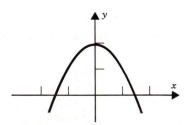

From $x^2 = 4py$, a general form of a parabola is $(x - h)^2 = 4p(y - k)$ where (h,k) is the vertex.

13. Graph the following parabolas and identify the vertex.
    (a) $y = x^2 - 2$ (d) $y = -(x + 3)^2$
    (b) $y = x^2 + 2x + 4$ (e) $y = 2x^2 - 4x$
    (c) $y = -x^2 - 2x - 4$ (f) $y = x^2 + 6x + 5$

14. Graph the following equations.
    (a) $y = 8x^2$ (d) $y = x^2 - 5x - 6$
    (b) $y = -x^2$ (e) $y = 3x^2 + x - 10$
    (c) $y = x^2 - 9$ (f) $y = x^2 + x + 1$

15. The profit equation of a certain company was found to be $y = x - .01x^2$ (the gross profit, y, in thousands of dollars, for producing x units of a commodity). Graph this equation and estimate the maximum profit.

16. The Jay Co. has found that the cost per unit, y (in dollars), for producing one TV tuner depends on the size x of the production run according to the equation $y = 35 + 4x - \frac{1}{2}x^2$ (x, in hundreds and $0 \leqslant x \leqslant 10$). Graph this cost equation.

17. A quadratic utility equation is $y = 2x - .04x^2$ (y, the utility; x, dollars). Estimate, by graphing the equation, the maximum value y reaches.

18. The growth of a controlled culture can be estimated by the equation $y = .2x^2 + 100$, where y is the quantity after x hours. Graph this equation and determine the quantity present initially ($x = 0$) and after one hour and two hours.

19. A cost-benefit equation relates cost to the benefit obtained. For example, a chemical company relates cost to the percentage of particulate matter that can be eliminated from its emissions. The company finds the cost-benefit equation to be $y = \dfrac{5x}{100 - x}$ where x is the percent of particulate matter removed ($0 \leqslant x < 100$) and y the cost in thousands of dollars. Find the cost of removing 0%, 25%, 50%, 75% and 90% of the particulate matter. Graph this equation.

20. A mathematics teacher has determined that grades on her tests depend on the number of homework problems she assigned that week. The average test score S is given by $S = -\frac{1}{2}x^2 + 15x$ where x is the number of problems assigned.
    (a) Graph this equation.
    (b) How many problems should she assign to attain the highest average test score?

21. A biologist determines that the temperature of the water has an effect on the number of fish caught in a day. He has found the relationship to be $N = -\frac{1}{2}t^2 + 60t - 1300(t \geqslant 25)$, where N represents the number of fish caught in a day, and t is the temperature (°F).
    (a) Graph this equation.
    (b) What is the ideal water temperature?

In Problems 22 to 27, find the distance between the pairs of points.

22. $(2,-4)$; $(2,3)$

23. $(-2,-3)$; $(3,-3)$

24. $(-5,2)$; $(4,-2)$

25. $(0,-3)$; $(-4,1)$

26. $(-5,2)$; $(-1,11)$

27. (c,d); $(-c,d)$, $c > 0$

In Problems 28 to 31, find the length of the sides of the triangle denoted by ABC. Are the triangles isosceles or right triangles or neither?

28. $A(2,1)$; $B(-4,1)$; $C(-4,-3)$

29. $A(0,0)$; $B(-5,-4)$; $C(-6,-2)$

30. $A(-1,4)$; $B(6,2)$; $C(4,-5)$

31. $A(-2,-1)$; $B(0,7)$; $C(3,2)$

32. A *circle* is defined as the set of all points whose distance from a fixed point is a constant. The fixed point is called the *center* of the circle and the constant distance is called the *radius* of the circle. Show that the equation of a circle with center at (h,k) and radius of r is $(x - h)^2 + (y - k)^2 = r^2$.
(*Hint:* Use the distance formula.)

As an example using this definition, the equation of a circle with center at $(-3,4)$ and radius equal to 6 is $(x + 3)^2 + (y - 4)^2 = 36$.

In Problems 33 to 37, find the equation of the circle and graph the equation (see Problem 32).

33. Center at the origin and radius 2.

34. Center at $(1,-2)$ and radius 5.

35. Center at $(1,-2)$ and passing through the point $(4,2)$.

36. Center at $(3,2)$ and passing through the origin.

37. Center at $(a,0)$ and passing through the point $(0,0)$.

38. Prove the following theorem:

If $P_1(x_1,y_1)$ and $P_2(x_2,y_2)$ are points in two-dimensional space, then the midpoint of the line segment P_1P_2, denoted by P, has coordinates:

$$x = \frac{x_1 + x_2}{2}, \; y = \frac{y_1 + y_2}{2}.$$

Hint. Construct the points P_3, R, S and note their coordinates. Recall the theorem—If a line bisects two sides of a triangle, then it is parallel to the third side and equal in length to one half of the length of the third side. Therefore, $PS = RP_3 = \frac{1}{2}P_1P_3$ and $RP = P_3S = \frac{1}{2}P_3P_2$.

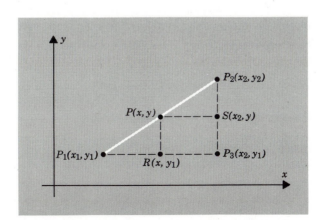

39. Name the coordinates of the midpoints of the line segments in Problems 22 to 27.

40. Find the y coordinate of the point $(2,y)$ that lies on the perpendicular bisector of the line segment formed by $(-4,2)$ and $(3,7)$.

41. Find the x coordinate of the point $(x,-3)$ that lies on the perpendicular bisector of the line segment formed by $(5,0)$ and $(-4,3)$.

1.6   SLOPE; STRAIGHT LINE

The so-called linear relationship plays an important role in all disciplines and is one that you have met before. For review purposes, the slope of a straight line, equations of straight lines, and parallel and perpendicular lines are now explored.

Given nonvertical straight lines L_1, L_2, L_3, we wish to compare the steepness or slant of these lines.

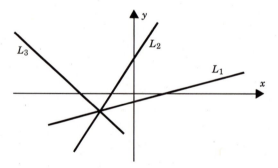

Let us concentrate on a single nonvertical line L and assume that we know the coordinates of two points of L: $P_1(x_1,y_1)$ and $P_2(x_2,y_2)$.

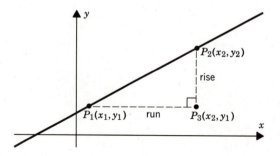

We will call the quantity $x_2 - x_1$ a run for line L and the quantity $y_2 - y_1$ the corresponding rise for L. Now, we will define the slope of line L as the ratio of the rise to the run. That is, the slope of line L is given by $m = \dfrac{\text{rise}}{\text{run}}$. Again, the rise is $y_2 - y_1$ and the run is $x_2 - x_1$. Therefore:

DEFINITION

Slope.   The slope of a line $= m = \dfrac{\text{rise}}{\text{run}} = \dfrac{y_2 - y_1}{x_2 - x_1}$, where (x_1,y_1) and (x_2,y_2) are any two points on the line and $x_1 \neq x_2$.

Notice that with the exception of vertical lines[1] every straight line will have a slope assigned to it.

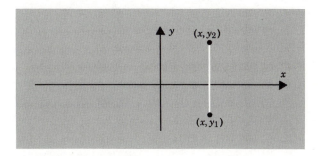

With a vertical line $x_1 = x_2 = x$. When we substitute in the formula,

$$m = \frac{y_2 - y_1}{x - x},$$

we are attempting to divide by zero. Recall that we can *never divide by zero.* Therefore, a vertical line does not have a slope.

Also, we should note that the slope of a given line does not depend on the choice of points P_1 and P_2. The same answer for m will be obtained as long as P_1 and P_2 are any two different points on the given line. To summarize, slope is a property of a nonvertical straight line that measures the slant of this line. The slope, m, is found by $m = \dfrac{\text{rise}}{\text{run}} = \dfrac{y_2 - y_1}{x_2 - x_1}$ and m may be positive, negative or zero. Different possible values of m are illustrated in the following example.

---

[1] A line is said to be vertical if it is parallel to the y axis.

Example 1

Purpose   To use the formula to find the slopes of lines:

Problem A   Find the slope of line L_A determined by $P_1(3,2)$ and $P_2(-3,-1)$.

Solution
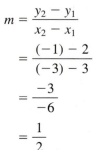

$$m = \frac{y_2 - y_1}{x_2 - x_1}$$

$$= \frac{(-1) - 2}{(-3) - 3}$$

$$= \frac{-3}{-6}$$

$$= \frac{1}{2}$$

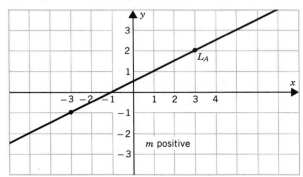

Problem B   Find the slope of line L_B determined by $P_1(3,2)$ and $P_3(4,0)$.

Solution

$$m = \frac{0 - 2}{4 - 3}$$

$$= -2$$

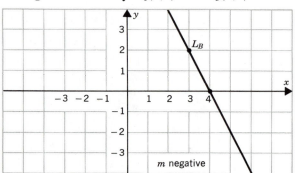

Problem C   Find the slope of line L_C determined by $P_1(3,2)$ and $P_4(-2,2)$.

Solution

$$m = \frac{2 - 2}{(-2) - 3}$$

$$= \frac{0}{-5}$$

$$= 0$$

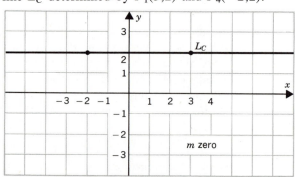

Problem D   Find the slope of line L_D determined by $P_1(3,2)$ and $P_5(3,-2)$.

Solution

$$m = \frac{(-2) - 2}{3 - 3}$$

Since, in the denominator, $3 - 3 = 0$, the slope does not exist.

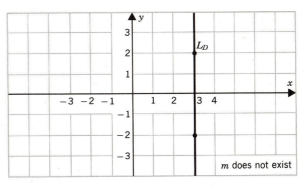

m does not exist

Note the special property of L_C whose slope is zero. Such a line is called a *horizontal line* and is parallel to the *x* axis.

Example 2

Purpose To use the slope formula in an applied problem:

Problem The growth of a biological culture, *y*, increases from 16 square centimeters (cm^2) to 20 square centimeters as time, *x*, increases from two hours to four hours. Find the average rate of growth, denoted by the slope between $(2,16)$ and $(4,20)$.

Solution

$$m = \frac{20 - 16}{4 - 2} = \frac{4}{2} = 2.$$

The average rate of growth is 2 square centimeters per hour.

One of the applications of slopes is to establish the relationship that must exist between the slopes of two nonvertical parallel lines and between the slopes of two perpendicular lines (neither vertical). These relationships are shown in the following diagrams.

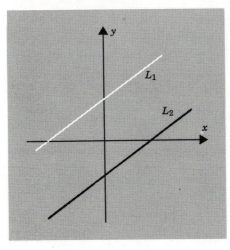

$L_1 \parallel L_2$ and $m_1 = m_2$.

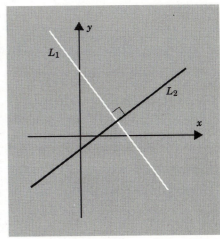

$L_1 \perp L_2$ and $m_1 = -\dfrac{1}{m_2}$.

In words, if two lines are parallel their slopes are equal. If two lines are perpendicular, the slope of one is equal to the negative reciprocal of the slope of the other line (or the product of the slopes equals -1).

Example 3

Purpose To illustrate relationships of slopes of nonvertical lines:

Problem A If line L_1 has a slope of $-\frac{2}{3}$ and line L_1 is parallel to L_2 ($L_2 \| L_1$), what is the slope of L_2?

Solution Since $L_1 \| L_2$, then $m_1 = m_2$ and $m_2 = -\frac{2}{3}$.

Problem B Is the triangle ABC with vertices $A(4,4)$, $B(3,-8)$, and $C(-1,2)$ a right triangle?

Solution
$$m_{AC} = \frac{4-2}{4-(-1)} = \frac{2}{5}; \; m_{BC} = \frac{(-8)-2}{3-(-1)} = \frac{-10}{4} = -\frac{5}{2}.$$

Since $m_{AC} = -\dfrac{1}{m_{BC}}$, it follows that side AC is perpendicular to side BC ($AC \perp BC$) and the triangle ABC is a right triangle. Plot the points and graph the triangle for further clarification.

In Exercise 1.6, Problems 1 to 18 will give you practice with the concept of slope.

Straight Line

Let us now formulate the standard equations of vertical and nonvertical straight lines. We will first consider the equation of a vertical line; then we will develop the two forms for the nonvertical line—commonly referred to as the point-slope form and the slope-intercept form.

The vertical line, L, passing through a point (x_1, y_1) has as its equation:

$$x = x_1.$$

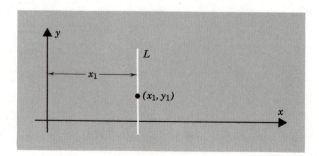

Since every point of the line L is the same distance from the y axis, $x = x_1$ for all points of L. Therefore, $x = x_1$ is the desired equation.

Example 4

Purpose To write the equation of the vertical line through a given point:

Problem The vertical line passes through (2,4). Write the equation of this vertical line.

Solution $x = 2$.

From the previous discussion we observe that every nonvertical line has a slope. When a point on this line and its slope are known, its equation may be written using point-slope form.

The *point-slope equation* for a line L passing through a point (x_1,y_1) with slope m is

$$y - y_1 = m(x - x_1).$$

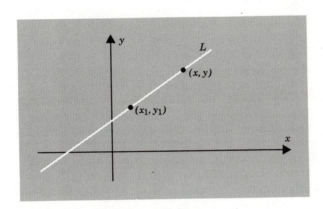

Let (x,y) be any point on line L such that $(x,y) \neq (x_1,y_1)$.
Then the slope of L is

$$\frac{y - y_1}{x - x_1} = m.$$

Multiply both sides of the equation by $x - x_1$ to obtain

$$y - y_1 = m(x - x_1).$$

Example 5

Purpose To illustrate the use of point-slope form in the writing of the equation of a straight line:

Problem A Write the equation of the straight line that passes through the points (5,2) and (−1,−2).

Solution

$$m = \frac{2 - (-2)}{5 - (-1)} = \frac{4}{6} = \frac{2}{3} \quad \text{or} \quad m = \frac{(-2) - 2}{(-1) - 5} = \frac{-4}{-6} = \frac{2}{3}$$

Using $y - y_1 = m(x - x_1)$:

$$m = \frac{2}{3} \qquad\qquad\qquad m = \frac{2}{3}$$

$$(x_1, y_1) = (5, 2) \quad \text{or} \quad (x_1, y_1) = (-1, -2)$$

$$y - 2 = \frac{2}{3}(x - 5) \qquad\qquad y - (-2) = \frac{2}{3}[x - (-1)]$$

$$3y - 6 = 2x - 10 \qquad\qquad 3y + 6 = 2x + 2$$

$$2x - 3y - 4 = 0 \qquad\qquad 2x - 3y - 4 = 0$$

Problem B The Clayton Cycle Shop has plotted sales for the last four years. The line passing through the points corresponding to the first and fourth years is used to determine future sales. Find this sales equation.

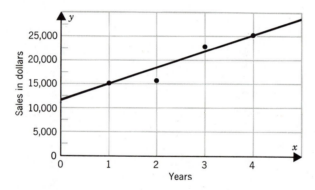

Solution Using the points (1, 15,000) and (4, 25,000), find the slope, m.

$$m = \frac{25,000 - 15,000}{4 - 1} = \frac{10,000}{3}.$$

Using $y - y_1 = m(x - x_1)$, find the equation:

$$y - 15,000 = \frac{10,000}{3}(x - 1)$$

$$3y - 45,000 = 10,000x - 10,000$$

$$10,000x - 3y + 35,000 = 0$$

or

$$y = \frac{10,000}{3}x + \frac{35,000}{3}.$$

Another useful form for the equation of a nonvertical line is the slope-intercept form. This form makes use of the y intercept, that is, the y coordinate of a point where the curve touches or crosses the y axis.

The *slope-intercept equation* of a line L with slope m and y intercept b is

$$y = mx + b.$$

This equation may be obtained from the point-slope equation. Since we know the slope of the line and a point on the line, that is, the y intercept, we use the point-slope form with slope m and point $(0,b)$.

$$y - y_1 = m(x - x_1)$$
$$y - b = m(x - 0)$$
$$y - b = mx$$
$$y = mx + b$$

Example 6

Purpose To illustrate the use of the slope-intercept form both in writing the equation of a line and as an aid in graphing:

Problem A Write the equation of the line whose slope is 2 and crosses the y axis at $(0,3)$.

Solution $y = mx + b$

$y = 2x + 3$

Problem B Graph $2x - y = 1$.

Solution $2x - y = 1$ or $y = 2x - 1$. The slope, m, is the coefficient of x or $m = 2$.

Therefore, $m = \dfrac{\text{rise}}{\text{run}} = \dfrac{2}{1}$. Also, b, the y intercept $= -1$. Starting at $(0,-1)$ rise 2 and run 1. Connect the determined points with a straight line.

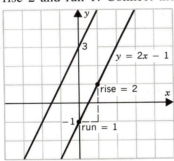

The graph from the equation of Problem A is also drawn. Notice that the slopes are the same, the y intercepts differ, and the lines are parallel.

The preceding discussion and examples lead us to conclude that there is a relationship between linear equations and straight lines.

1. Every straight line has an equation of the type $Ax + By = C$ where A and B are not both zero; and

2. Every linear equation, $Ax + By = C$ (A and B not both zero) has a straight line for a graph.

Familiarity with these ideas is particularly helpful when you are graphing. If you recognize that the equation is linear, then you know the graph of the equation is a straight line and two points will suffice to graph the equation. The following example illustrates this idea and also reviews the method for solving a system of simultaneous equations.

Example 7

Purpose To identify the graphs of two equations and to solve the system of simultaneous equations both algebraically and by utilizing graphs:

Problem Identify the graphs of $x + y = 3$ and $y = x^2 - 2x - 3$. Graph $x + y = 3$ and $y = x^2 - 2x - 3$ on the same coordinate system and determine graphically their points of intersection. Also, solve the system algebraically.

$$\begin{cases} x + y = 3 \\ y = x^2 - 2x - 3 \end{cases}$$

Solution From the discussion above, $x + y = 3$ has a straight line for its graph. $y = x^2 - 2x - 3$ is not a straight line, since it is not of the form $Ax + By = C$. Recall that $y = x^2 - 2x - 3$ is of the form $y = Ax^2 + Bx + C$, which is the equation of a curve called a parabola.
Graphically:

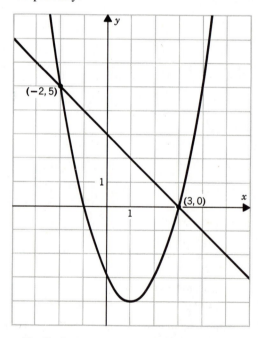

To find the points of intersection algebraically, substitute $y = 3 - x$ for y in the equation $y = x^2 - 2x - 3$ and solve for x. Then find the corresponding y values. Thus

$$3 - x = x^2 - 2x - 3$$
$$0 = x^2 - x - 6$$
$$0 = (x - 3)(x + 2)$$

$$x = 3 \qquad\qquad\qquad x = -2$$

$$y = 3 - x = 3 - 3 = 0 \qquad y = 3 - x = 3 - (-2) = 5$$

Therefore, the points of intersection are $(3,0)$ and $(-2,5)$.

Linear Demand and Supply Equations, Equilibrium Point

It is not unusual for economists to develop their demand equations as linear relationships, since these equations give a fairly close approximation and it is far easier to work with linear forms. The demand equation can be of the form $y = ax + b$, $a < 0$ (as price, y, increases, the demand for the commodity, x, will fall). A supply equation, also linear, $y = cx + d$, $c > 0$, represents the opposite effect (the higher the price, y, the supply of the commodity, x, will increase). The intersection of these two lines, where demand equals supply, is the equilibrium quantity and price. A manufacturer of raincoats has established for his product the demand equation, $y = -\frac{5}{3}x + \frac{230}{3}$, and the supply equation, $y = \frac{2}{3}x + \frac{55}{3}$ where x is the number of raincoats in hundreds and y is the price per coat. Find the equilibrium price and quantity.
Graphically:

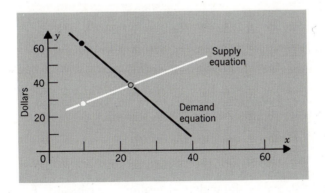

Algebraically, substitute $y = -\frac{5}{3}x + \frac{230}{3}$ for y in the equation $y = \frac{2}{3}x + \frac{55}{3}$ and solve for x. Then find the corresponding y value. Thus,

$$-\frac{5}{3}x + \frac{230}{3} = \frac{2}{3}x + \frac{55}{3}$$

$$-7x = -175$$

$$x = 25$$

$$y = \frac{2}{3}(25) + \frac{55}{3}$$

$$y = 35.$$

The equilibrium coordinates are $x = 25$, $y = 35$. Thus, when 2500 raincoats are supplied at a price of $35 each, the supply and demand for this product are equal.

Break-even Point

Another business situation involving the solution of simultaneous linear equations is to determine the break-even point. The *break-even point* is the level of production where the revenue from the sales is equal to the cost of production. At the break-even point the company is neither making a profit nor losing money. In other words, profit is equal to zero at this point.

Graphically, the break-even point is the intersection of the total revenue and the total cost curves. Total cost is equal to the fixed costs plus the variable costs. Fixed costs include items such as taxes, upkeep of the building, and interest on loans. Variable costs depend on the number of units produced. Labor and material costs are examples of variable costs.

The break-even point can also be found by the simultaneous solution of the total revenue and total cost equations.

These concepts are illustrated in the following example.

Example 8

Purpose To use methods of solving simultaneous linear equations in solving a break-even analysis problem:

Problem A company has fixed costs of $20,000, and the cost of producing one unit of their product is $5. If each unit sells for $9, find the break-even point.

Solution To solve this problem we must first form the total cost and total revenue equations. Let x equal the number of items produced and sold. Then,

$$\text{Total cost} = \text{fixed costs} + \text{variable cost} = 20{,}000 + 5x$$

and

$$\text{Total revenue} = 9x$$

The break-even point may be determined graphically. From this graph it can be seen that the break-even point occurs when the level of production is 5000.

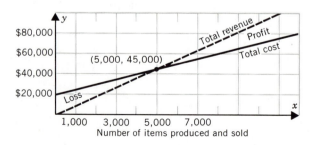

Number of items produced and sold

To find the break-even point algebraically, solve the total revenue and total cost equations simultaneously. At the break-even point, total revenue equals total cost. Let us call this common value y. Then we have

$$y = 5x + 20{,}000$$
$$y = 9x$$

Substituting $\quad 9x = 5x + 20{,}000$

Solve for $x \quad 4x = 20{,}000$

$$x = 5000$$

We find the break-even point to be 5000 items.

Exercise 1.6

In Problems 1 to 6, find the slopes (if they exist) that are determined by the following pairs of points.

1. $(-5,2)$; $(4,-2)$
2. $(0,-3)$; $(-4,1)$
3. $(3,4)$; $(-3,-4)$
4. $(-2,-3)$; $(3,-3)$
5. $(2,-4)$; $(2,3)$
6. (c,d); $(-c,d)$; $c > 0$

7. Determine from the following graphs whether the slope of each line is positive, negative, zero, or undefined.

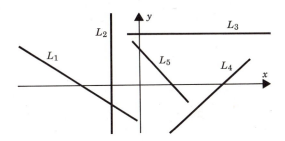

8. In the graphs above (Problem 7) list in order of magnitude the slopes of L_1, L_3, L_4, and L_5 stating the largest slope first. Assume L_1 has slope m_1, L_3 has slope m_3, etc.

9. A scientist doing an experiment on the relationship between the pressure and volume of a gas finds that when the pressure is 1 atmosphere (atm), the volume is 30 cubic feet (ft^3). When the pressure is 10 atm, the volume is 5 ft^3. Plot these two points and determine the slope of the line that they form $(x, \text{pressure})$.

10. A psychologist studies the effects of stimulation on the maze-solving ability of a mouse. She finds that if one piece of cheese is placed in a corner, the mouse takes 12 minutes to find his way through the maze. If five pieces of cheese are used, the mouse requires only 3 minutes. Plot these

two points and find the slope of the straight line through the points. (Let x = number of pieces of cheese.)

11. When the number of items a company sells, x, increases from 100 to 125, the cost for producing each item, y, decreases from $80 to $75. Determine the average change in cost by finding the slope between (100,80) and (125,75).

12. A Lorenz curve provides a way of visualizing the distribution of a resource by relating a fraction of the population to the fraction of the resource accounted for. If the fraction of the population being considered is always equal to the fraction of the resource accounted for, then a completely equal distribution is said to exist. The slope between two points on the curve may be used as an indicator of distribution. If the slope between two points is 1, then there is an equal distribution of the resource for this fraction of the population. If the slope is greater than 1, then the fraction of the population between the two points is receiving more than an equal share of the resource. If the slope is less than 1, then the fraction of the population between the two points is receiving less than an equal share of the resource.

   Use the following data to graph the Lorenz curve.

Fraction of Farmers (x)	Fraction of Land (y)
0.0	0.000
0.2	0.008
0.4	0.064
0.6	0.216
0.8	0.512
1.0	1.000

   Determine whether the following fraction of farmers are receiving more than or less than an equal share of land.
   (a) Fraction of farmers between .2 and .4.
   (b) Fraction of farmers between .6 and .8.

13. What is the slope of a line parallel to the line determined by $(-2,-4)$ and $(3,-1)$?

14. What is the slope of a line perpendicular to the line determined by $(-2,-4)$ and $(3,-1)$?

15. If line L_1, with points $(-2,-3)$ and $(2,3)$, is parallel to line L_2, with points $(1,5)$ and $(x,8)$, find x.

16. If line L_1, with points $(-2,-3)$ and $(2,3)$, is perpendicular to line L_2, with points $(1,5)$ and $(4,y)$, find y.

17. Each pair of points determines a line. Which lines are parallel? Which lines are perpendicular?

(a) L_1: $(-5,2)$; $(4,-2)$ (d) L_4: $(4,-2)$; $(8,7)$
(b) L_2: $(-5,2)$; $(-1,11)$ (e) L_5: $(5,2)$; $(14,-2)$
(c) L_3: $(-5,2)$; $(1,2)$ (f) L_6: $(-3,2)$; $(1,3)$

18. Indicate which pair of lines are parallel and which are perpendicular.
(a) $x + y = 7$, $x + y = 9$ (c) $x + 2y = 5$, $2x - y = 5$
(b) $x + y = 7$, $x - y = 9$ (d) $3x - y = 0$, $3x - y = 1$

19. Find the equation of the line and graph the line:
(a) through $(2,-5)$ and $(-4,-1)$
(b) with $m = \frac{5}{3}$ and through $(-1,-2)$
(c) through $(3,0)$ and parallel to the y axis
(d) through $(0,3)$ and parallel to the x axis
(e) with $m = \frac{1}{3}$ and y intercept -2.

20. Find the equation of the line and graph the line:
(a) through $(-2,3)$ and $(1,-1)$
(b) through $(2,2)$ with slope 3
(c) parallel to the y axis and through $(-2,5)$
(d) perpendicular to the y axis and through $(-2,5)$
(e) with slope $-\frac{5}{4}$ and y intercept $\frac{3}{4}$.

21. What is the equation whose graph is the x axis?

22. What is the equation whose graph is the y axis?

23. True or False
(a) An equation whose graph is parallel to the x axis is of the form $y = c$, where c is a constant.
(b) An equation whose graph is parallel to the y axis is of the form $x = c$, where c is a constant.

24. Find the equation of a line through $(-3,2)$ and parallel to the line $3x + 4y = 4$.

25. Find the equation of a line through $(-3,2)$ and perpendicular to the line $3x + 4y = 4$.

26. If $6x + cy = c$, find a value for c such that the line will have a slope of 2. What is the y intercept?

27. Given a triangle with vertices $A(5,2)$; $B(1,-3)$; $C(-3,4)$. Find:
(a) the equation of side AB
(b) the equation of a line through C and parallel to AB
(c) the equation of the perpendicular bisector of side AB
(d) the equation of the altitude from C to side AB
(e) the equation of the median from side AB to C.

28. Given the triangle with vertices $A(-3,-1)$; $B(3,-4)$; $C(5,2)$. Find:
(a) the equation of side BC
(b) the equation of a line through A and parallel to side BC
(c) the equation of the median from side AB to C
(d) the equation of the altitude from B to side AC.

29. Find the equation of the straight line with intercepts $(3,0)$ and $(0,2)$. Show, in general, with intercepts $(a,0)$ and $(0,b)$, the equation is

$$\frac{x}{a} + \frac{y}{b} = 1.$$

30. Given the equation $2x - 3y = 6$:
    (a) graph by transforming the equation into slope-intercept form.
    (b) graph by transforming the equation into intercept form (see Problem 29).

31. Which of the following are equations of a straight line? Why?
    (a) $2x^2 + 3y = 4$ (d) $2x = 4$
    (b) $2x + 3y^2 = 4$ (e) $3y = 4$
    (c) $2x + 3y = 4$ (f) $2x^2 + 3y^2 = 4$

32. A factory bought a machine in 1979 for $10,000. Due to depreciation the machine was worth $5000 in 1984. If depreciation is considered to be linear, find the equation describing the value of the machine after x years. (Let y be the value of the machine and x be the number of years the machine has been owned.) Find the value of the machine in 1986.

33. A fundamental principle of spectrophotometry is that the concentration and absorbance of a solution are related in a linear fashion. A student finds that for a 5-grams per liter solution, the absorbance is .5. For a 15-grams per liter solution, the absorbance turns out to be 1.5. Using these two points, $(5, .5)$ and $(15, 1.5)$, determine the equation of the line with x representing concentration and y being the absorbance. If a solution of unknown concentration is found to have an absorbance of 1.1, what must be its concentration?

34. A local supermarket compiled the following data for the total number of loaves of bread sold by a given hour in an average workday.

Total number of loaves sold	13	25	38	51	66	78
Hour	10:00	11:00	12:00	1:00	2:00	3:00

    (a) Plot the total number of loaves, y, versus the hour, x, with 10:00 as 1, 11:00 as 2, etc.
    (b) Graph a line through the points corresponding to 10:00 and 3:00.
    (c) Find the sales equation.

35. The following data have been compiled by the Brandon Electric Co. for the last five years.

Total sales in thousands of dollars	9	12	14	21	22
Year	1	2	3	4	5

(a) Plot the total sales (y axis) versus the year (x axis).

(b) To determine future sales a line is passed through the points corresponding to the first and fifth years. Find this sales equation.

(c) From this equation what would be the sales prediction for the seventh year?

36. A manufacturer finds the cost, y, in dollars, of producing x units of a commodity is given to be $y = \frac{1}{3}x + 212$. This equation was formulated from the data:

x, units	0	30	60	90	120
y, dollars	212	222	232	242	252

(a) Graph this equation.

(b) What is the cost of producing 48 units?

(*Note.* To be meaningful, both x and y must be greater than or equal to zero.)

37. A laboratory pigeon has been taught to peck a disk for a reward of food. The researcher determined the following data:

Responses per minute, y	1	2	3	4
Units of food, x	5	12	15	18

(a) Plot the points (5,1), (12,2), (15,3), and (18,4).

(b) Determine the response equation through (5,1) and (15,3). What would the frequency of response be for 30 units of food?

(c) Determine the response equation through (12,2) and (18,4). What would the frequency of response be for 30 units of food?

38. To gain data on the demand for rented cars, a rent-a-car agency ran advertised specials each week in the local newspaper. They obtained the following data:

$$x = \text{number of cars rented per week,}$$
$$y = \text{price per week for renting the car (gas excluded).}$$

Week	1	2	3	4
x	270	324	378	432
y	54	47	40	33

(a) What is the linear demand equation?

(b) Graph this equation.

39. A manufacturer wants to predict the sales of an item if it is sold at a certain price. The demand for the item at a given price is diagramed below.

Price, y	$2.00	$2.50	$3.00	$3.50
Number of items sold, x (in hundreds)	50	42	27	20

(a) Plot these points.

(b) Determine the demand equation using the points corresponding to $2 and $3.

(c) What would the predicted quantity demanded be if the items were sold at $1.50? At $4?

In Problems 40 to 42, graph the two equations on the same coordinate system. What are their points of intersection? Check your results algebraically.

40. $y = x^2 - 4$ and $x - y = 2$.

41. $y = x^2 - 5x + 4$ and $3x + y = 7$.

42. $x^2 + y^2 = 25$ and $3x + 4y = 0$.

43. Assume that in a biology laboratory one technician is assigned the task of supplying a certain bacteria for a scientist who is using them for experimental purposes. Find the equilibrium quantity given the demand equation, $y = -\frac{2}{3}x + \frac{16}{3}$ and supply equation, $8y = 10x - 11$ (x in thousands of bacteria and y in hundreds of dollars). Also, graph these equations.

44. The Bestarox Company faces a monthly demand for their production of pocket calculators represented by the equation $y = 30 - x$ and has established a supply equation of $y = \frac{2}{3}x + 5$; where x stands for the number of calculators in hundreds and y is the selling price per calculator. Determine the equilibrium price and quantity (the point at which the demand and supply curves intersect).

45. In Problem 44, suppose that consumer tastes change and the demand for pocket calculators decreases. At each possible price, consumers now demand 500 fewer calculators than before.

(a) What is the new demand equation?

(b) Does the demand curve shift to the right or to the left?

(c) What is the new equilibrium price and quantity?

46. In a competitive market, equilibrium occurs at the price where market demand equals market supply. Suppose that the total demand for corn and the supply of corn in the Verdeen market are as follows:

Bushels demanded (in thousands)	110	105	100	95	90	85	80	75	70
Price per bushel	3.00	3.25	3.50	3.75	4.00	4.25	4.50	4.75	5.00
Bushels supplied (in thousands)	70	73	76	79	82	85	88	91	94

(a) Form the demand and supply equations.

(b) What is the equilibrium price and quantity?

47. The Armatron Company faces a monthly demand for their production of radios represented by the equation $y = -\frac{2}{3}x + 60$. Armatron has established the following supply equation of $y = \frac{1}{3}x + 30$; where y is the selling price per radio and x is the number of radios in hundreds. Graph the supply and demand curves on the same axes. What is the equilibrium price and quantity?

48. A company finds its total cost equation, TC, to be $TC = 1200 + 40x$ and its total revenue equation, TR, to be $TR = 50x$. Find the break-even point.

49. A company has fixed costs of \$15,000, and the cost of producing one unit of their product is \$10. The units sell for \$15 each.
(a) Write an equation for the total cost, TC.
(b) Write an equation for the total revenue, TR.
(c) Graph these equations to find the break-even point.
(d) Solve algebraically to find the break-even point.

50. A company has fixed costs of \$26,000. The cost of producing one item is \$30. If this item sells for \$43, find the break-even point.

51. The Alber Company sells its product at \$4 per unit.
(a) Write the total revenue, TR, equation and graph this equation.
(b) Fixed costs for the Alber Company are \$2800, and variable costs are estimated to run 30% of the total revenue. Form the total cost, TC, equation.
(c) What is the break-even point?
(d) What quantity must the Alber Company sell to cover their fixed cost?

52. Using the same production process, a company manufactures two different types, A and B, of automobile tires. Thus, x is the number of type A and y is the number of type B produced per hour, and the relationship between x and y is found to be $-(y - 81) = (x + 2)^2$, $(x \geqslant 0, y \geqslant 0)$.
(a) What is the largest number of type A that can be produced?
(b) What is the largest number of type B that can be produced?
(c) What amounts of x and y should be produced to satisfy the additional condition $4x - y + 12 = 0$?

1.7 FUNCTIONS

To continue our prelude to the study of the calculus, the concept of a function is now developed. Indeed, it is this concept that is exceptionally convenient for expressing the correspondence between quantities such as salary and hours worked, revenue and the number of units sold, the rate of growth of cancerous tissue and radioactive treatments, and the like.

Consider this correspondence between hours worked and weekly take-home pay.

Hours Worked	Weekly Pay
30	\$150
40	\$200
55	\$275

Thus, to each value of hours worked there corresponds one and only one take-home pay. This type of correspondence is essentially what we mean by a function.

DEFINITION

Function. A function is a correspondence that exists between two sets, X and Y, such that to each element x belonging to set X there corresponds one and only one element y belonging to set Y. X is called the *domain* of the function. The set of y values in Y that are associated with the x values of the domain is called the *range*. The elements x and y are called the *independent* and *dependent* variables, respectively, associated with the function. An alternate definition is that a function is a set of ordered pairs in which no two of its ordered pairs have the same first member.

Example 1

Purpose   To illustrate the concept of a function:

Illustration A   Let x represent the time and y represent the corresponding voltage reading on a patient's electrocardiograph (EKG).

EKG Readings							
x (time)	0	1	2	3	5	7	8
y (voltage)	0	4	0	-4	0	2	4

Illustration B   On a given day, a small factory can produce zero through four units. The daily operating cost of the factory is given by:

Cost of Operating Factory					
x (number of units)	0	1	2	3	4
y (daily cost)	$500	$700	$900	$1100	$1300

Illustration C   Let $C = \{(1,2),\ (1,-2),\ (2,4),\ (3,6)\}$.

Illustration D   Let $D = \{(1,3),\ (2,3),\ (-2,3),\ (3,3)\}$.

In each illustration, test to see if the definition for a function is satisfied.

Solution   (A)   To each x (time) there corresponds one and only one y (voltage). Thus, this correspondence is a function.

(B)   To each x (number of units) there corresponds one and only one y (daily cost). Again we have a function. This could be illustrated by the following mapping.

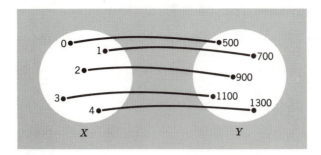

(C) Set *C* is not a function. Notice that $(1,2)$ and $(1,-2)$ mean that to 1 there corresponds both 2 and -2; or two ordered pairs have the same first member.

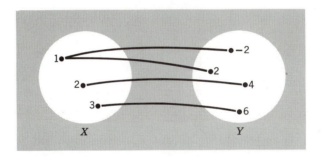

(D) Set *D* represents a function. To each *x* there corresponds one and only one *y*. It is irrelevant that the second members of the ordered pairs are the same.

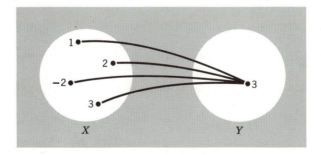

Notice the important feature: To each element of *X* there is associated one and only one element of *Y*. Contrast this picture with *C*, which is not a function. Also the function shown in Illustration *D* is a particular type of function, called a constant function. A *constant function* has only a single element in its range.

Representation of a Function

A function may be represented by four common ways. A function may be represented (1) by a *table*.

Example

x (Number of Units)	y (Daily Operating Cost)
0	$500
1	$700
2	$900
3	$1100
4	$1300

This table is equivalent to

$$(0,500), \ (1,700), \ (2,900), \ (3,1100), \ (4,1300).$$

A function may be represented (2) by a *rule*.

Example.   To obtain the daily operating cost in the above example for 0, 1, 2, 3, or 4 units, add $500 to the number of units multiplied by $200.

A function may be represented (3) by a *graph*.

Example.   (same as above).

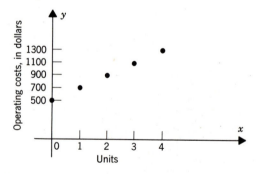

Caution.   Just as all tables are not representations of a function and all rules are not representations of a function, all graphs are not representations of a function.

An example of a graph that does not represent a function follows:

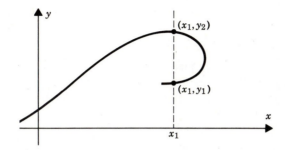

This graph does not represent a function, since to the value x_1 is associated two different y values: y_1 and y_2. Correspondingly, by the *vertical line test,* a graph represents a function if any vertical line crosses or touches the graph in at most one point.

A function may be represented (4) by an *equation.*

Example.   To obtain the daily operating cost in the above example, $y = \$200x + \$500$, where x (number of units) $= 0, 1, 2, 3, 4$, and y is the daily operating cost.

Although an equation may be used to represent a function, not all equations represent functions. This fact is illustrated in Example 2.

Example 2

Purpose   To identify equations that represent functions:

Problem A   Does $y = 2x$ represent a function?

Solution $y = 2x$ is an equation that represents a function, since to each value of the independent variable x is associated one and only one value of the dependent variable y.

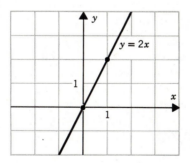

In fact, if we have the equation of any nonvertical straight line, we will always be dealing with a function. This is true, since to each value of x there will be associated one and only one value of y. However, if the equation is of the type $x = x_1$, the equation of a vertical line, we do not have a function. For a vertical line, to the value $x = x_1$, there is associated an unlimited number of values for y.

Problem B   Does $y^2 = x$ represent a function?

Solution $y^2 = x$ or $y = \pm \sqrt{x}$ is an equation that does not represent a function. For example, to the value $x = 4$ there are associated two values of y, namely $+2$ and -2. Also, the vertical line test would readily guarantee that $y^2 = x$ does not represent a function of x.

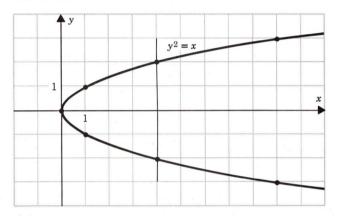

The *f* Notation

Often we wish to give a function a particular name so that when this name is called we will know to which function we are referring. The most popular name for a function is f. It is the usual practice to let f represent the function and to have $f(x)$ equal the value of the function for a specific value of x.

Example 3

Purpose   To illustrate the use of the f notation and to show that it is equivalent to the previously discussed concept of function:

Illustration A   The function f is defined by $y = f(x) = x^2$. This names the function f and specifies that to each member x is associated one and only one value y, namely, the square of x. That is, $f(x) = x^2$ is simply another way to indicate the function whose mapping is shown below.

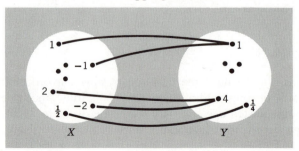

Another way to name the same function is by the set of ordered pairs $\{(1,1),$ $(-1,1), (\frac{1}{2},\frac{1}{4}), (2,4), (-2,4) \ldots \}$. Obviously, $f(x) = x^2$ is preferred to the two previous ways of identifying this function. This is true, since we never could give all the ordered pairs that are elements of this function.

Illustration B   Now, we wish to name another function, call it g, where $g(x) = 2x + 4$. We have used a letter, g, different from f so that we can distinguish between the f and g functions, which certainly do represent different mappings.

Example 4

Purpose   To further illustrate the use of the f notation:

Problem   The functions f and g are defined by $f(x) = x^2 + 2$; $g(x) = 3x - 1$. What is meant by (A) $f(-1)$, (B) $g(1)$, (C) $f(-1) + g(1)$?

Solution   (A) $f(-1)$ means use the function f to find the value of the dependent variable corresponding to a value of $x = -1$.

$$f(x) = x^2 + 2; \qquad \text{replace } x \text{ by } (-1)$$
$$f(-1) = (-1)^2 + 2$$
$$= 1 + 2$$
$$= 3$$

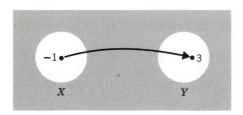

-1 is mapped into 3 by the function f.

This function could have also been given by $y = x^2 + 2$. When $x = -1$, $y = 3$. Therefore, the point $(-1,3)$ is an element of the graph of $y = x^2 + 2$. Also, note that the point identified by $(-1, f(-1))$ or $(-1,3)$ is a point on the graph of $f(x) = x^2 + 2$. Thus, $f(-1)$ is simply another name for the y coordinate of the point where $x = -1$.

(B)   If $g(x) = 3x - 1$, find $g(1)$.

$$g(x) = 3x - 1$$
$$g(1) = 3(1) - 1$$
$$= 2$$

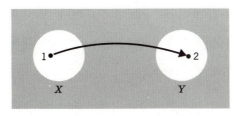

1 is mapped into 2 by the function g.

Also, $g(1)$ is the y coordinate of the point $(1, g(1))$ of the graph of $g(x) = 3x - 1$.

(C)   If $f(x) = x^2 + 2$ and $g(x) = 3x - 1$, find $[f(-1) + g(1)]$.
From parts (A) and (B), $f(-1) = 3$ and $g(1) = 2$. We have,

$$f(-1) + g(1) = 3 + 2 = 5.$$

Example 5

Purpose   To use the f notation to determine certain functional values:

Problem   Given the function f defined by $f(x) = 2x^2 - 3x + 7$. Find $f(0)$, $f(-2)$, $f(x_1)$, $f(x_1 + \Delta x)$ and $f(x_1 + \Delta x) - f(x_1)$. Δx is a variable that represents a change in x.

Solution
$$f(x) = 2x^2 - 3x + 7$$
$$f(0) = 2(0)^2 - 3(0) + 7 = 7$$
$$f(-2) = 2(-2)^2 - 3(-2) + 7 = 21$$
$$f(x_1) = 2x_1{}^2 - 3x_1 + 7$$
$$f(x_1 + \Delta x) = 2(x_1 + \Delta x)^2 - 3(x_1 + \Delta x) + 7$$
$$f(x_1 + \Delta x) - f(x_1) = [2(x_1 + \Delta x)^2 - 3(x_1 + \Delta x) + 7] - [2x_1{}^2 - 3x_1 + 7]$$
$$= 2x_1{}^2 + 4x_1\Delta x + 2(\Delta x)^2 - 3x_1 - 3\Delta x + 7 -$$
$$2x_1{}^2 + 3x_1 - 7$$
$$= 4x_1\Delta x + 2(\Delta x)^2 - 3\Delta x.$$

Example 6

Purpose   To illustrate the method for finding the domain of a function when x is the independent variable:

Problem   Find the domain for $(x^2 - 1)y = 4$ and $g(x) = \sqrt{9 - x^2}$.

Solution   To obtain the domain, solve for y in terms of x, if possible, and then delete from the set of real numbers those values of x that cause us to attempt division by zero. Also, delete those values of x that cause $h(x)$ to be negative in $\sqrt[n]{h(x)}$, when n is an even positive integer.

Let D_f represent the domain of the function f.

Let D_g represent the domain of the function g.
$$(x^2 - 1)y = 4 \qquad g(x) = \sqrt{9 - x^2}$$

$$y = f(x) = \frac{4}{x^2 - 1}$$

$D_f = \{x \mid x \neq \pm 1\}$ $\qquad$ $D_g = \{x \mid -3 \leqslant x \leqslant 3\}$
or all real numbers $\qquad$ or all real numbers
x except $\qquad$ x such that x is greater
$+1$ and -1. $\qquad$ than or equal to -3 and
$\qquad\qquad\qquad\qquad\qquad$ less than or equal to 3.

For D_f we exclude $x = \pm 1$, since $x = \pm 1$ would yield zero in the denominator of $\dfrac{4}{x^2 - 1}$.

For D_g we exclude $x > 3$ and $x < -3$, since for these values of x it follows that $9 - x^2$ is negative, and the square root of a negative number is not a real number.

Throughout this book it should be understood that we shall deal only with values of x that will cause the corresponding functional value to be a real number. Also, in practical applied situations, the domain of a function is often restricted by the realities of the condition under study. For example, if the independent variable x represents the number of units manufactured by a company, then x is restricted to values greater than or equal to zero.

Cost, Revenue, and Profit Functions

We use Problems A, B, and C that follow to illustrate three very important functions. These functions are extremely useful in the business world and appear frequently throughout the remainder of this book.

Problem A   The cost function C, defined by $C(x)$, is the total cost of producing and marketing x units of a commodity. Using this in an example, a company has carefully studied its cost structure, compiled the following data

x (units)	0	10	20	30	40
$C(x)$ (dollars)	50	350	650	950	1250

and determined the daily operating cost to be a linear relation that is a fixed cost of \$50 plus \$30 times the number of units produced and marketed. Thus, the function $C(x) = \$(50 + 30x)$ represents the company's costs. It is understood that $x \geqslant 0$. Graph this function. Find $C(30)$.

Solution

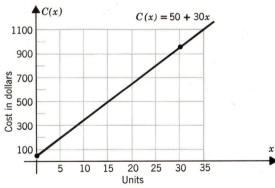

$$C(x) = 50 + 30x$$
$$C(30) = 50 + (30)(30)$$
$$C(30) = \$950$$

Problem B   The revenue function R, defined by $R(x)$, is the total revenue from the sale of x units. As an example, the same company finds its revenue structure can be determined by the function $R(x) = \$(90x - x^2)$. Graph this function. Find $R(30)$.

Solution

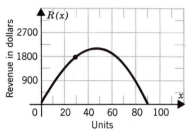

$$R(x) = 90x - x^2$$
$$R(30) = (90)(30) - (30)^2$$
$$R(30) = \$1800$$

Problem C   The profit function P, defined by $P(x)$, is the profit from producing, marketing, and selling x units of a commodity. $P(x)$ is the revenue function minus the cost function; $P(x) = R(x) - C(x)$. State the profit function for this same company. Graph the profit function. Find $P(20)$, $P(30)$, $P(50)$. Estimate from the graph the company's maximum profit.

Solution $P(x) = R(x) - C(x)$

$P(x) = (90x - x^2) - (50 + 30x)$

$P(x) = -x^2 + 60x - 50$

$$P(20) = -(20)^2 + 60(20) - 50$$
$$= \$750$$
$$P(30) = -(30)^2 + 60(30) - 50$$
$$= \$850$$
$$P(50) = -(50)^2 + 60(50) - 50$$
$$= \$450$$

Estimate of maximum profit is $850.

Exercise 1.7

1. Given the function f, defined by $f(x) = 7 - x$.
   Find:
   (a) $f(3)$; $f(-4)$; $f(7)$

   (b) $f(x_1)$; $f(x_1 + \Delta x)$; $f(x_1 + \Delta x) - f(x_1)$; $\dfrac{f(x_1 + \Delta x) - f(x_1)}{\Delta x}$

   ($\Delta x \neq 0$). *Note:* Δx is a variable that represents the change in x.

2. Given $g(x) = 2x + 1$.
   Find:
   (a) $g(-5)$; $g(-\frac{1}{2})$; $g(2)$

   (b) $g(x_1)$; $g(x_1 + \Delta x)$; $g(x_1 + \Delta x) - g(x_1)$; $\dfrac{g(x_1 + \Delta x) - g(x_1)}{\Delta x}$

3. Given $h(x) = 2x^2 + 1$.
   Find:
   (a) $h(3)$; $h(-1)$; $h(0)$

   (b) $h(x_1)$; $h(x_1 + \Delta x)$; $h(x_1 + \Delta x) - h(x_1)$; $\dfrac{h(x_1 + \Delta x) - h(x_1)}{\Delta x}$

4. Given $g(x) = x^2 - x + 1$.
   Find:
   (a) $g(-2)$; $g(1)$; $g(3)$

   (b) $g(x_1)$; $g(x_1 + \Delta x)$; $g(x_1 + \Delta x) - g(x_1)$; $\dfrac{g(x_1 + \Delta x) - g(x_1)}{\Delta x}$

5. Given $f(x) = 2x^2 + x - 3$.
   Find:
   (a) $f(1)$; $f(-2)$; $f(2)$

   (b) $f(2 + \Delta x)$; $f(2 + \Delta x) - f(2)$; $\dfrac{f(2 + \Delta x) - f(2)}{\Delta x}$

6. Given $g(x) = \dfrac{1}{x + 1}$.
   Find:
   (a) $g(0)$; $g(1)$; $g(-1)$

   (b) $g(x_1)$; $g(x_1 + \Delta x)$; $g(x_1 + \Delta x) - g(x_1)$; $\dfrac{g(x_1 + \Delta x) - g(x_1)}{\Delta x}$

7. Given $f(t) = \dfrac{2}{t - 2}$.
   Find:
   (a) $f(4)$; $f(0)$; $f(2)$

   (b) $f(t_1)$; $f(t_1 + \Delta t)$; $f(t_1 + \Delta t) - f(t_1)$; $\dfrac{f(t_1 + \Delta t) - f(t_1)}{\Delta t}$

8. Given $h(x) = \sqrt{2x - 3}$.
   Find:
   (a) $h(2)$; $h(\frac{3}{2})$; $h(5)$

   (b) $h(x_1)$; $h(x_1 + \Delta x)$; $h(x_1 + \Delta x) - h(x_1)$; $\dfrac{h(x_1 + \Delta x) - h(x_1)}{\Delta x}$

9. Given $s = f(t) = \sqrt{2 - t}$.
   Find:
   (a) $f(0)$; $f(2)$; $f(-7)$

   (b) $f(t_1)$; $f(t_1 + \Delta t)$; $f(t_1 + \Delta t) - f(t_1)$; $\dfrac{f(t_1 + \Delta t) - f(t_1)}{\Delta t}$

10. Given $f(x) = 2^x$.
    Find:
    $f(3)$; $f(1)$; $f(0)$; $f(\frac{1}{2})$; $f(-2)$.

11. In Pennsylvania the sales tax is 6% of the purchasing price. Therefore, the tax may be calculated by the function $T(x) = .06x$ where $T(x)$ is the tax in dollars and x is the purchasing price in dollars.
    (a) Find the tax on an item costing one dollar, $T(1)$.
    (b) Find $T(10)$.
    (c) Find $T(100)$.

12. The equation used to convert pounds, P, to kilograms, K, is $K = .454P$. Does this equation represent a function? If a student weighs 120 pounds, how many kilograms does she weigh?

13. Due to depreciation, the value of factory equipment decreases with time. The equipment is assumed to depreciate linearly over a period of 15 years according to the equation $V = 15,000 - 1000x$ where x is the number of years since the purchase and V is the value of the equipment after x years.
    (a) Graph this function.
    (b) What is the value of the equipment after 5 years? 10 years?

14. A 4-mg dosage of a drug is given to a patient. The concentration of the drug in the patient's bloodstream after t hours can be found by
    $$K(t) = \frac{4}{1 + t^3}.$$
    (a) What is the concentration at the end of one hour, $K(1)$? After two hours?
    (b) Graph the function to exhibit the correspondence between drug concentration and elapsed time.

15. The time, in minutes, it takes a person being tested to complete a task may be found by $T(x) = \dfrac{440}{\sqrt{x}}$, where x is the person's IQ.
    (a) How long does it take a person to complete the task if his or her IQ is 100, $T(100)$?
    (b) What is the IQ of a person who completes the task in 40 minutes?

16. Does $y = x - 3$ define a function?
    (a) If so, represent the function by a table and show a mapping for $x = -1, 0, 2, 4$.
    (b) Represent the function by a rule.
    (c) Represent the function by a graph.

17. Does $y = 2 - 5x$ define a function?
    (a) If so, represent the function by a table and show a mapping for $x = -2, 0, \frac{2}{5}, 1$.
    (b) Represent the function by a rule.
    (c) Represent the function by a graph.

Determine which of the equations in Problems 18 to 30 are functions. If the equation represents a function, determine its domain.

18. $y = x + 3$

19. $y = \frac{1}{2}x + \frac{1}{4}$

20. $x = 7$

21. $y^2 = x - 3$

22. $y = x^2 - x - 6$

23. $y = \sqrt{x - 3}$

24. $y = \dfrac{1}{x}$

25. $y^2 = \dfrac{1}{x^2 + 1}$

26. $y = \dfrac{1}{x^2 + 1}$

27. $y = \dfrac{1}{x^2 - 1}$

28. $y = \sqrt{4 - x^2}$

29. $y = \sqrt{4 + x^2}$

30. $y^2 = 4 - x^2$

31. Graph each of these equations: $y = \sqrt{4x}$, $y = -\sqrt{4x}$, and $y^2 = 4x$. Which of these represent functions?

32. Product-exchange functions are functions that relate the production of two items produced by the same factory. For example, a steel mill that can produce stainless steel for tablewear, rolled steel for cars, or a combination of both has the product-exchange function $y = \dfrac{5,000 - 250x}{100 + 25x}$ where

    x represents the tons of stainless steel and y represents the tons of rolled steel produced per day. Graph this function. From the graph estimate the maximum number of tons of stainless steel that can be produced. From the graph estimate the maximum number of tons of rolled steel that can be produced per day. (Note that for the function to be meaningful both x and y must be greater than or equal to zero.)

33. The Ajax Electric Company currently has a working force of 185 men. They expect to hire five employees per year, thus increasing their working force. State a function to indicate the number of employees as a function of the year.

34. (a) Determine the linear cost function for a small manufacturing company, given the following data:

x (units)	0	5	10	15	20
$C(x)$ (dollars)	29	74	119	164	209

Notice that the cost to produce and sell x units of a product is \$29 added to \$9 times x (units), $x \geq 0$.

(b) The revenue from the sales is $R(x) = \$(39x - x^2)$. Find the profit function; graph this function, and estimate from the graph the company's maximum profit.

35. A tractor manufacturer finds the weekly total cost for x tractors to be $\$(18000 + 26000x)$. The total revenue from the sale of x tractors is $\$(38000x - 300x^2)$.

(a) Graph the cost function.

(b) Graph the revenue function.

(c) Find the profit function, and from the graph of the profit function estimate the company's maximum profit.

36. A function of particular interest is the absolute value function, $f(x) = |x|$. The absolute value of a real number x is defined as

$$|x| = \begin{cases} x, & \text{if } x \geq 0 \\ -x, & \text{if } x < 0 \end{cases}$$

The graph of $f(x) = |x|$ is

x	$f(x)$
0	0
1	1
2	2
-1	1
-2	2

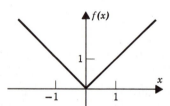

Graph the following absolute value functions.

(a) $f(x) = |-x|$

(b) $f(x) = |x + 2|$

(c) $f(x) = |x| + 2$

1.8   INEQUALITIES

In some of the early work in the calculus and in the establishment of confidence intervals in statistics, it is necessary to work with inequalities. For these reasons, we now give a brief summary of the mechanics of inequalities. Assume that a and b represent real numbers.

> **DEFINITION**
>
> $a > b$. $a > b$, read a is greater than b (or b is less than a), if $a - b$ is some positive number.

This definition is known as the algebraic definition of "greater than." Note that the larger open end of the inequality sign points toward the larger number. There is also an equivalent geometric definition of "greater than" which is, $a > b$ if a lies to the right of b on the number line.

Example 1

Purpose   To illustrate the two definitions of "greater than":

Problem A   Verify $4 > -2$.

Solution $4 > -2$ since $4 - (-2) = +6$

$\qquad\qquad\qquad\qquad$ a positive number.

or

$4 > -2$, since 4 lies to the right of -2 on the number line.

Problem B   Verify $-6 < -3$.

Solution $-6 < -3$, or $-3 > -6$ since $-3 - (-6) = +3$

$\qquad\qquad\qquad\qquad\qquad\qquad$ a positive number

or

$-6 < -3$, since -3 lies to the right of -6 on the number line.

We shall agree that $a \geqslant b$ will be true if a is greater than b or if a is equal to b. Thus, $7 \geqslant 7$ and $5 \geqslant 2$ are both "true" statements.

Some of the theorems on inequalities are now stated. Additional theorems are found in Exercise 1.8, Problem 35.

Theorem	*Examples*
1. If $a > b$, then $a + c > b + c$.	$2 > -3$ and $2 + 4 > -3 + 4$ or $6 > 1$ $2 > -3$ and $2 + (-3) > -3 + (-3)$ or $-1 > -6$

2. If $a > b$ and $c > 0$, $2 > -3$ and $2(4) > -3(4)$ or
    then $ac > bc$. $8 > -12$

3. If $a > b$ and $c < 0$, $2 > -3$ and $2(-3) < -3(-3)$ or
    then $ac < bc$. $-6 < 9$

The proofs for Theorems 1 and 2 follow. The proof for Theorem 3 is outlined in Exercise 1.8, Problem 36.

TO PROVE.   If $a > b$, then $a + c > b + c$.

Here we are proving that we have a right to add the same real number c to both sides of the inequality and that the sense of the inequality will be preserved. $a + c > b + c$ is true if we can show that it meets the definition for "greater than."

Proof

1. $a > b$. Given.

2. $a - b = k$ where $k > 0$. By the definition of "greater than."

3. $a = b + k$

4. $(a + c) = (b + c) + k$
    We may add the same number c to both sides of the equality.

5. $(a + c) - (b + c) = k$
    We may subtract the same number, $b + c$, from both sides of the equality.

6. $a + c > b + c$, since $(a + c) - (b + c)$ is equal to a positive number k.

Example 2

Purpose   To illustrate the use of adding the same number to both sides of an inequality in order to solve the inequality:

Problem   Solve $2x - 7 > x + 2$.

Solution
$$2x - 7 > x + 2$$
$$2x - 7 + 7 > x + 2 + 7$$
$$2x > x + 9$$
$$2x - x > x + 9 - x$$
$$x > 9$$

Now consider the second theorem: if $a > b$ and $c > 0$, then $ac > bc$.

Here we are proving that we have a right to multiply both sides of an inequality by a *positive number* and that the sense, or direction, of the inequality sign will be preserved. $ac > bc$ is true if it meets the definition for "greater than."

Proof

1. $a > b$ and $c > 0$ Given.

2. $a - b = k$ where $k > 0$ Why?

3. $ac - bc = kc$ where $kc > 0$ Why?
4. Therefore, $ac > bc$.   Why?

Caution. We may multiply both members of an inequality by a positive number and the sense or direction of the inequality sign is preserved. However, it is obvious by the following example that if the multiplier is a negative number, the sense of the inequality sign is not preserved. For example, we have $-2 < 8$. Multiply both sides by -1, obtaining $2 < -8$, which is a false statement! Therefore, Theorem 3 is a companion theorem for the multiplication of both sides of an inequality by a negative number. That is: if $a > b$ and $c < 0$, then $ac < bc$. Notice that the sense of the inequality is changed.

Example 3

Purpose   To use inequalities to solve an applied problem:

Problem   Mr. Samuels is paid a monthly commission of 20 percent on all sales over $1000. If he wishes his monthly commission to be greater than $500, what must his monthly sales be?

Solution   Let x represent his monthly sales in dollars.

$$.20(x - 1000) > 500$$
$$.20x - .20(1000) > 500$$
$$.20x - 200 > 500$$
$$.20x - 200 + 200 > 500 + 200$$
$$.20x > 700$$
$$\frac{1}{.20}(.20x) > \frac{1}{.20}(700)$$
$$x > 3500$$

Mr. Samuels' sales for the month must be greater than $3500 for his monthly commission to be greater than $500.

Example 4

Purpose   To illustrate the use of the previous theorems in the solution of linear and quadratic inequalities:

Problem A   Solve $3x - 4 < x - 8$.

Solution $3x - 4 < x - 8$
$$2x < -4$$
$$x < -2$$

Geometrically:

Problem B   Solve $x - 7 < 3x - 2$.

Solution $x - 7 < 3x - 2$

$-2x < 5$

$x > -\dfrac{5}{2}$

Geometrically:

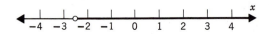

Problem C   Solve $(x - 3)(x - 7) < 0$.

Solution   This problem may be solved by making use of the values of x that cause the factors in the inequality to take on the value zero.

$$(x - 3)(x - 7) = 0 \text{ if } x = 3 \quad \text{or} \quad x = 7.$$

The "key values" for obtaining the solution are 3 and 7.

If $x < 3$,	then	$\begin{array}{l} x - 3 < 0 \\ x - 7 < 0 \end{array}$	and	$(x - 3)(x - 7) > 0$
If $3 < x < 7$,	then	$\begin{array}{l} x - 3 > 0 \\ x - 7 < 0 \end{array}$	and	$(x - 3)(x - 7) < 0$
If $x > 7$,	then	$\begin{array}{l} x - 3 > 0 \\ x - 7 > 0 \end{array}$	and	$(x - 3)(x - 7) < 0$

Therefore, $(x - 3)(x - 7) < 0$ if $3 < x < 7$

Problem D   Solve $x^2 - 5 > 4x$.

Solution $x^2 - 5 > 4x$

$x^2 - 4x - 5 > 0$

$(x + 1)(x - 5) > 0$

This problem may be solved by making use of the values of x that cause the factors in the inequality to take on the value zero.

$$(x + 1)(x - 5) = 0 \text{ if } x = -1 \text{ or } x = 5.$$

The "key values" for obtaining the solution are -1 and 5.

If $x < -1$,	then	$\begin{array}{l} x + 1 < 0 \\ x - 5 < 0 \end{array}$	and	$(x + 1)(x - 5) > 0$
If $-1 < x < 5$,	then	$\begin{array}{l} x + 1 > 0 \\ x - 5 < 0 \end{array}$	and	$(x + 1)(x - 5) < 0$

If $x > 5$, then $\begin{aligned} x + 1 &> 0 \\ x - 5 &> 0 \end{aligned}$ and $(x + 1)(x - 5) > 0$

Therefore, $(x + 1)(x - 5) > 0$ if $x < -1$ or $x > 5$.

Linear Programming (Optional)

Among the applications of inequalities is a type of problem from economics called linear programming. The graphing of inequalities in two variables, x and y, aids us in solving this type problem.

A graph of an equation in x and y describes a set of points on the Cartesian Coordinate System, which is the representation of a plane. An inequality in x and y describes a region of a plane.

Example 5

Purpose To illustrate the graph of a region of a plane:

Problem (A) Graph $2x + y = 1$.

(B) Graph $2x + y > 1$.

(C) Graph $2x + y < 1$.

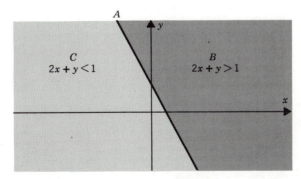

Solution The graph of the straight line is $2x + y = 1$, and this line divides the plane into two regions or half planes. Region B satisfies all points (x,y) such that $2x + y > 1$ and region C satisfies all points (x,y) such that $2x + y < 1$.

The graphical interpretation of inequalities has many applications, one of which is utilized in linear programming. Linear programming develops a procedure for determining the maximum or minimum value of a dependent variable (for instance, profit or cost) that is a function of several independent variables (for instance, hours of labor available, material required, or the length of time a particular piece of machinery is used). These independent variables are subject to various constraints or limitations that require the independent variables to satisfy simultaneous linear inequalities. The simultaneous linear inequalities are the constraints placed on the linear programming problem. The idea just discussed can best be understood by an example.

Example 6

Purpose   To find the maximum profit by linear programming:

Problem   A manufacturer produces two different models, X and Y, of the same product. The raw materials M_1 and M_2 are required for production. At least 18 pounds of M_1 and 12 pounds of M_2 must be used daily. Also, at most, 34 hours of labor are to be utilized. 2 pounds of M_1 are needed for each model X and 1 pound of M_1 for each model Y. For each model of X and Y, 1 pound of M_2 is required. It takes 3 hours to manufacture a model X and 2 hours to manufacture a model Y. The profit is $8 for each model X and $5 for each model Y. How many of each model should be produced to maximize the profit?

Solution   Let x represent the number of model X produced and y represent the number of model Y produced. Notice the question asked. Find x and y to maximize the profit; or for the equation, $P = 8x + 5y$, find an x and y that will maximize P and at the same time satisfy the constraints:

$$2x + y \geqslant 18 \qquad \text{Constraint for material } M_1.$$
$$x + y \geqslant 12 \qquad \text{Constraint for material } M_2.$$
$$3x + 2y \leqslant 34 \qquad \text{Constraint on labor.}$$

Graph the constraints and find the points of intersection: (2,14), (6,6), and (10,2).

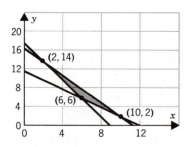

The shaded area is the region, called the ''feasible'' region, that contains our possible solutions, and it can be proven that the solution is at a point of intersection. Check these points to determine maximum profit.

$$P = 8x + 5y$$
$$(2,14) \qquad P = 16 + 70 = \$86$$
$$(6,6) \qquad P = 48 + 30 = \$78$$
$$(10,2) \qquad P = 80 + 10 = \$90$$

The maximum profit of $90 will be realized if 10 units of model X and 2 units of model Y are produced.

Exercise 1.8

In Problems 1 to 25, solve the inequalities and show their solution sets on a number line.

1. $2x - 3 < x + 5$

2. $3x + 4 > x - 7$

3. $\frac{1}{2}x + 1 < 2x - 3$

4. $7 - x \geqslant \frac{1}{3}x - 5$

5. $-2(x - 3) \leqslant 7$

6. $1 \leqslant 4x + 3 \leqslant 8$

7. $-1 < 5 - 2x < 3$

8. $x^2 - 9 > 0$

9. $x^2 - 9 < 0$

10. $x^2 + 9 > 0$

11. $(x - 8)(x + 1) > 0$

12. $(x + 3)(2x - 1) < 0$

13. $(x + 1)(x - 1) > 0$

14. $(2x + 7)(x - 7) < 0$

15. $x^2 - x - 6 < 0$

16. $2x^2 - 5x - 3 > 0$

17. $2x^2 + x - 1 < 0$

18. $3x^2 - 10x + 3 > 0$

19. $3x^2 + 8x + 6 < 2$

20. $\dfrac{x + 3}{x - 4} > 0$

21. $\dfrac{2x - 1}{x + 2} < 0$

22. $(x - 2)^2(x + 3) > 0$

23. $(x + 1)(x - 4)^2 < 0$

24. $(x - 1)(x - 2)(x + 3) < 0$

25. $(x - 1)^3 > 0$

26. Determine the values of x such that the values of $f(x)$ will be real numbers.

   (a) $f(x) = \sqrt{x^2 - 25}$ (b) $f(x) = \sqrt{x^2 - 5x + 6}$

In Problems 27 to 34, graph:

27. $2x - y = 3$

28. $2x - y > 3$

29. $2x - y < 3$

30. $3x + 4y = 9$

31. $3x + 4y < 9$

32. $3x + 4y > 9$

33. $y = x^2 + x - 2$

34. $y < x^2 + x - 2$

35. Additional theorems on inequalities. Prove:
   (a) If $a > b$ and $b > c$, then $a > c$ (transitive principle).
   (b) If $a > b$ and if a and b have the same sign, both positive or both negative, then $\dfrac{1}{a} < \dfrac{1}{b}$.

36. Given $a > b$ and $c < 0$. Prove $ac < bc$. The steps of this proof are given below. Give the reason why each statement can be made.

   1. $a > b$.

   2. $a - b = k$ where $k > 0$.

   3. $ac - bc = kc$ where kc is a negative number.

   4. $bc - ac = -kc$ where $-kc$ is a positive number.

   5. Therefore, $bc > ac$ or $ac < bc$.

37. A car manufacturer gives the inequality $3x - 5 < 2x + 25$, where x is the maximum speed that one can drive and keep a mileage rating of 25 miles per gallon. By solving the inequality for x, determine this maximum average driving speed.

38. A salesman's commission per week is 5% on all sales over $350. If he realizes a commission of at least $137.50, what must be his weekly sales?

39. A local department store budgets at least $3670 per month for advertising. Of this amount, $1500 is spent on television presentations. If a half-page newspaper advertisement costs $45, state an inequality that must be satisfied by x, the number of half-page advertisements taken. What would be the minimum number of half-page advertisements?

40. A crane with a maximum carrying load of 1500 pounds is being used to lift steel beams in the construction of an apartment building. If each beam weighs 235 pounds, set up an inequality to show the number of beams, x, that can be lifted at one time. What is the largest number of beams the crane can lift?

41. A stock advisory service predicts that a stock, with a current price of $57 per share will rise at least $5 within the next three months. State the predicted price, x, in terms of an inequality. If you buy 75 shares of this stock, state your predicted investment, y, in terms of an inequality. What would be your predicted profit? (Disregard commissions.)

42. Since cigarette smoke is known to be injurious to one's health, a certain convention center restricts the number of people allowed in a conference room. If the conference room is 45 feet by 30 feet by 10 feet and each person needs 120 cubic feet of air, what is the greatest number of people permitted in the conference room?

43. (Optional) The Thermo Company produces two models of electric ovens. Each unit of model A requires 3 hours of work on assembly line 1 and 3 hours of work on assembly line 2. Model B requires 10 hours of work on assembly line 1 and 2 hours of work on assembly line 2. During one work week assembly line 1 has up to 150 hours and line 2 has up to 54 hours to devote to electric oven production. How should production be allotted to maximize total profit if Model A nets a profit of $40 per oven and Model B nets a profit of $60 per oven? All models of both ovens that are produced are sold.

44. (Optional) A biologist wishes to make two chemical solutions. There are 36 grams each of chemical I and chemical II available and 66 grams of chemical III. Each liter of solution A requires 1 gram of chemical I, 4 grams of chemical II, and 6 grams of chemical III. Each liter of solution B requires 3 grams of chemical I, 1 gram of chemical II, and 3 grams of chemical III. If the biologist wishes to make a maximum number of liters, how many liters of each solution should she make?

45. (Optional) A dog food company makes two kinds of dog food, mixture A

and mixture B. A grinder can take up to 390 ounces of horsemeat and 240 ounces of meal per hour. Each can of mixture A contains 10 ounces of horsemeat and 5 ounces of meal. Mixture B contains 7 ounces of horsemeat and 8 ounces of meal per can. How many cans of each mixture should be produced hourly to maximize profit if the profit on mixture A is 10 cents a can and if the profit on mixture B is 8 cents a can?

46. (Optional) A company must ship items produced in Branch A and Branch B to a warehouse. The warehouse needs at least 2000 items. It costs $5 to ship an item from Branch A to the warehouse, while it costs $3 to ship an item from Branch B to the warehouse. Branch A can supply at most 1500 items, and Branch B can supply at most 1000 items. How should the company plan its shipping schedule to minimize cost?

47. (Optional) Before building a new factory the following information (based on 1 hour of operation) is considered.

	No. of Type I Articles Produced	No. of Type II Articles Produced	Amount of Sulfur Oxides Released
Production line A	10	20	3
Production line B	10	30	5

The factory is to produce at least 50 articles of Type I and 120 of Type II per hour. How many of each type of production lines should be installed to minimize the amount of sulfur oxides released?

1.9 FORMING ALGEBRAIC FUNCTIONS

In the modeling of real world situations it is often necessary to form an equation that accurately describes the relationship between certain quantities. For example, one may wish to know how profit corresponds to production cost, how sales volume relates to advertising expenditures, or how levels of a certain drug in the bloodstream relate to the elapsed time since injection. After the correspondence is represented by a function, several alternate approaches are available. The function may be graphed or mathematical techniques may be applied or both to arrive at an intelligent course of action. Subsequent chapters of this book show the many uses of the calculus in this ongoing analysis and decision process. Therefore, we must practice the formation of equations, a step of paramount importance in the building of mathematical models. The following problems illustrate this process.

Practice in Forming Algebraic Functions

Problem A After much research a drug company has determined that the concentration of the drug in the bloodstream may be calculated as twice the amount of time elapsed since injection divided by the elapsed time cubed.

Solution Concentration, $C = \dfrac{2(\text{time elapsed, } t)}{(\text{time elapsed, } t)^3}$

$$C = \frac{2t}{t^3}$$

$$C = \frac{2}{t^2}$$

Problem B A company has observed that it can sell 50 items when 5 advertisements a week are placed in the newspaper. For each additional weekly advertisement, sales increase by 7 items. If the number of advertisements is greater than 5, express the number of sales as a function of the number of advertisements.

Solution Let x be the number of advertisements greater than 5. The number of sales, y, would equal $50 + 7(x)$; therefore, $y = 50 + 7x$.

Problem C In business models, the gross profit, $P(x)$, is equal to the total revenue, $R(x)$, minus the total costs, $C(x)$; or $P(x) = R(x) - C(x)$. Express the gross profit for an industrial firm if the total production cost is $\frac{4}{5}$ of their sales.

Solution Let x represent the firm's sales (revenue); then $\frac{4}{5}x$ will represent the production costs. Therefore, the gross profit function is $P(x) = R(x) - C(x)$ or $P(x) = x - \frac{4}{5}x$.

Problem D In the preceding problem it is stated that $P(x) = R(x) - C(x)$. $R(x)$ is often determined by the product of x (the number of units) and $D(x)$, the demand function. The demand function represents the price per unit at which x units are sold. Thus, $R(x) = x \cdot D(x)$.

An automobile manufacturer can sell x cars per week at $(2460 - .2x^2)$ dollars per car. It costs the manufacturer $(2.4x + 1200)$ dollars to produce x cars per week. Form the profit function.

Solution $D(x) = (2460 - .2x^2)$

$R(x) = x \cdot D(x)$ or

$R(x) = x(2460 - .2x^2)$

$C(x) = (2.4x + 1200)$

$P(x) = R(x) - C(x)$ or

$P(x) = x(2460 - .2x^2) - (2.4x + 1200)$

$P(x) = 2460x - .2x^3 - 2.4x - 1200$

$P(x) = -.2x^3 + 2457.6x - 1200$

Exercise 1.9

1. A keypunch operator for a special project gets paid 30 cents per card punched plus a base salary of $100 a week. Write an equation that gives the amount of money, M, she has earned at the end of three weeks.

2. A newspaper charges x cents per word to print an advertisement for one week and $(x + 3)$ cents per word for two weeks. If a man has a 50-word advertisement that he wants published for one week and a 32-word advertisement that he wants in the paper for two weeks, write an expression showing his total cost.

3. A car salesman receives a monthly salary of $550 plus a bonus of $75 for each new car he sells in excess of 15 per month. Write a mathematical statement to indicate his salary if he sells more than 15 cars. What is his salary if he sells exactly 15 cars? Less than 15 cars?

4. The total value of sales (total revenue) for the Allen Company amounted to x dollars, and their costs amounted to $\frac{5}{2}$ of their profit.
    (a) Let y represent the profit and express the cost in terms of y.
    (b) Since profit equals revenue minus cost, find the profit function for the Allen Company.
    (c) If sales are $14,000 what is the profit?

5. A store can sell 50 items of a commodity when the sale price is $10. When the items are sold at $20 each, only 30 items are sold. What is the linear demand function?

6. The monthly cost of manufacturing sets of china plates is given by the equation $C(x) = 2000 + 10x - .01x^2$. If the company can sell x sets a month at $\$(300 - .1x)$ per set, what is the profit function?

7. A restaurant manager needs 10 waitresses at 8 a.m. For each hour after 8 a.m. until 10 p.m. she needs one additional waitress. State a function to indicate the number of employees needed as a function of the hour.

8. A furniture store can purchase 50 easy chairs at $260 each. For each chair over 50 purchased, the manufacturer will reduce the price of each chair by $2. Assuming that over 50 chairs are purchased:
    (a) Show that the price per chair is $\$(360 - 2x)$ where x is the number of chairs.
    (b) State the manufacturer's total revenue as a function of the number of chairs sold.

9. It has been found that the amount of relief that an aspirin gives is equal to four times the number of hours elapsed since taking the medication minus the square of the hours elapsed. State the amount of relief as a function of time.

10. The cost of renting a hauling truck is $5 per hour. The cost of fuel is equal to $\frac{1}{10}$ times the sum of the velocity and the velocity squared. Find the total hourly cost of using the truck as a function of velocity.

11. Fifty apple trees are grown on a plot. Each tree will yield 2 bushels of apples. For every additional tree planted and grown to maturity, the yield of each tree decreases by $\frac{1}{12}$ of a bushel. Find an expression that gives the total number of bushels produced when the number of trees is greater than 50.

12. A company must pay $500 a day to maintain its factory. There are 40 production employees, and each worker is paid $5 for each item produced. The cost of the materials for each item is $3. Express cost as a function of the number produced.

13. A firm's revenue can be found by subtracting the square of the number of units of output sold from 500 times the number of units. Also, the cost of producing a given number of units of output is one half the square of the number of units plus 150. If x represents the number of units of output,
    (a) What is the revenue function?
    (b) What is the cost function?
    (c) What is the profit function?

14. Find the area of a square if its perimeter is known by:
    (a) Perimeter is equal to four times the length of a side, find the length of one side in terms of the perimeter.
    (b) Area is equal to the length of a side squared, so express area in terms of perimeter.

15. A rectangular box is constructed such that its length is twice its width, and its height is one and one half times its length. To express the volume in terms of width:
    (a) If the width is w, then the length, l, is $2w$. State the height in terms of width.
    (b) Given volume = lwh, state the volume in terms of the width.

16. How could the surface area of the closed rectangular box in Problem 15 be expressed in terms of its width? [Surface area = $2(lw) + 2(wh) + 2(lh)$.]

17. To find the area of the shaded portion in the diagram:
    (a) Find the area of the circle in terms of r.
    (b) Express the length of the side of the square in terms of r.
    (c) Find the area of the square in terms of r.
    (d) Write the expression for finding the area of the shaded portion.

18. How could the perimeter of an isosceles right triangle be written as a function of one of its legs?

19. The hypotenuse of a right triangle is 25. Write an expression for determining the area of this triangle.

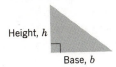

Height, h

Base, b

20. A sphere and a right circular cylinder have equal radii. The height of the cylinder is equal to the diameter of the sphere. If the radius is known, what would be an expression for the sum of the volumes of the given cylinder and sphere?

$$V_{\text{sph}} = \tfrac{4}{3}\pi \, (\text{radius})^3; \; V_{\text{cyl}} = \pi(\text{radius})^2(\text{height})$$

21. A painting is hanging in a frame made from wood 1 inch in width. The dimensions of the painting are v by $v + 3$. What would be an expression for the dimensions of the picture frame?

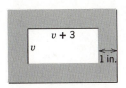

$v + 3$

v

1 in.

22. An outdoor swimming pool is twice as long as it is wide. A cement walk 3 feet wide is to be poured around the pool. How many square feet of cement will be poured if the width of the pool is known?

23. The sum of two numbers is 96. How would the product of their squares be expressed? How would the square of their product be expressed? Would these expressions yield the same answer?

24. A right circular cone is inscribed within a cylinder having the same radius and height. What would be the volume of the portion of the cylinder not enclosed within the cone? (See diagram.)

(a) The volume of a cylinder can be found by $\pi(\text{radius})^2(\text{height})$ or $V = \pi r^2 h$. Express the volume of the cylinder in terms of the radius.

(b) The volume of a cone is found by $\tfrac{1}{3}\pi(\text{radius})^2(\text{height})$ or $V = \tfrac{1}{3}\pi r^2 h$. Find the volume of the cone in terms of the radius.

(c) The portion of the cylinder not enclosed within the cone may be found by $V_{\text{cyl}} - V_{\text{cone}}$. Express the volume of the portion in terms of radius.

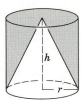

h

r

25. The shape of a tin can is a right circular cylinder whose radius is equal to its height. What would the expression for the total cost of producing one tin can be if it costs 3 cents per square inch to make the top and bottom and 5 cents per square inch to manufacture the side? [The surface area for the side may be found by using the equation 2π(radius)(height). The surface area of both the top and the bottom is given by 2π(radius)^2.]

■ 1.10 CHAPTER REVIEW

Important Ideas

Note. Ideas enclosed in parentheses are optional topics.
real number system
number line
rectangular coordinate system
graphs of equations
distance between two points
slope of a line
nonvertical parallel lines have equal slopes
for nonvertical perpendicular lines, the product of their slopes is -1
the forms of equations of a straight line
function
domain of a function
range of a function
f notation
$a > b$
(linear programming)
forming algebraic functions

■ REVIEW EXERCISE (optional)

1. (a) What is meant by slope?
   (b) What is the slope of a horizontal line?
   (c) Does a vertical line have a slope? Why?
   (d) In the graphs below which line(s) have a positive slope? A negative slope? Zero slope?

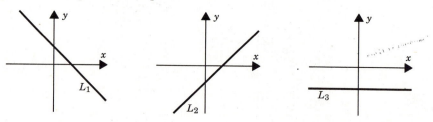

2. The officials of the soapbox derby must make certain that the slope of all the racing hills around the country are equal. If the slope is required to be .63, and the horizontal measure of one hill is 300 feet, what is the required drop in height for the hill?

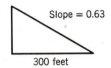

Slope = 0.63

300 feet

3. Describe the set of lines $y = mx + b$
   (a) when b is fixed and m varies;
   (b) when m is fixed and b varies.

4. State the equation $3x - 4y = 16$ in slope-intercept form. What is the slope? What is the y intercept? Graph the equation.

5. Find the equation of the line and graph the line:
   (a) through $(0,3)$ and $(4,-2)$.
   (b) through $(-2,1)$ with slope $\frac{1}{3}$.
   (c) parallel to the x axis and through $(-4,4)$.
   (d) perpendicular to the y axis and through $(4,-4)$.
   (e) with slope $\frac{3}{4}$ and y intercept -1.
   (f) with x intercept 2 and y intercept -5.

6. Line L_1 has a slope of $\frac{1}{3}$. What is the slope of L_2 if
   (a) L_2 is parallel to L_1?
   (b) L_2 is perpendicular to L_1?

7. Find the equation of the line through $(-1,1)$ and parallel to the line $3x + 2y = 6$.

8. Find the equation of the line through $(-1,1)$ and perpendicular to the line $3x + 2y = 6$.

9. Given the triangle with vertices $A(-1,4)$; $B(4,-1)$; $C(3,6)$. Find:
   (a) the slope of side AC.
   (b) the equation of side AC.
   (c) the equation of a line through B and parallel to side AC.
   (d) the midpoint of AC.
   (e) the equation of the perpendicular bisector of side AC.
   (f) the equation of the altitude from B to side AC.
   (g) the equation of the median from side AC to B.
   (h) the length of side AB.
   (i) the length of side BC.
   (j) if the triangle is isosceles. Why?

10. A manufacturer of nylon all-weather backpacks did a study on the demand for his product. He observed the following data, where x represents the

number of backpacks he can sell and y, in dollars, the selling price per backpack.

x	150	180	210	240
y	30	26	22	18

(a) Find the linear demand equation.
(b) Graph this equation.
(Assume all backpacks produced are sold.)

11. Form a linear cost equation for a fixed cost of $2000 and $720 for each item produced. [Fixed costs do not change with changes in the quantity sold. Such expenses as heat, light, and depreciation, and the like remain the same regardless of how many units of the product are sold. Therefore, the cost equation is written as $C = F + V(x)$ where F is the fixed cost and $V(x)$ represents the variable cost. Here, $V(x)$ does depend on the number x of the quantity produced.]

12. Form a linear cost equation for the following data: A cost of $525 occurs when 150 units are produced and a cost of $700 occurs when 400 units are produced. What is the fixed cost? (See Problem 11.)

13. Graph $x^2 + y^2 = 25$ and $3y = 2x^2 - 6$ on the same coordinate axis. What are their points of intersection? Check your results algebraically.

14. A dairy farmer has found that the demand equation for his unprocessed milk is $200y = -2x + 300$ while the supply equation is $600y = 2x + 100$ where x is in gallons and y is the price per gallon in dollars. Find the equilibrium price and quantity.

15. A company has fixed costs of $25,000, and the cost of producing one unit of their product is $25. If the units sell for $35 each, find the break-even point.

16. A circle has its center at $(3,1)$ and is tangent to the line $2x + 3y = 22$. Find the equation of the circle.

17. The end points of the diameter of a circle are $(-1,3)$ and $(7,-3)$. Find the equation of the circle.

18. The function f is defined by $f(x) = \dfrac{1}{\sqrt{x + 2}}$.

(a) Find $f(7)$.
(b) Does $f(-2)$ exist?
(c) What is the domain of f?

19. The function g is defined by $g(x) = 3 + \sqrt{x - 3}$.
(a) Find $g(3)$.
(b) Does $g(0)$ exist?
(c) What is the domain of this function?

20. The function f is defined by $f(x) = 4 - 3x$. Find:
(a) $f(-2)$; $f(0)$; $f(2)$.

(b) $f(x_1)$; $f(x_1 + \Delta x)$; $f(x_1 + \Delta x) - f(x_1)$; $\dfrac{f(x_1 + \Delta x) - f(x_1)}{\Delta x}$.

21.   Given $g(x) = x^2 - 3x - 4$. Find:
(a) $g(-1)$; $g(1)$; $g(3)$.

(b) $g(x_1)$; $g(x_1 + \Delta x)$; $g(x_1 + \Delta x) - g(x_1)$; $\dfrac{g(x_1 + \Delta x) - g(x_1)}{\Delta x}$.

22.   Examine the graphs and determine if the graph represents a function.

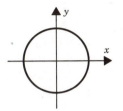 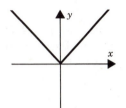 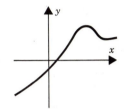

23.   Graph the following. Which of them are functions?
(a) $y^2 - x + 2 = 0$
(b) $y = |x|$
(c) $y = x + 2$
(d) $y > x + 2$
(e) $y - x^2 = 2$

(f) $y = \begin{cases} 2x + 1, & x > 1 \\ x - 2, & x \leqslant 1 \end{cases}$

(g) $y = \begin{cases} x^2 + 1, & x \geqslant 2 \\ 3x - 1, & x < 2 \end{cases}$

24.   The amount of time in minutes, T, it takes to complete a jigsaw puzzle for the xth time is given by the function $T = 90 + \dfrac{c}{x}$, where c is dependent on the individual being tested. What is the learning function for the following cases?
(a) It takes Diane two and one-half hours to build the puzzle for the first time.
(b) Janet completes the puzzle for the fifth time taking two hours and 10 minutes.
(c) Bruce finishes the jigsaw for the second time, taking two hours.

25.   How is the sense of an inequality affected if we: (a) Add the same number to both sides? (b) Multiply both sides by the same positive number? (c) Multiply both sides by the same negative number?

26.   Represent the following region(s) using inequalities.

In Problems 27 to 38, solve the inequalities and show their solution sets on a number line.

27. $3x + 2 < 4$

28. $3 - x \leq 2$

29. $x + \frac{1}{2} < 3x - \frac{5}{6}$

30. $3 - 2x > 2x - 5$

31. $2 < 2 - 5x < 7$

32. $x^2 > 4$

33. $x^2 \leq 4$

34. $(3x + 2)(x + 2) < 0$

35. $(x + 1)(3x - 8) > 0$

36. $3x^2 - 11x + 6 > 0$

37. $2 + x - x^2 > 0$

38. $x(3x + 1)(x - 3) > 0$

39. (Optional) The Amber Company specializes in two styles of decorator lampshades: modern and traditional. Each style requires the craftsmanship of both the frame and the shade departments. The total time available weekly in each department is 100 hours. Three hours from the frame department and 1 hour from the shade department are needed for a modern style shade; whereas 1 hour and 2 hours, respectively, are needed for a traditional style shade. What quantity of each shade should the company produce to maximize its profit, if its profit on the modern style is $5 and its profit on the traditional style is $4?

40. A gas company has found that natural gas produces more heat if it is burned at a higher temperature. At temperatures greater than 1000° F, 5000 Btu per lb of gas are produced, whereas at temperatures between 500 and 1000° F, 4500 Btu per lb of gas are produced. Make a graph of the number of British thermal units produced per pound of gas versus the temperature.

41. The charges for a checking account at City Bank are a monthly charge of $1.50 plus $.10 for each check written. There is no charge for deposits. If x represents the number of checks written, express the total monthly charges, $T(x)$, as a function of x.

42. An object is falling freely from a height of 50 feet. The equation for computing the distance traveled as a function of time, t, is sixteen times t squared subtracted from the original distance. What would an appropriate equation be for computing the distance, d, using a given time?

43. To manufacture x pairs of skis, a company has a cost of $(80 + .02x^2)$. The revenue function is $150x - .1x^2$. Find the profit function.

44. A grain silo is in the shape of a right circular cylinder topped by a hemisphere. The radius of the cylinder is a fourth of the total height of the silo. Find an equation for the volume of the silo in terms of the radius of the cylinder by:
(a) If the radius of the cylinder is r, what is the height of the cylinder in terms of r?
(b) What is the volume of the hemisphere in terms of r?
(c) What is the volume of the cylinder in terms of r?
(d) What is the total volume of the silo?

45. Prove analytically that line segments joining consecutive midpoints of the sides of any quadrilateral form a parallelogram.

■ **IN RETROSPECT**

Chapter 1 is a review of techniques that you will need in later sections of this book. Indeed, not only does every topic of this chapter appear later in this book but many of these topics will also appear in subsequent courses in statistics, probability, biology, and management.

The amount of time that you and your instructor devoted to this chapter has depended on the recency of your previous mathematical preparation and the confidence with which you have approached these topics. We suggest that you take the opportunity to review this chapter, whether or not it is considered a part of your course.

With the preliminaries concluded, we proceed with the main objective of this book: the development of the calculus and its use as a powerful tool in many diverse academic fields.

We repeat: you are dealing with the ideas and the thinking of some of the most brilliant minds of the past 300 years. If at first you do not understand a topic, keep returning to it. Also, use the books in your library to review any other needed algebraic techniques or to extend the concepts of the calculus. Good luck!

TWO

Limits, Derivatives, and Continuity

2.1 OVERVIEW OF CHAPTER

A favorite topic of commencement speakers concerns the fact that we live in a world of change—our values, ideals, hopes, and institutions are undergoing constant change. Usually the speaker's subtopic is a warning that certain changes are happening too rapidly, whereas other changes are not occurring fast enough.

This illustrates that, although the topic of change is important, often the concept of rate of change is more relevant. For example, in the study of population growth it is not sufficient to know that the population changed by doubling. We need to know the rate at which this doubling took place. It is significant that at one time the doubling of the world population took a thousand years, but that now the doubling takes only 35 years. In a similar manner, knowledge of increased industrial productivity is not very meaningful unless one knows at what rate this increase in productivity took place. This idea is particularly important when one considers the rate at which the change in food productivity compares with the rate of change in world population.

The theme of this chapter is change and rate of change. The mathematical tools for measuring rates of change are developed. You will notice that this chapter is developmental in nature—an idea will be carried along until we need another concept. Then this concept will be developed and fitted into its proper place.

Further evidence of the power of calculus is given in this chapter: the tools developed nearly 300 years ago to solve Basic Problem I, the slope of the tangent line to a curve, are still relevant. These tools are the major devices used today in the building of mathematical models for measuring rates of change.

2.2 INCREMENT NOTATION AND AVERAGE RATE OF CHANGE

Δx and Δy

When x changes from an initial value of 3.0 to a final value of 4.5 we say that:

$$\text{the change in } x = \text{final } x \text{ value} - \text{initial } x \text{ value}$$

or

$$\text{the change in } x = 4.5 - 3.0 = 1.5.$$

DEFINITION

Increment in x. The mathematical symbol for the change in x, called the *increment in x,* is Δx (read ''delta x'').

Therefore, in our example:

$$\Delta x = \text{final } x \text{ value} - \text{initial } x \text{ value}$$
$$\Delta x = 4.5 - 3.0 = 1.5$$

Correspondingly, the increment in y, Δy, is defined.

DEFINITION

Increment in y. The change in y, called the *increment in* y, is Δy; and $\Delta y = \text{final } y \text{ value} - \text{initial } y \text{ value}$.

It is noted that the values of Δx and Δy depend on the problem and, when $y = f(x)$, the value of Δy usually depends on the values of x and Δx. The value for Δx is never zero ($\Delta x \neq 0$); however, Δy may be positive, negative, or zero.

Example 1

Purpose To find the increments Δx and Δy under specified conditions:

Problem $y = x^2$. Let x change from a value of 1 to a value of 2. Find Δx and Δy and represent these quantities on a properly labeled graph.

Solution $\Delta x = \text{final } x \text{ value} - \text{initial } x \text{ value}$

$\Delta x = 2 - 1 = 1$

$\Delta y = \text{final } y \text{ value} - \text{initial } y \text{ value}$

$\Delta y = 4 - 1 = 3$ (when $x = 1$, $y = 1$ and when $x = 2$, $y = 4$)

Notice that the value of Δy depends on the values of x and Δx. Graphically:

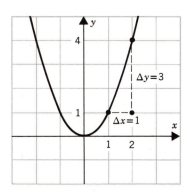

<table>
<tr><td></td><td>In General</td><td>For Example 1</td></tr>
</table>

	In General	*For Example 1*
Initial x value $= x_1$		$x_1 = 1$
Final x value $= x_1 + \Delta x$		$x_1 + \Delta x = 2$
		(Therefore, $\Delta x = 1$.)
Initial y value $= f(x_1) = y_1$		$f(x_1) = f(1) = 1$
Final y value $= f(x_1 + \Delta x) = y_1 + \Delta y$		$f(x_1 + \Delta x) = f(2) = 4$
		(Therefore, $\Delta y = 3$.)

Example 2

Purpose To find the increments ΔN and Δt:

Problem In 1965 the world population was 3,280,000,000. In 1990 it is projected that the world population will be in excess of 5,189,000,000. Find ΔN, the change in population, and Δt, the change in years.

Solution $\Delta N = 5{,}189{,}000{,}000 - 3{,}280{,}000{,}000 = 1{,}909{,}000{,}000$

$\Delta t = 1990 - 1965 = 25$ years

Example 3

Purpose To illustrate that increments are not always positive:

Problem Consider the constant function f, defined by $y = f(x) = 5$. Let x change from a value of 2.5 to a value of 1.3. Find Δx and Δy.

Solution $\Delta x = $ final x value $-$ initial x value

$\Delta x = 1.3 - 2.5 = -1.2$

$\Delta y = $ final y value $-$ initial y value

$\Delta y = 5 - 5 = 0$

From our examples we notice that it would be possible to define Δx and Δy by using more elegant notation:

$$\Delta x = (x_1 + \Delta x) - x_1$$
$$\Delta y = f(x_1 + \Delta x) - f(x_1)$$

or

$$\Delta y = (y_1 + \Delta y) - y_1$$

Example 4

Purpose To find expressions using the new notation for x_1, $f(x_1)$, $x_1 + \Delta x$, $f(x_1 + \Delta x)$, Δx and Δy:

Problem $y = f(x) = 2x + 1$. The initial value of x is 1 and the final value of x is 3. Find expressions for x_1, $f(x_1)$, $x_1 + \Delta x$, and $f(x_1 + \Delta x)$. Then find Δx and Δy and represent these quantities on a graph.

Solution

$$x_1 = \text{initial value of } x = 1$$
$$f(x_1) = y \text{ value corresponding to } x = 1, f(1) = 2(1) + 1 = 3$$
$$x_1 + \Delta x = \text{final value of } x = 3$$
$$f(x_1 + \Delta x) = y \text{ value corresponding to the final } x \text{ value of } 3$$
$$= f(3) = 2(3) + 1 = 7$$
$$\Delta x = (x_1 + \Delta x) - x_1 = 3 - 1 = 2$$
$$\Delta y = f(x_1 + \Delta x) - f(x_1) = 7 - 3 = 4$$

Graphically:

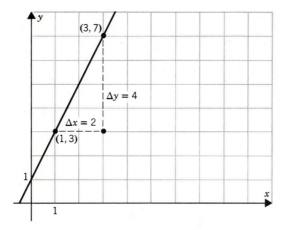

In Example 4 the slope of the straight line $y = 2x + 1$ can be recognized from the equation as being equal to 2. (The equation is in slope-intercept form, $y = mx + b$, where $m = 2$.) However, we could have found the slope of this straight line by noting that

$$\text{slope} = m = \frac{\text{rise}}{\text{run}} = \frac{\Delta y}{\Delta x} = \frac{4}{2} = 2.$$

Thus, it is possible to express the slope of a straight line as the quotient of two increments:

$$m = \frac{\Delta y}{\Delta x} = \frac{f(x_1 + \Delta x) - f(x_1)}{\Delta x}.$$

This consideration of the slope, as the quotient of two increments, $m = \dfrac{\Delta y}{\Delta x}$ is really considering the slope as an *average rate of change*. That is, the slope of a straight line is the average rate of change of y with respect to x.

To summarize:

$$m = \frac{y_2 - y_1}{x_2 - x_1} = \frac{\Delta y}{\Delta x} = \frac{f(x_1 + \Delta x) - f(x_1)}{\Delta x} = \begin{array}{l} \text{average rate of change} \\ \text{of } y \text{ with respect to } x \end{array}$$

are all equivalent ways of identifying the same concept, *slope of a straight line*.

One particular slope that is of extreme importance to us is the slope of the secant line to a curve. First, we define the secant line and then consider its slope.

DEFINITION

Secant Line. Let P_1 and P_2 be two points on a curve. The straight line that passes through P_1 and P_2 is said to be a *secant line* to the curve at P_1.

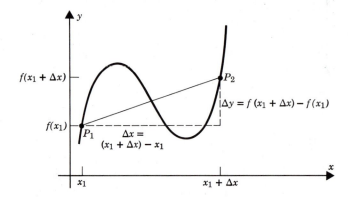

This drawing suggests that there exists an unlimited number of secant lines to the curve C at P_1. As soon as P_2 is shifted to another location on C, a different secant line is determined. From the previous comments on slope, it follows that the slope of secant line P_1P_2 equals

$$\frac{\Delta y}{\Delta x} = \frac{f(x_1 + \Delta x) - f(x_1)}{\Delta x}.$$

Difference Quotients

Previously, we mentioned the average rate of change of y with respect to the corresponding change in x. We formally state this definition.

DEFINITION

Average Rate of Change. The function f is defined by $y = f(x)$. When x changes from an initial x_1 to a final value $x_1 + \Delta x$, the corresponding values of y change from $y_1 = f(x_1)$ to $y_1 + \Delta y = f(x_1 + \Delta x)$. The *average rate of change* of y with respect to x, as x changes from x_1 to $x_1 + \Delta x$, is the change in y divided by the change in x or

$$\frac{\Delta y}{\Delta x} = \frac{(y_1 + \Delta y) - y_1}{\Delta x} = \frac{f(x_1 + \Delta x) - f(x_1)}{\Delta x}.$$

This fraction is also called a Difference Quotient or Newton's Quotient. Therefore, we have

$$\begin{array}{l} \text{Average rate of change} \\ \text{of } y \text{ with respect to} \\ x, \text{ as } x \text{ changes from} \\ x_1 \text{ to } x_1 + \Delta x \end{array} = \begin{array}{l} \text{A} \\ \text{Difference} \\ \text{Quotient} \end{array} = \frac{f(x_1 + \Delta x) - f(x_1)}{\Delta x}.$$

The average rate of change of y with respect to x is the quotient of Δy divided by Δx. Also, $\dfrac{\Delta y}{\Delta x}$ is the slope of secant line P_1P_2.

Example 5

Purpose To find an average rate of change:

Problem If the function f is defined by $y = f(x) = \dfrac{1}{x}$, find the average rate of change of y with respect to x, as x changes from x_1 to $x_1 + \Delta x$.

Solution Form the difference quotient:

$$\frac{\Delta y}{\Delta x} = \frac{f(x_1 + \Delta x) - f(x_1)}{\Delta x}$$

$$= \frac{\dfrac{1}{x_1 + \Delta x} - \dfrac{1}{x_1}}{\Delta x}$$

$$= \frac{\dfrac{x_1 - (x_1 + \Delta x)}{(x_1)(x_1 + \Delta x)}}{\Delta x}$$

$$= \frac{-\Delta x}{(\Delta x)(x_1)(x_1 + \Delta x)}$$

$$= \frac{-1}{(x_1)(x_1 + \Delta x)}$$

Graphically (if $x > 0$ and $\Delta x > 0$):

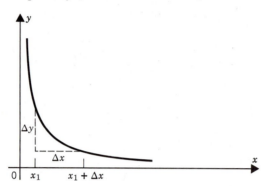

Example 6

Purpose To apply an average rate of change:

Problem A $C(x) = 2x^2 + .5x + 10$ is a cost function that, in dollars, gives the total cost of producing x units of a commodity. Find the average rate of change of total cost with respect to x, as x changes from x_1 to $x_1 + \Delta x$.

Solution $\dfrac{C(x_1 + \Delta x) - C(x_1)}{\Delta x}$

$$= \frac{[2(x_1 + \Delta x)^2 + .5(x_1 + \Delta x) + 10] - [2x_1^2 + .5x_1 + 10]}{\Delta x}$$

$$= \frac{2x_1^2 + 4x_1 \Delta x + 2(\Delta x)^2 + .5x_1 + .5 \Delta x + 10 - 2x_1^2 - .5x_1 - 10}{\Delta x}$$

$$= \frac{4x_1 \Delta x + 2(\Delta x)^2 + .5 \Delta x}{\Delta x}$$

$$= 4x_1 + 2 \Delta x + .5$$

Problem B In a controlled experiment the total area utilized by an algae growth is noted after each hour. A student obtains the following data.

Area of algae growth (cm^2)	320	500	600	540	504
Time (hours)	0	12	24	36	48

(a) Determine the average rate of change in area from $t = 0$ to $t = 12$.
(b) Determine the average rate of change in area from $t = 24$ to $t = 36$.
(c) Determine the average rate of change in area from $t = 24$ to $t = 48$.
(d) What does a negative rate of change mean?

Solution	Δ area	Δt	$\dfrac{\Delta \text{ area}}{\Delta t}$
(a)	$500 - 320 = 180$	$12 - 0 = 12$	$\dfrac{180}{12} = 15$
(b)	$540 - 600 = -60$	$36 - 24 = 12$	$\dfrac{-60}{12} = -5$
(c)	$504 - 600 = -96$	$48 - 24 = 24$	$\dfrac{-96}{24} = -4$

(d) A negative rate of change indicates a decrease in the amount of algae present.

Example 7

Purpose   To find an average velocity:

Problem   A particle of blood moves along an artery in such a manner that its distance, s, from a reference point is given by $s = f(t) = 8t$ (where t is the time expressed in seconds). Find the average velocity of the particle of blood in the time interval from t_1 to $t_1 + \Delta t$ seconds.

Solution   The average velocity is a special case of the difference quotient. The average velocity is the average rate of change of distance with respect to time.

$$\frac{\Delta s}{\Delta t} = \frac{f(t_1 + \Delta t) - f(t_1)}{\Delta t}$$

$$= \frac{8(t_1 + \Delta t) - 8t_1}{\Delta t}$$

$$= \frac{8t_1 + 8\,\Delta t - 8t_1}{\Delta t} = 8$$

There are many instances in mathematics where we use the concept of average rate of change. For example, J. Doe began a journey at 8 a.m. and arrived at 5 p.m. the same day. Initially, his car's speedometer read 30,000 miles. On arriving, the reading was 30,450 miles. What was his average velocity?

$$\Delta s = \text{change in distance} = 30{,}450 - 30{,}000 = 450 \text{ miles}$$

$$\Delta t = \text{change in time} = 5 \text{ p.m.} - 8 \text{ a.m.} = 9 \text{ hours}$$

$$\text{Average velocity} = \frac{\Delta s}{\Delta t} = \frac{450 \text{ miles}}{9 \text{ hours}} = 50 \text{ miles per hour}$$

It is evident that we don't mean that J. Doe was always traveling at 50 miles per hour. He had to accelerate and brake. His velocity varied but, on the average, it was 50 miles per hour.

Here, we see one of the problems that the calculus will help us solve. That is, although averages are useful, they do not give us the exact behavior at a particular instant. For example, knowing an individual's average velocity is 50 miles per hour does not tell us his exact velocity at 1 p.m. What we need is not an average velocity but an instantaneous velocity. In other applications, it will turn out that average rates of change will not be useful; what we need is an instantaneous rate of change. To progress from average rate of change to exact or instantaneous rate of change, we must take the *limit* of the average rate of change as the change in the independent variable approaches zero.

Example 8

Purpose To motivate the need for a "limit" process to find an instantaneous rate of change:

Problem The function f defined by $s = f(t) = 10t^2$ ($0 \leq t \leq 4$) gives the straight line distance in miles of a trucker from his terminal after t hours have elapsed. Find the trucker's average velocity for the time intervals 2.8 to 3.0 hours, 2.9 to 3.0 hours, 3.0 to 3.1 hours, and 3.0 to 3.2 hours. From these results, estimate the truck's exact velocity (instantaneous velocity) at $t = 3$ hours.

Solution

t (hours)	2.8	2.9	3.0	3.1	3.2
s (miles)	78.4	84.1	90.0	96.1	102.4

Δt (hours)	Δs (miles)	$\dfrac{\Delta s}{\Delta t}$ (average velocity)
$3.0 - 2.8 = .2$	$90.0 - 78.4 = 11.6$	$\dfrac{11.6}{.2} = 58$ miles per hour
$3.0 - 2.9 = .1$	$90.0 - 84.1 = 5.9$	$\dfrac{5.9}{.1} = 59$ miles per hour
$3.1 - 3.0 = .1$	$96.1 - 90.0 = 6.1$	$\dfrac{6.1}{.1} = 61$ miles per hour
$3.2 - 3.0 = .2$	$102.4 - 90.0 = 12.4$	$\dfrac{12.4}{.2} = 62$ miles per hour

It appears that at $t = 3$ the exact velocity should be approximately 60 miles per hour. To obtain the exact velocity at $t = 3$ we need to observe the magnitudes of the average velocities, about 3, as the time interval, Δt, becomes shorter, that is, as Δt approaches zero, $\Delta t \rightarrow 0$. In the next section we will see that exact velocity at $t = 3$ is the $\lim\limits_{\Delta t \to 0} \dfrac{\Delta s}{\Delta t}$ where lim represents a new concept called the limit.

Exercise 2.2

1. Given the function f defined by $y = f(x) = 3x + 1$.
    (a) Find Δy if x changes from 3 to 3.5.
    (b) Find Δy if x changes from 3 to 2.5.
    (c) Find Δy if x changes from -2 to -1.
    (d) Find Δy if x changes from -2 to -2.5.
    (e) Find $\dfrac{\Delta y}{\Delta x}$ for part (a).
    (f) What is $\dfrac{\Delta y}{\Delta x}$ for parts (b), (c), (d)? Are you surprised?
    (g) What is x_1, Δx, and $x_1 + \Delta x$ for parts (b) and (c)?
    (h) Find the slope of this straight line by using the quotient of the incre-
    ments $\dfrac{\Delta y}{\Delta x} = \dfrac{f(x_1 + \Delta x) - f(x_1)}{\Delta x}$.
    (i) Graph the function.

2. Given the function f defined by $y = f(x) = 3 - 2x$.
    (a) Find Δy if x changes from 0 to $\frac{1}{2}$.
    (b) Find Δy if x changes from 0 to $-\frac{1}{2}$.
    (c) Find $\dfrac{\Delta y}{\Delta x}$ for parts (a) and (b).
    (d) What is x_1, Δx, and $x_1 + \Delta x$ for parts (a) and (b)?
    (e) Find the slope of this straight line by using the difference quotient.
    (f) Graph the function.

3. Given the function f defined by $f(x) = 6 + 5x - x^2$.
    (a) Find Δy if x changes from -1 to 1.
    (b) Find $\dfrac{\Delta y}{\Delta x}$ for part (a).
    (c) Construct the secant line from $P_1(-1,0)$ to $P_2(1,10)$. Find the slope of
    the secant line using $\dfrac{f(x_1 + \Delta x) - f(x_1)}{\Delta x}$.
    (d) Compare parts (b) and (c).
    (e) Find Δy if x changes from 2 to 4.
    (f) Find the slope of the secant line determined by $P_1(x_1,y_1)$ and $P_2(x_2,y_2)$
    where $x_1 = 2$ and $x_2 = 4$.

4. Given the function $y = f(x) = 4x + 7$.
    (a) Find Δy if x changes from 1 to 1.001.
    (b) Find Δy if x changes from 1 to 0.999.
    (c) Find $\dfrac{\Delta y}{\Delta x}$ for parts (a) and (b).

5. (Optional calculator problem) Given the function f defined by $y = f(x) = 5 - 6x^2$.
   (a) Find Δy if x changes from 2 to 2.05.
   (b) Find Δy if x changes from 2 to 1.95.
   (c) Find $\dfrac{\Delta y}{\Delta x}$ for parts (a) and (b).
   (d) Find Δy if x changes from 2 to 2.0005.
   (e) Find Δy if x changes from 2 to 1.9995.
   (f) Find $\dfrac{\Delta y}{\Delta x}$ for parts (d) and (e).
   (g) Approximate the exact change at $x = 2$.

6. Graph $y = f(x) = x^3 + 1$. Construct the secant line connecting the points $P_1(-2,-7)$ and $P_2(0,1)$. Find the slope of this secant line using $\dfrac{f(x_1 + \Delta x) - f(x_1)}{\Delta x}$.

In Problems 7 to 18, find the average rate of change of y with respect to x (or form the difference quotient) for the function f.

7. $y = f(x) = \frac{1}{2}x + 1$

8. $y = f(x) = \frac{1}{3}x + 4$

9. $y = f(x) = 5 - 4x$

10. $y = f(x) = x^2 - 7x + 16$.

11. $y = f(x) = x^2 + 6x + 5$

12. $y = f(x) = 3 + x - x^2$

13. $y = f(x) = \dfrac{3}{x}$

14. $y = f(x) = \dfrac{5}{x + 2}$

15. $y = f(x) = \dfrac{1}{x + 4}$

16. $y = f(x) = \sqrt{3 - 2x}$

17. $y = f(x) = \sqrt{2x + 5}$

18. $y = f(x) = \sqrt{1 - x}$

19. On June 12 the Dow-Jones hourly averages for 30 industrial stocks were:

11 a.m.	1179.70
12 noon	1180.24
1 p.m.	1178.05
2 p.m.	1179.22
3 p.m.	1181.07

   (a) Find the average rate of change between 12 noon and 1 p.m.
   (b) Find the average rate of change between 11 a.m. and 3 p.m.

20. A specialty company has determined the total daily cost for producing x units of their commodity to be $\$(\frac{1}{2}x^2 + x + 2)$. Find the average rate of change of cost with respect to x.

21. Given a linear cost function, $C(x) = ax + b$, $(a > 0, b > 0)$, find the

average rate of change of cost with respect to x, as x changes from x_1 to $x_1 + \Delta x$.

22. Given a quadratic cost function $C(x) = ax^2 + bx + c$, $(a > 0, b > 0, c > 0)$, find the average rate of change of cost with respect to x, as x changes from x_1 to $x_1 + \Delta x$.

23. The growth of a biological culture, y, increases from 4 square centimeters (cm^2) to 24 cm^2 as time, t, increases from 1 hour to 11 hours. Find the average rate of growth, $\dfrac{\Delta y}{\Delta t}$.

24. The distance an oxygen-enriched red blood cell travels from the heart after t seconds can be found by the equation $s = kt^2$. Find the expression for the average velocity using t_1 and $t_1 + \Delta t$. What is the average velocity when t changes from .01 to .02 seconds? When t changes from .02 to .04 seconds?

25. A trucker, starting at station A, travels along a straight turnpike. The distance from the starting point is given by $s = f(t) = 32t + 6$ (s, miles; t, hours). Find his average velocity (average velocity is the average rate of change of distance with respect to time).

26. To relate distance s to time t, the function $s = f(t) = 16t^2$ is given. Find $\dfrac{\Delta s}{\Delta t}$ when
    (a) t changes from t_1 to $t_1 + \Delta t$.
    (b) t changes from $t_1 = 2$ to $t_1 + \Delta t = 2.5$.

27. A moving van leaves Pittsburgh to make the five-hour trip to Detroit. After each hour the driver makes note of his distance. The total distance after a given hour can be found by the equation $s = \frac{1}{2}(-5t^2 + 115t)$.
    (a) Make a table listing all of the distances after each hour interval, from $t = 0$ to $t = 5$.
    (b) Find $\dfrac{\Delta s}{\Delta t}$ for each of these hourly intervals.
    (c) What is the van's average velocity for the entire trip?

28. A circle has a given radius r. When the radius changes from r_1 to $r_1 + \Delta r$, find (a) the change in area, (b) the change in circumference, (c) the average rate of change in area with respect to r, and (d) the average rate of change in circumference with respect to r.

29. If the radius of a circle is given to be 2 feet and the radius changes by .5 feet ($\Delta r = .5$), find (a) the change in area, and (b) the change in circumference. Also find (c) the average rate of change in area with respect to r, and (d) the average rate of change in circumference with respect to r.

30. A record and tapes store charted the following graph to relate total sales,

x, in thousands of dollars, to advertising costs, y, in thousands of dollars. Find the average rate of change of advertising costs with respect to sales when x changes from:

(a) 20 to 40   (b) 40 to 60   (c) 60 to 80.

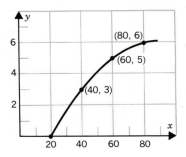

2.3   LIMIT OF A FUNCTION

We mentioned at the end of the last section that the limit is needed to pass from the average rate of change to the more useful concept of an exact or instantaneous rate of change. Indeed, it is this concept of the limit, imposed on the then existing mathematical system, that resulted in the "invention" of the calculus. Moreover, the limit is often called the glue that holds all of the calculus together.

Our approach to the limit will proceed according to the way that it was developed historically. The approach is intuitive, and we begin the development of the limit by the consideration of specific illustrations.

Consider the function f defined by $f(x) = 2x + 1$. Calculate the values of $f(x)$ for some values of x that are very close to but unequal to 3.

From the following table and the graph, it appears that if x is very close to 3 in value, then $2x + 1$ is very close to 7. We represent this statement in mathematical shorthand by the following.

If x is very close to 3 in value, then

$2x + 1$ is very close to 7 in value

or

limit of $(2x + 1)$, as x approaches 3, is 7

or

$$\lim_{x \to 3} (2x + 1) = 7.$$

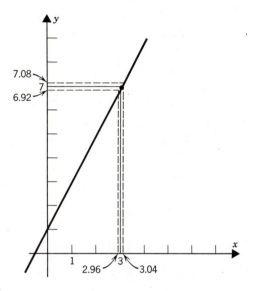

x	f(x) = 2x + 1
2.96	6.92
2.98	6.96
2.99	6.98
3.01	7.02
3.02	7.04
3.04	7.08

Such a limit as $\lim_{x \to 3} (2x + 1) = 7$ does not seem to have much practical significance, but this as well as the illustrations that follow are used to introduce the limit notion. As we progress, the practical significance will become clearer.

As an illustration of the limit of a rational function, consider $\lim_{x \to 2} \dfrac{x^2 - 1}{x^3 - 4}$.

Here we are concerned with the behavior of the fraction as x takes on values nearly equal to 2. Construct a table and examine the graph of this function near x equal to 2.

x	$f(x) = \dfrac{x^2 - 1}{x^3 - 4}$
1.98	$\dfrac{2.92}{3.76} \simeq .78$
1.99	$\dfrac{2.96}{3.88} \simeq .76$
2.01	$\dfrac{3.04}{4.12} \simeq .74$
2.02	$\dfrac{3.08}{4.24} \simeq .73$

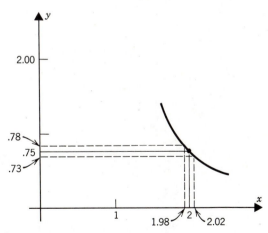

We should strongly suspect that if x is very near 2 in value, $f(x) = \dfrac{x^2 - 1}{x^3 - 4}$

will be very near to .75 or $\frac{3}{4}$ in value. Indeed, it does turn out that to be consistent, with the later definition of limit, we must say that

$$\lim_{x \to 2} \frac{x^2 - 1}{x^3 - 4} = \frac{3}{4}.$$

In the first limit problem, $\lim_{x \to 3} (2x + 1)$, the fact that the function does exist at 3 is entirely irrelevant from the limit standpoint. However, since $f(3)$ does exist, it gives us a convenient way to find this limit, without resorting to a table. That is, since $f(3) = 2(3) + 1 = 7$,

$$\lim_{x \to 3} (2x + 1) = 7.$$

In a similar manner for the $\lim_{x \to 2} \frac{x^2 - 1}{x^3 - 4}$, the fact that $\frac{x^2 - 1}{x^3 - 4}$ does exist at 2 is of no importance to the limit concept. However, let us take advantage of our good fortune and note that $\lim_{x \to 2} \frac{x^2 - 1}{x^3 - 4}$ can be found in the most expedient manner by substituting 2 for x. Thus,

$$\lim_{x \to 2} \frac{x^2 - 1}{x^3 - 4} = \frac{2^2 - 1}{2^3 - 4} = \frac{4 - 1}{8 - 4} = \frac{3}{4}.$$

Unfortunately, we are not always so lucky. Let us consider as another limit problem, $\lim_{x \to 2} \frac{x^2 - 4}{x - 2}$. The constructed table and graph are:

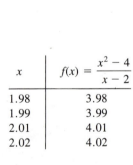

x	$f(x) = \dfrac{x^2 - 4}{x - 2}$
1.98	3.98
1.99	3.99
2.01	4.01
2.02	4.02

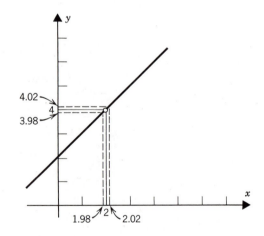

From the table and graph it appears that if x is very near 2 in value then $f(x) = \dfrac{x^2 - 4}{x - 2}$ is very near 4 in value. We cannot take the previous shortcut of substituting 2 for x. To do so would introduce $2 - 2$ or zero into the denominator, and we remember to *never divide by zero*. However, we do note that

$$\frac{x^2 - 4}{x - 2} = \frac{(x - 2)(x + 2)}{x - 2}.$$

Now, we can divide numerator and denominator of this fraction by $x - 2$, providing we are not attempting to divide by zero. Since x is only approaching 2, and is therefore unequal to 2, we are safe in doing this division. We then state that when $x \neq 2$:

$$\frac{x^2 - 4}{x - 2} = \frac{\cancel{(x - 2)}(x + 2)}{\cancel{x - 2}} = x + 2.$$

Organizing our work:

$$\lim_{x \to 2} \frac{x^2 - 4}{x - 2} = \lim_{x \to 2} \frac{\cancel{(x - 2)}(x + 2)}{\cancel{x - 2}} = \lim_{x \to 2} (x + 2) = 4$$

Notice that when we reach $\lim_{x \to 2} (x + 2)$ we are safe to substitute 2 for x. Notice also that the "lim" symbol is carried in the problem until the last step where the limit is actually found.

Example 1

Purpose To illustrate the mechanics of finding the limit and to give an intuitive meaning to limit:

Problem A Find $\lim_{x \to 10} (2x - 1)$ and give an intuitive meaning to your answer.

Solution $\lim_{x \to 10} (2x - 1) = 2(10) - 1 = 19$

When x is very close to 10, then $2x - 1$ has a value very close to 19.

Problem B Find $\lim_{x \to 3} \frac{x^2 - 5x + 6}{x - 3}$ and give an intuitive meaning to your answer.

Solution $\lim_{x \to 3} \frac{x^2 - 5x + 6}{x - 3} = \lim_{x \to 3} \frac{\cancel{(x - 3)}(x - 2)}{\cancel{x - 3}} = \lim_{x \to 3} (x - 2) = 1$

When x is very close to 3, but $x \neq 3$, then $\frac{x^2 - 5x + 6}{x - 3}$ has a value very close to 1.

We repeat for emphasis: The fact that $\frac{x^2 - 5x + 6}{x - 3}$ does not exist when $x = 3$ has no influence on whether the $\lim_{x \to 3} \frac{(x^2 - 5x + 6)}{x - 3}$ exists or does not exist.

Problem C Find $\lim_{\Delta x \to 0} \frac{2(\Delta x)^2 + 3 \Delta x}{\Delta x}$ and give an intuitive meaning to your answer.

Solution $\lim\limits_{\Delta x \to 0} \dfrac{2(\Delta x)^2 + 3\,\Delta x}{\Delta x} = \lim\limits_{\Delta x \to 0} \dfrac{\cancel{\Delta x}(2\,\Delta x + 3)}{\cancel{\Delta x}} = \lim\limits_{\Delta x \to 0} (2\,\Delta x + 3) = 3.$

When Δx is very close to 0, but $\Delta x \neq 0$, then $\dfrac{2(\Delta x)^2 + 3\,\Delta x}{\Delta x}$ has a value very close to 3.

In general, the limit problem is stated by $\lim\limits_{x \to a} f(x) = L$ where L represents a real number. Cauchy's definition of the limit is given in Appendix II. Here we state a working "definition" for the limit.

WORKING DEFINITION

Limit of a Function.   We say that $\lim\limits_{x \to a} f(x) = L$ if the numbers $f(x)$ remain arbitrarily close to the real number L, whenever x is very near to a in value but $x \neq a$.

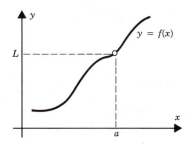

Whenever x is very near to a in value, $f(x)$ is close to L in value. Whether $f(x)$ does or does not exist at a has no influence on the $\lim\limits_{x \to a} f(x) = L$.

It should be noted that a limit does not always exist. Consider the following example.

Example 2

Purpose   To give an example of a function which has no limit as $x \to a$:

Problem A   Given the function f represented by

$$f(x) = \begin{cases} \frac{1}{2}x + 2, & x < 2 \\ x + 4, & x \geq 2. \end{cases}$$

The graph of f is given. Investigate to see if $\lim\limits_{x \to 2} f(x)$ exists.

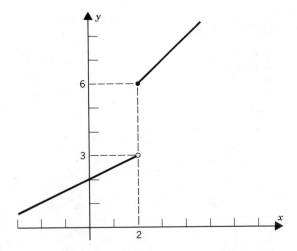

Solution As x approaches 2 from the right side of 2 (that is, $x > 2$ but approaching 2), the values of $f(x)$ approach 6. We say that the right-hand limit does exist and write

$$\lim_{x \to 2^+} f(x) = 6.$$

As x approaches 2 from the left side of $2(x < 2)$, the values of $f(x)$ approach 3. We say that the left-hand limit does exist and write

$$\lim_{x \to 2^-} f(x) = 3.$$

Since $\lim_{x \to 2^+} f(x) \neq \lim_{x \to 2^-} f(x)$, we will agree that $\lim_{x \to 2} f(x)$ does not exist. To generalize:

$$\lim_{x \to a} f(x) = L \text{ means } \lim_{x \to a^+} f(x) = L \text{ and } \lim_{x \to a^-} f(x) = L.$$

Problem B Given the function defined by $g(x) = \sqrt{x} + 3$. Find, if possible, $\lim_{x \to 0} g(x)$.

Solution As x approaches 0 and $x > 0$, $\sqrt{x}$ approaches 0, and $\lim_{x \to 0^+} (\sqrt{x} + 3) = 0 + 3 = 3$. However, as x approaches 0 and $x < 0$, $\sqrt{x}$ is not defined, and $\lim_{x \to 0^-} (\sqrt{x} + 3)$ does not exist. Therefore, $\lim_{x \to 0} g(x)$ does not exist, since $\lim_{x \to 0^+} g(x) \neq \lim_{x \to 0^-} g(x)$.

The Limit Theorems

There exist several theorems that we wish to use in our work with limits. Each of these theorems can be proved, using as a starting point the formal definition of limit. However, we will state the theorems without proofs and illustrate their use in finding limits.

THEOREMS ON LIMITS

Let f and g be functions defined by $y = f(x)$ and $y = g(x)$, respectively. Also, let $\lim\limits_{x \to a} f(x) = L$ and $\lim\limits_{x \to a} g(x) = M$.

1. $\lim\limits_{x \to a} c = c$. In words, the limit of a constant is the given constant.

   Examples. $\lim\limits_{x \to 4} 7 = 7$

   $\qquad\qquad \lim\limits_{x \to 15} 7 = 7$

2. $\lim\limits_{x \to a} [f(x) \pm g(x)] = \lim\limits_{x \to a} f(x) \pm \lim\limits_{x \to a} g(x) = L \pm M$. In words, the limit of a sum or difference is equal to the sum or difference of its limits.

   Examples. $\lim\limits_{x \to 4} (x + 7) = \lim\limits_{x \to 4} x + \lim\limits_{x \to 4} 7 = 4 + 7 = 11$

   $\qquad\qquad \lim\limits_{x \to 2} (x^2 + x - 11) = \lim\limits_{x \to 2} x^2 + \lim\limits_{x \to 2} x - \lim\limits_{x \to 2} 11$

   $\qquad\qquad\qquad\qquad\qquad = 4 + 2 - 11 = -5$

   Note. Theorem 2 is valid for the sum or difference of more than two functions.

3. $\lim\limits_{x \to a} [f(x) \cdot g(x)] = \lim\limits_{x \to a} f(x) \cdot \lim\limits_{x \to a} g(x) = L \cdot M$. In words, the limit of a product is equal to the product of the limits.

   Example. $\lim\limits_{x \to 4} (x + 7)(6 - x) = \lim\limits_{x \to 4} (x + 7) \cdot \lim\limits_{x \to 4} (6 - x)$

   $\qquad\qquad\qquad\qquad\qquad = 11 \cdot 2 = 22$

   As a special case of this theorem, let $f(x) = c$ (c, a constant), then

3a. $\lim\limits_{x \to a} c \cdot g(x) = \lim\limits_{x \to a} c \cdot \lim\limits_{x \to a} g(x) = c \cdot \lim\limits_{x \to a} g(x) = c \cdot M$.

   Example. $\lim\limits_{x \to 4} 3(x + 7) = 3 \cdot \lim\limits_{x \to 4} (x + 7) = 3 \cdot 11 = 33$

4. $\lim\limits_{x \to a} \left[\dfrac{f(x)}{g(x)} \right] = \dfrac{\lim\limits_{x \to a} f(x)}{\lim\limits_{x \to a} g(x)} = \dfrac{L}{M}$, $(M \neq 0)$. In words, the limit of a quotient is the quotient of the limits.

   Example. $\lim\limits_{x \to 3} \dfrac{1 - x^2}{8 + x} = \dfrac{\lim\limits_{x \to 3} (1 - x^2)}{\lim\limits_{x \to 3} (8 + x)} = \dfrac{-8}{11}$

It may be surprising to learn that Newton did not have a complete understanding of the limit. Many years later, Cauchy (1789–1857) put the concept of limit on a sound mathematical basis. In this chapter the approach is intuitive. Those of you who wish to consider the mathematically elegant Cauchy epsilon-delta definition of limit should now turn to Appendix II.

Exercise 2.3

1. Given the function f defined by $f(x) = x + 5$. Find $f(x)$ when $x = 1.9$, 1.99, 2.01, 2.1. When x is very close to 2, you would conclude that $f(x)$ is very close to what number?

2. Given the function f defined by $f(x) = 7 - 3x$. Find $f(x)$ when $x = 3.97$, 3.99, 4.01, 4.03. When x is very close to 4, you would conclude that $f(x)$ is very close to what number?

3. Given the function f defined by $f(x) = x^2$. Find $f(x)$ when $x = -3.2$, -3.1, -2.9, -2.8. When x is very close to -3, you would conclude that $f(x)$ is very close to what number?

4. The function f is defined by $f(x) = \dfrac{3x^2 - x - 2}{x - 1}$. Find $f(x)$ when $x = .8$, $.9$, 1.1, 1.2. When x is very close to 1, you would conclude that $f(x)$ is very close to what number?

5. (Optional calculator problem) Consider $f(x) = (1 + x)^{1/x}$. Let $x = 1$, $.1$, $.01$, $.001$, $.0001$, $.00001$, ... In this manner estimate $\lim\limits_{x \to 0^+} (1 + x)^{1/x}$.

6. Given the function f defined by $f(x) = \begin{cases} 4 - 3x, & x \neq 1 \\ 4 & , x = 1 \end{cases}$

   (a) Sketch the graph of f.
   (b) Find $\lim\limits_{x \to 1} f(x)$.
   (c) Show that $\lim\limits_{x \to 1} f(x) \neq f(1)$.

7. The function g is defined by $g(x) = \begin{cases} 1 + x^2, & x \leq 2 \\ 7 - x, & x > 2 \end{cases}$

   (a) Sketch the graph of g.
   (b) Find $\lim\limits_{x \to 2} g(x)$.
   (c) Does $\lim\limits_{x \to 2} g(x) = g(2)$?

8. Let F be the number of furnaces operating in a steel mill and $P(F)$ be the profit in thousands of dollars.

$$P(F) = \begin{cases} 2F + 1 \text{ if } 1 \leq F \leq 3 \\ 3F - 3 \text{ if } 3 < F \leq 5 \\ 4F - 10 \text{ if } 5 < F \leq 6 \end{cases}$$

   (a) Plot the graph of $P(F)$.
   (b) Determine: $\lim\limits_{F \to 2} P(F)$, $\quad \lim\limits_{F \to 3^+} P(F)$, $\quad \lim\limits_{F \to 3^-} P(F)$, $\quad \lim\limits_{F \to 3} P(F)$.

9. Let $P(t) = 3t - 6$ measure the profits a company makes if its production is t tons of goods per day. This equation applies only when $4 \leqslant t \leqslant 15$ tons. To produce more than 15 tons per day, the company must add a night shift; and $P(t) = 2t + 5$ when $15 < t \leqslant 30$ tons.
   (a) Plot the graph of $P(t)$ from $t = 4$ to $t = 25$ tons. $P(t)$ is measured in thousands of dollars.
   (b) Find (if possible): $\lim_{t \to 15^+} P(t)$, $\lim_{t \to 15^-} P(t)$, $\lim_{t \to 15} P(t)$.

10. The final height of a plant is dependent on the amount of sunlight it receives. The final height of a certain plant may be expressed as $H = \dfrac{t^2 - 22t - 48}{t - 24}$ where t is the hours of sunlight received each day. If the plant is given 24 hours of sunlight, what is the limit of the plant's height?

In Problems 11 to 42, find the limit of the function.

11. $\lim_{x \to 3} (x + 4)$

12. $\lim_{x \to 1} (5 - 9x)$

13. $\lim_{x \to -2} (3x + 10)$

14. $\lim_{x \to 0} 4x^2$

15. $\lim_{x \to 3} (x^2 - x - 7)$

16. $\lim_{x \to -1} (2x^2 - x + 4)$

17. $\lim_{x \to 4} (16 + x - x^3)$

18. $\lim_{x \to 0} (x^4 - 4x^3 + 3x^2 - x + 2)$

19. $\lim_{x \to 1/2} (2x^2 + 4x - 3)$

20. $\lim_{x \to 5} [(x + 5)(3x - 4)]$

21. $\lim_{x \to 1} [(x)(x^2 + 4)(3x - 2)]$

22. $\lim_{x \to 3} \dfrac{1}{x}$

23. $\lim_{x \to -3} \dfrac{1}{x}$

24. $\lim_{x \to 5} \sqrt{x + 4}$

25. $\lim_{x \to -4} \sqrt[3]{x + 3}$

26. $\lim_{x \to 3} (3x - 7)^3$

27. $\lim_{x \to 2} \dfrac{5x - 3}{x^2 + 1}$

28. $\lim_{x \to 3} \dfrac{x^2 - 9}{x + 3}$

29. $\lim_{x \to -3} \dfrac{x^2 - 9}{x + 3}$

30. $\lim_{\Delta x \to 0} \dfrac{-1}{(x_1)(x_1 + \Delta x)}$

31. $\lim_{\Delta t \to 0} \dfrac{16}{\sqrt{4 + \Delta t}}$

32. $\lim_{\Delta x \to 0} \dfrac{2x(\Delta x) - 7(\Delta x)^2}{\Delta x}$

33. $\lim_{x \to 2} \dfrac{3x^2 - 4x - 4}{x - 2}$

34. $\lim_{x \to -1/2} \dfrac{4x^2 - 4x - 3}{2x + 1}$

35. $\lim_{x \to 5} \dfrac{x^2 - 10x + 25}{x^2 - 25}$

36. $\lim_{x \to 0} \dfrac{x^3 - x^2 + x - 1}{x^2 + 1}$

37. $\lim_{x \to 1} \dfrac{x^3 - x^2 + x - 1}{x^2 - 1}$

38. $\lim\limits_{x\to 0} \dfrac{\sqrt{1+x}-1}{x}$

(*Hint.* Multiply the numerator and denominator by $\sqrt{1+x}+1$.)

39. $\lim\limits_{x\to 4} \dfrac{\sqrt{x}-2}{x^2-16}$

40. $\lim\limits_{x\to 1} \dfrac{x+2\sqrt{x}-3}{x^2+x-2}$

41. $\lim\limits_{x\to 0^+} (\sqrt{x}-4)$

42. $\lim\limits_{x\to 0} (\sqrt{x}-4)$

43. Given the function f graphed as:

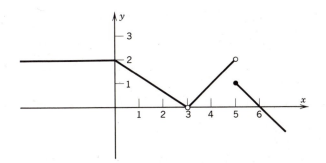

Find the limit, if it exists:

(a) $\lim\limits_{x\to 3^-} f(x) =$

(b) $\lim\limits_{x\to 3^+} f(x) =$

(c) $\lim\limits_{x\to 3} f(x) =$

(d) $\lim\limits_{x\to 5^-} f(x) =$

(e) $\lim\limits_{x\to 5^+} f(x) =$

(f) $\lim\limits_{x\to 5} f(x) =$

2.4 ADDITIONAL LIMITS

There are three other types of fractions whose limits we must agree on. These limits are different from the types just mentioned in that they involve the concept of a quantity growing or diminishing without bound. Therefore, it would take additional definitions, found in Appendix II, to establish these limits. We will use examples to study the behavior of these functions and to intuitively decide their limits.

Recall that $x \to a^+$ means x approaches a from the right-hand side of a, and $x \to a^-$ means x approaches a from the left-hand side of a.

Consider the behavior of $y = \dfrac{1}{x}$.

Graphically:

x	$y = \dfrac{1}{x}$
1000	.001
10	.1
.01	100
.001	1000
-1000	$-.001$
-10	$-.1$
$-.01$	-100
$-.001$	-1000

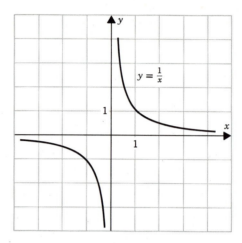

Determine $\displaystyle\lim_{x \to 0^+} \frac{1}{x}$.

Analysis: As x approaches zero from the right side of zero, y grows larger and larger. The closer x is to zero, the larger y is. This limit does not exist in the previous sense where y approaches some number L and remains near to L. However, we would like to describe the phenomenon by saying that y increases without bound or $y \to +\infty$ or $y \to \infty$. Therefore, we will agree

$$\lim_{x \to 0^+} \frac{1}{x} = +\infty.$$

Also,

$$\lim_{x \to 0^-} \frac{1}{x} = -\infty.$$

Notice that the symbol ∞ is not a number. ∞ or $+\infty$ is a symbol for "increasing without bound." As such, the symbol ∞, instead of being a point on the number line, represents a happening, the phenomenon of growing larger and larger. $-\infty$ is a symbol for "decreasing without bound."

By using the same graph, we can also determine answers for $\displaystyle\lim_{x \to \infty} \frac{1}{x}$ and $\displaystyle\lim_{x \to -\infty} \frac{1}{x}$.

Determine $\displaystyle\lim_{x \to \infty} \frac{1}{x}$.

Analysis: As x grows "larger and larger," the values of x are farther from the origin and the graph is nearly touching the x axis or $y \to 0$. Therefore, we will agree that

$$\lim_{x \to \infty} \frac{1}{x} = 0.$$

Also, as $x \to -\infty$, $y \to 0$ or $\frac{1}{x} \to 0$.

Therefore,

$$\lim_{x \to -\infty} \frac{1}{x} = 0.$$

In general, for $c =$ any constant, $\lim\limits_{x \to \infty} \dfrac{c}{x} = 0$ and $\lim\limits_{x \to -\infty} \dfrac{c}{x} = 0$.

The third new type of limit to consider is when the numerator and the denominator both $\to \pm\infty$.

Determine $\lim\limits_{x \to \infty} \dfrac{3x^2 - 2x + 7}{5x^2 - 8}$.

Analysis. Divide numerator and denominator by the highest power of x present. In this problem divide by x^2.

$$\lim_{x \to \infty} \frac{3x^2 - 2x + 7}{5x^2 - 8} = \lim_{x \to \infty} \frac{3 - \dfrac{2}{x} + \dfrac{7}{x^2}}{5 - \dfrac{8}{x^2}} = \frac{3}{5}.$$

This is true, since $-\dfrac{2}{x}$, $\dfrac{7}{x^2}$, and $-\dfrac{8}{x^2}$ each $\to 0$ as $x \to \infty$.

Example 1

Purpose To find limits by the use of these additional techniques:

Problem A Find $\lim\limits_{x \to 0^+} \dfrac{3}{x^2}$ and $\lim\limits_{x \to 0^-} \dfrac{3}{x^2}$.

Solution As $x \to 0^+$, $\dfrac{3}{x^2}$ increases without bound; or $\lim\limits_{x \to 0^+} \dfrac{3}{x^2} = \infty$.

As $x \to 0^-$, $\dfrac{3}{x^2}$ increases without bound; or $\lim\limits_{x \to 0^-} \dfrac{3}{x^2} = \infty$.

Problem B Find $\lim\limits_{x \to \pm\infty} \dfrac{-6}{x + 1}$.

Solution   As x increases without bound, the denominator increases and the fraction approaches 0.

$$\lim_{x \to +\infty} \frac{-6}{x + 1} = 0$$

$$\left(\text{if } x = 1000, \text{ then } \frac{-6}{x + 1} = \frac{-6}{1001}, \text{ a number near zero}\right)$$

$$\lim_{x \to -\infty} \frac{-6}{x + 1} = 0$$

$$\left(\text{if } x = -1000, \frac{-6}{-1000 + 1} = \frac{6}{999}, \text{ a number near zero}\right)$$

Problem C   Find $\lim\limits_{x \to \infty} \dfrac{2x^3 + x + 7}{6x^3 - 5x^2 + 11}$.

Solution   Divide the numerator and denominator by the highest power of x present, x^3.

$$\lim_{x \to \infty} \frac{2 + \dfrac{1}{x^2} + \dfrac{7}{x^3}}{6 - \dfrac{5}{x} + \dfrac{11}{x^3}} = \frac{\lim\limits_{x \to \infty}\left(2 + \dfrac{1}{x^2} + \dfrac{7}{x^3}\right)}{\lim\limits_{x \to \infty}\left(6 - \dfrac{5}{x} + \dfrac{11}{x^3}\right)} = \frac{2 + 0 + 0}{6 - 0 + 0} = \frac{1}{3}$$

Problem D   Find $\lim\limits_{x \to \infty} \dfrac{2x^2 + x + 7}{6x^3 - 5x^2 + 11}$.

Solution   Divide both numerator and denominator by the highest power of x present, x^3.

$$\lim_{x \to \infty} \frac{\dfrac{2}{x} + \dfrac{1}{x^2} + \dfrac{7}{x^3}}{6 - \dfrac{5}{x} + \dfrac{11}{x^3}} = \frac{0}{6} = 0$$

Review of Five Types of Limit

Let us review the five types of limits by stating the usual method of solution and showing another example for each type.

1. Numerator approaches a real number and denominator approaches a non-zero real number.
    Method of solution—substitution.

    Example. $\lim\limits_{x \to 4} \dfrac{2x}{x^2 - 3} = \dfrac{8}{13}$

2. Numerator and denominator both approach zero.
    Method of solution—factor numerator and denominator, divide by the

common factor, and then choose the appropriate method to take the limit, if possible.

Example. $\lim\limits_{x\to 2} \dfrac{5x^2 - 8x - 4}{x - 2} = \lim\limits_{x\to 2} \dfrac{(5x + 2)(x - 2)}{x - 2}$

$$= \lim\limits_{x\to 2} (5x + 2) = 12$$

3. Numerator approaches a nonzero number and denominator approaches zero.

   Method of solution—since the denominator is approaching zero, the fraction will either increase or decrease without bound and the answer is $+\infty$ or $-\infty$. To determine the correct sign, $+$ or $-$, try an appropriate value of x.

   Example. $\lim\limits_{x\to -2^+} \dfrac{x}{4 - x^2} = -\infty$

   $\underset{-2}{\xleftarrow{}}\left(\text{Let } x = -1.9, \text{ then } \dfrac{x}{4 - x^2} < 0.\right)$

4. Numerator approaches a real number and denominator approaches $+\infty$ or $-\infty$.

   Method of solution—answer is always zero.

   Example. $\lim\limits_{x\to\infty} \dfrac{3}{x^2 - 4} = 0$.

5. Numerator and denominator both approach $+\infty$ or $-\infty$.

   Method of solution—divide numerator and denominator by the highest power of the variable that appears and then proceed by choosing the appropriate technique to take the limit, if possible.

   Example. $\lim\limits_{x\to\infty} \dfrac{7x^2 - 2}{3x^2 + 10x - 100} = \lim\limits_{x\to\infty} \dfrac{7 - \dfrac{2}{x^2}}{3 + \dfrac{10}{x} - \dfrac{100}{x^2}} = \dfrac{7}{3}$.

Problems that give additional practice in these five types of limits are found in Exercise 2.4–2.5. These problems may be considered either before or after the concept of continuity.

2.5 CONTINUOUS FUNCTIONS

An important use of limits is in the study of continuous functions. Many of the theorems of the calculus are simply not true unless we are dealing with a function that may be considered as continuous. Therefore, we must be able to readily identify a continuous function.

Consider the graph of a function as shown in Graph I.

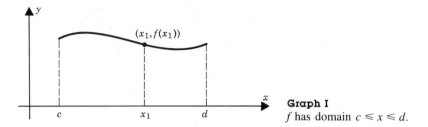

Graph I
f has domain $c \leq x \leq d$.

This is a well-behaved function and, since it is a function, to each x in its domain is associated one and only one value $f(x)$. Also, it can be graphed without ever removing your pencil from the paper.

Now consider two other graphs of functions: Graph II and Graph III.

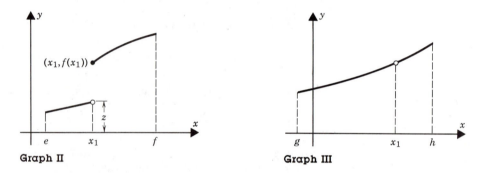

Graph II   **Graph III**

In graph II, we have a function whose domain is $e \leq x \leq f$. This function is defined at $x = x_1$. In Graph III, we have a function whose domain excludes $x = x_1$. The function is not defined at $x = x_1$. This means that $f(x_1)$ does not exist. Also, notice that in Graphs II and III we cannot draw the graph without removing our pencil from the paper.

In all three cases (Graphs I, II, and III), we are dealing with functions. However, Graph I behaves well, whereas Graphs II and III misbehave. We would like a way to partition functions into two sets: one having members that behave as Graph I, and the other having members that behave as Graph II or Graph III. This is done by defining a continuous function.

DEFINITION

Continuous Function at $x = x_1$. Let the function f have a value at $x = x_1$ and at all points in some interval containing x_1. Then f is said to be *continuous at $x = x_1$ if* $\lim\limits_{x \to x_1} f(x) = f(x_1)$.

Comment on this definition—For f to be continuous at $x = x_1$, three criteria must be met:

1. $f(x_1)$ must exist, and

2. $\lim\limits_{x \to x_1} f(x)$ must exist, and

3. $\lim\limits_{x \to x_1} f(x) = f(x_1)$.

If any one of these three criteria is not met, then f is *discontinuous at* $x = x_1$.

In addition, the $\lim\limits_{x \to x_1} f(x) = f(x_1)$ is equivalent to the requirement that the $\lim\limits_{\Delta x \to 0} f(x_1 + \Delta x) = f(x_1)$. From the definition, let $x = x_1 + \Delta x$ and then $x \to x_1$ is equivalent to $x_1 + \Delta x \to x_1$ or $\Delta x \to 0$. These substitutions yield an alternate definition.

ALTERNATE DEFINITION

Continuous Function at $x = x_1$. Let f be defined at $x = x_1$ and at all points in some interval containing x_1. Then f is said to be *continuous at* $x = x_1$ if $\lim\limits_{\Delta x \to 0} f(x_1 + \Delta x) = f(x_1)$.

DEFINITION

Continuous Function on an Interval. If f is continuous at every point in an interval, then f is said to be *continuous on that interval*.

Now, let us return to Graphs I, II, and III.

I. f is continuous at $x = x_1$, since $\lim\limits_{x \to x_1} f(x) = f(x_1)$. Indeed, f is continuous over $c \leqslant x \leqslant d$.

II. f is discontinuous at $x = x_1$, since $\lim\limits_{x \to x_1} f(x)$ does not exist. Notice why this limit does not exist. $\lim\limits_{x \to x_1^-} f(x) = z$ while $\lim\limits_{x \to x_1^+} f(x) = f(x_1)$ and since the left-hand limit, z, does not equal the right-hand limit, $f(x_1)$, then $\lim\limits_{x \to x_1} f(x)$ cannot exist. However, f is continuous at all other points of $e \leqslant x \leqslant f$. Such a discontinuity, as occurs in Graph II at x_1, is called a *jump discontinuity*.

III. f is discontinuous at $x = x_1$, since $f(x_1)$ does not exist. However, f would be continuous at all other points in $g \leqslant x \leqslant h$. Such a discontinuity, as occurs in Graph III at x_1, is called a *point discontinuity*.

Example 1

Purpose To classify a function as being continuous or discontinuous at $x = x_1$ by making use of definitions:

Problem A If $f(x) = \dfrac{x^2 - 9}{x - 3}$, is f continuous at $x = 1$? Is f continuous at $x = 3$?

Solution The function is continuous at $x = 1$.

$$f(1) = \frac{1^2 - 9}{1 - 3} = 4$$

$$\lim_{x \to 1} f(x) = \lim_{x \to 1} \frac{x^2 - 9}{x - 3} = \lim_{x \to 1} \frac{(x + 3)(x - 3)}{x - 3} = 4$$

And, $\lim\limits_{x \to 1} f(x) = f(1) = 4$.

The function is not continuous at $x = 3$, since $f(3)$ does not exist. Remember, division by zero is impossible.

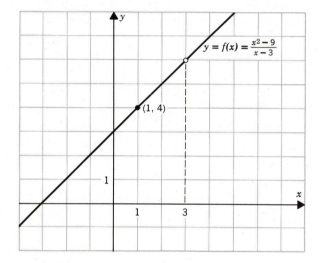

Problem B If $g(x) = \begin{cases} \dfrac{x^2 - 9}{x - 3} & \text{if } x \neq 3 \\ 6 & \text{if } x = 3 \end{cases}$, is g continuous at $x = 3$?

Solution Yes!

$g(3) = 6$

$$\lim_{x \to 3} g(x) = \lim_{x \to 3} \frac{x^2 - 9}{x - 3} = \lim_{x \to 3} \frac{(x + 3)(x - 3)}{x - 3} = 6$$

Therefore, $\lim\limits_{x \to 3} g(x) = g(3) = 6$.

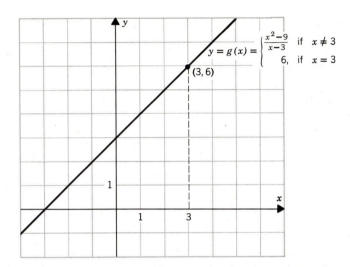

$$y = g(x) = \begin{cases} \dfrac{x^2-9}{x-3} & \text{if } x \neq 3 \\ 6, & \text{if } x = 3 \end{cases}$$

Problem C Let a function f be specified by the statement: in 1985, the postage for first-class mail was 22 cents for the first ounce or fraction thereof and 17 cents for each additional ounce or fraction thereof. Is this function continuous at $x = 3$? At $x = .5$? Here x represents the weight in ounces for $0 \leq x \leq 4$.

Solution

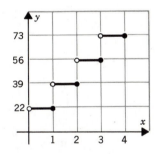

At $x = 3$, $f(3) = 56$. However, $\lim\limits_{x \to 3} f(x)$ does not exist, since $\lim\limits_{x \to 3^+} f(x) = 73$ and $\lim\limits_{x \to 3^-} f(x) = 56$. Therefore, f is discontinuous at $x = 3$. At $x = .5$, $f(.5) = 22$. Moreover, $\lim\limits_{x \to .5} f(x)$ exists and is equal to 22. Since $\lim\limits_{x \to .5} f(x) = f(.5)$, f is continuous at $x = .5$.

At what other values of x will f be continuous? Discontinuous?

Problem D If $h(x) = \begin{cases} 1 \text{ if } x \text{ is an integer} \\ -1 \text{ if } x \text{ is not an integer} \end{cases}$, is h continuous at $x = 1$? at $x = -\frac{5}{2}$?

Solution At $x = 1$,

$$h(1) = 1, \lim_{x \to 1} h(x) = -1 \text{ and } \lim_{x \to 1} h(x) \neq h(1).$$

Therefore, h is not continuous at $x = 1$.

At $x = -\frac{5}{2}$,

$h(-\frac{5}{2}) = -1$, $\lim\limits_{x \to -5/2} h(x) = -1$, and $\lim\limits_{x \to -5/2} h(x) = h(-\frac{5}{2})$

Therefore, h is continuous at $x = -\frac{5}{2}$.
Graph of h:

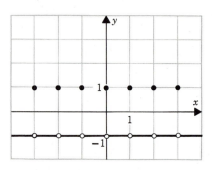

Exercise 2.4–2.5

In Problems 1 to 26, evaluate the limit.

1. $\lim\limits_{x \to 2} \dfrac{3x^2 + 1}{4x - 3}$

2. $\lim\limits_{t \to -2} \dfrac{2t - 3}{t^3 + 2}$

3. $\lim\limits_{w \to 3} \dfrac{w^2 - 9}{w - 3}$

4. $\lim\limits_{x \to -1} \dfrac{x^2 - x - 2}{x + 1}$

5. $\lim\limits_{\Delta x \to 0} (2x_1 + 3\,\Delta x)$

6. $\lim\limits_{x \to 2} \sqrt{\dfrac{x^2 + 7x - 6}{x + 1}}$

7. $\lim\limits_{\Delta x \to 0} \dfrac{4\,\Delta x + (\Delta x)^2}{\Delta x}$

8. $\lim\limits_{x \to 3^+} \dfrac{x + 4}{x - 3}$

9. $\lim\limits_{x \to 1^+} \dfrac{7}{1 - x^2}$

10. $\lim\limits_{x \to 3^-} \dfrac{2}{(x - 3)^4}$

11. $\lim\limits_{v \to \infty} \dfrac{17}{v^3 + 11}$

12. $\lim\limits_{v \to -\infty} \dfrac{17}{v^3 + 11}$

13. $\lim\limits_{t \to \infty} \dfrac{2t - 9}{t + 3}$

14. $\lim\limits_{x \to \infty} \dfrac{4x^3 - 2x + 5}{3x^3 + 6x^2 - 1}$

15. $\lim\limits_{x \to \infty} \dfrac{x + 9}{x^2 + 9}$

16. $\lim\limits_{x \to -\infty} \dfrac{5x^2 + 1}{3x^3 - 4}$

17. $\lim\limits_{x \to 0} \dfrac{4x^2}{x - 4}$

18. $\lim\limits_{x \to \infty} \dfrac{4x^2}{x - 4}$

19. $\lim\limits_{x \to 0} \dfrac{4x}{x^2 - 4}$

20. $\lim\limits_{x \to \infty} \dfrac{4x}{x^2 - 4}$

21. $\lim\limits_{x \to 2} \dfrac{x - 2}{x^2 - 2x}$

22. $\lim\limits_{x \to 3} \dfrac{x - 3}{x^2 - 2x - 3}$

23. $\lim\limits_{x \to 2} \dfrac{\dfrac{1}{x} - \dfrac{1}{2}}{x - 2}$

24. $\lim\limits_{\Delta x \to 0} \dfrac{\dfrac{1}{1 + \Delta x} - 1}{\Delta x}$

25. $\lim\limits_{x \to 5} \dfrac{\sqrt{x} - \sqrt{5}}{x - 5}$

26. $\lim\limits_{x \to 3} \dfrac{\sqrt{x} - \sqrt{3}}{x - 3}$

27. Given $g(x) = \dfrac{x^2 + x - 6}{x^2 - 4}$.

Find:

(a) $\lim\limits_{x \to 0} g(x)$

(b) $\lim\limits_{x \to 1} g(x)$

(c) $\lim\limits_{x \to 2} g(x)$

(d) $\lim\limits_{x \to -3} g(x)$

(e) $\lim\limits_{x \to -2^+} g(x)$

(f) $\lim\limits_{x \to -2^-} g(x)$

(g) $\lim\limits_{x \to \infty} g(x)$

28. Given $h(x) = \dfrac{2x^2 - 7x + 3}{x^2 - 2x - 3}$.

Find:

(a) $\lim\limits_{x \to 0} h(x)$

(b) $\lim\limits_{x \to 3} h(x)$

(c) $\lim\limits_{x \to \infty} h(x)$

(d) $\lim\limits_{x \to 1} h(x)$

(e) $\lim\limits_{x \to -1^-} h(x)$

(f) $\lim\limits_{x \to 1/2} h(x)$

29. Given $f(x) = \dfrac{(3x - 4)^2(x - 1)}{(x + 7)(x - 1)}$.

Find:

(a) $\lim\limits_{x \to -1} f(x)$

(b) $\lim\limits_{x \to 1} f(x)$

(c) $\lim\limits_{x \to -7^+} f(x)$

(d) $\lim\limits_{x \to -7^-} f(x)$

(e) $\lim\limits_{x \to \infty} f(x)$

(f) $\lim\limits_{x \to 4/3} f(x)$

(g) $\lim\limits_{x \to 7} f(x)$

(h) $\lim\limits_{x \to 0} f(x)$

30. Consider the graph of the function f shown on the following page. Find, if possible, $\lim\limits_{x \to 2} f(x)$.

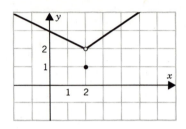

31. Given $f(x) = \begin{cases} x^2 + 1, & x < 1 \\ 2, & x = 1 \\ x + 2, & x > 1 \end{cases}$

Find, if possible, (a) $\lim\limits_{x\to 2} f(x)$, and (b) $\lim\limits_{x\to 1} f(x)$.

32. If $g(x) = \begin{cases} x, & x < 0 \\ -1, & x = 0, \\ -x, & x > 0 \end{cases}$ $\lim\limits_{x\to 0} g(x) =$

33. A psychologist is studying the effects of alcohol on a mouse that is trying to find its way through a maze leading to a piece of cheese. The mouse is awarded a score of one to ten depending on how close to the cheese the mouse comes in a certain period of time. One experiment gives the following results:

$t = cm^3$ of Alcohol	$s =$ Score
0.1 to 3.0	8
3.0 to 6.0	6
6.0 to 8.4	5

Plot score versus cubic centimeters of alcohol and determine the following limits (if possible).

(a) $\lim\limits_{t\to 3^-} s,$ $\lim\limits_{t\to 3^+} s,$ $\lim\limits_{t\to 3} s$

(b) $\lim\limits_{t\to 7^-} s,$ $\lim\limits_{t\to 7^+} s,$ $\lim\limits_{t\to 7} s$

34. A company finds its profit function to be $P(x) = \dfrac{26x - 25}{x}$ where profit is in thousands of dollars and x is the number of items in hundreds sold. What is the limit of the company's profit as the number of items sold increases without bound (approaches infinity)?

35. The time it takes a rat to run through a maze may be estimated by the equation $t = 3 + \dfrac{20}{n + 1}$ where t is the time in minutes and n is the number of previous trials. As the number of trials get very large, what time limit does the rat's performance reach?

36. By use of the following two problems show that if both the numerator and denominator of a fraction approach infinity the answer is not necessarily one.

(a) $\lim\limits_{x\to\infty} \dfrac{3x^2 - 2}{7x^2 - 1}$

(b) $\lim\limits_{x\to\infty} \dfrac{2x^3 + 3}{9x^5 - 4}$

In Problems 37 to 46, state the value(s) (if any) of x where the function is discontinuous.

37. $f(x) = 3x^2 - 4x + 5$

38. $f(x) = \dfrac{2}{x - 2}$

39. $g(x) = \dfrac{x}{4 - x^2}$

40. $h(x) = \dfrac{9 - x^2}{x + 3}$

41. $k(x) = \dfrac{x + 3}{x^2 + 9}$

42. $g(x) = \begin{cases} x & \text{if } x > 2 \\ x - 1 & \text{if } x \leqslant 2 \end{cases}$

43. $f(x) = \begin{cases} -2x + 3 & \text{if } x \geqslant 0 \\ \frac{1}{2}x + 1 & \text{if } x < 0 \end{cases}$

44. $g(x) = \begin{cases} -2x + 3 & \text{if } x \geqslant 0 \\ \frac{1}{2}x + 3 & \text{if } x < 0 \end{cases}$

45. $k(x) = \dfrac{3x^2 - 4x + 3}{x^2}$

46. $f(x) = \sqrt{x^2 + 7}$

47. The Ohio Highway Patrol uses the following function to relate the weight of a truck traveling on the turnpike to the taxes charged per mile of travel.

W = weight (tons)	10 to 15	15 to 18	18 to 22	22-up
T = taxes (cents per mile)	4	5	6	7

(a) Draw the graph of this function.

(b) For what values of W is this function discontinuous?

48. A certain commodity is sold by the gallon, x. The total cost, in dollars, of purchasing x gallons may be calculated by using the cost function, $C(x)$.

$$C(x) = \begin{cases} 3x & \text{if} & 0 \leqslant x \leqslant 100 \\ 100 + 2x & \text{if} & 100 < x \leqslant 500 \\ 500 + x & \text{if} & x > 500 \end{cases}$$

Graph this cost function. Is the cost function continuous at $x = 100$? At $x = 500$?

49. The cost function for a company is $C(x) = \dfrac{x^3 - 5x^2 + 13x - 14}{x - 2}$. For what values of x is this function continuous?

50. Let a function f be given by: A supply company sells buffered iodine by volume according to the following table.

	Less than		Greater than
Volume purchased (pint), x	5	5 to 10	10
Price per pint, y	$3.00	$2.50	$2.00

(a) Graph the function.

(b) Is this function continuous at $x = 7$? At $x = 10$?

51. (a) If $f(x) = \dfrac{x^2 - 25}{x - 5}$, is f continuous at $x = 5$?

(b) If $g(x) = \begin{cases} \dfrac{x^2 - 25}{x - 5} & \text{if } x \neq 5 \\ -10 & \text{if } x = 5 \end{cases}$, is g continuous at $x = 5$?

(c) If $h(x) = \begin{cases} \dfrac{x^2 - 25}{x - 5} & \text{if } x \neq 5 \\ 10 & \text{if } x = 5 \end{cases}$, is h continuous at $x = 5$?

If your answer for (a), (b), or (c) is no, state which of the continuity conditions is violated. Graph each function.

52. (a) If $f(x) = \dfrac{x^2 + 2x - 3}{x - 1}$, is f continuous at $x = 1$?

(b) If $g(x) = \begin{cases} \dfrac{x^2 + 2x - 3}{x - 1} & \text{if } x \neq 1 \\ 0 & \text{if } x = 1 \end{cases}$, is g continuous at $x = 1$?

(c) If $h(x) = \begin{cases} \dfrac{x^2 + 2x - 3}{x - 1} & \text{if } x \neq 1 \\ 4 & \text{if } x = 1 \end{cases}$, is h continuous at $x = 1$?

If your answer to (a), (b), or (c) is no, state which of the continuity conditions is violated. Graph each function.

53. $y = [x]$, the greatest integer function. It reads, y equals the greatest integer not greater than x.

For example:

$$x = 2, \qquad y = [2] = 2$$
$$x = \tfrac{3}{2}, \qquad y = [\tfrac{3}{2}] = 1$$
$$x = -\tfrac{1}{3}, \qquad y = [-\tfrac{1}{3}] = -1$$

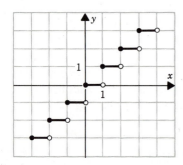

Where is the function discontinuous? Why?

This type of function is called a step function. Many cost functions are of this type.

54. Which of the following functions has a limit at $x = x_1$? Which of the following functions is continuous at $x = x_1$?

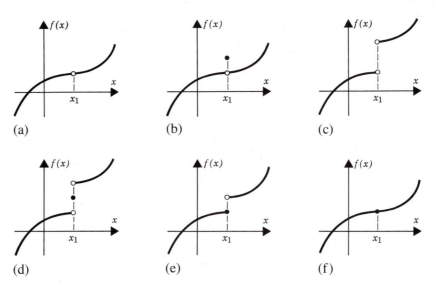

(a)   (b)   (c)

(d)   (e)   (f)

55. Consider the function $f(x) = a_0x^n + a_1x^{n-1} + a_2x^{n-2} + \cdots + a_n,$ where n is a positive integer and $a_1, a_2, \ldots, a_n$ are rational numbers. This type of function is called a polynomial.

(a) Does $\lim f(x)$ exist for all x? Discuss.

(b) Is this function continuous for all x? Discuss.

2.6 TANGENT LINE TO A CURVE AND THE DERIVATIVE

Now that we have investigated limits, we are ready to return to our basic problem of finding an exact rate of change. You recall that some of our reasons for interest in exact rates include wishing to find the slope of the tangent line to a curve, the exact rates of population growth, and instantaneous velocities. Before we find exact rates of change, let us briefly review the important concepts with which we will be working.

Thus far in this chapter the following topics and their mathematical symbols have been considered:

	Symbol
Change in x or increment in x	$\Delta x = (x_1 + \Delta x) - x_1$
Change in y or increment in y	$\Delta y = f(x_1 + \Delta x) - f(x_1)$
Average rate of change in y with respect to x, as x changes from x_1 to $x_1 + \Delta x$.	$\dfrac{\Delta y}{\Delta x} = \dfrac{f(x_1 + \Delta x) - f(x_1)}{\Delta x}$

This is a Difference Quotient

We have mentioned that the average rate of change, $\dfrac{\Delta y}{\Delta x}$, is often not adequate for solving a particular problem. Instead of an average rate of change over an interval of x, the exact rate of change at a particular value $x = x_1$ is needed. Let us return to Basic Problem I and show that the average rate of change is not sufficient to solve this problem. You recall Basic Problem I—Given any curve C, find the tangent line, t, to C at point P.

To find the equation of the tangent line, its slope is needed. The concept of the tangent line to the curve at a point P on the curve and a method for determining its slope are now developed.

Given a curve C with $P_1(x_1, f(x_1))$ and $P_2(x_1 + \Delta x, f(x_1 + \Delta x))$ distinct points of C, draw a secant line P_1P_2. Let P_2 approach P_1 along C. Every time P_2 moves to a new position draw a new secant line (P_2 could also lie to the left of P_1 on C). The slope of the secant line P_1P_2 is

$$\frac{\Delta y}{\Delta x} = \frac{f(x_1 + \Delta x) - f(x_1)}{\Delta x}.$$

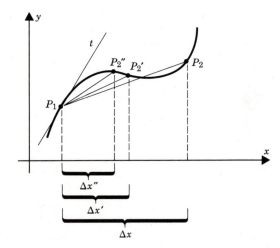

Suppose that as P_2 approaches P_1 or $\Delta x \to 0$ the secant line P_1P_2 approaches a limiting position. It is this limiting position that we wish to use as the tangent line to C at P_1. Therefore, we want the slope of the tangent line to C at P_1 to be $\lim\limits_{\Delta x \to 0} \dfrac{\Delta y}{\Delta x} = \lim\limits_{\Delta x \to 0} \dfrac{f(x_1 + \Delta x) - f(x_1)}{\Delta x}$. This motivates the following definition.

DEFINITION

Tangent Line to a Curve. If a function f has curve C for its graph then the *tangent line* to C at $P_1(x_1, f(x_1))$ is the straight line through P_1, which has slope equal to

$$\lim_{\Delta x \to 0} \frac{f(x_1 + \Delta x) - f(x_1)}{\Delta x}$$

(if this limit is a real number).[1]

Some notes on the above:

1. Students often confuse the tangent line and its slope. The tangent line to C at P_1 is a unique straight line, which is the limiting position of secant line $P_1 P_2$ as $P_2 \to P_1$ or as $\Delta x \to 0$. The slope is a property or feature of this tangent line. The slope is a number that gives us a measure of the slant of the tangent line. Therefore, we must be cautious not to use *tangent line* and *slope of the tangent line* synonymously, just as we do not say that Mary Jones and her blue eyes mean the same thing.

2. No longer is the definition of tangent line to a circle, that is, a line that touches the circle in one and only one point, of any use to us. A tangent line to C at P_1 may touch C at many points (see the following curve).

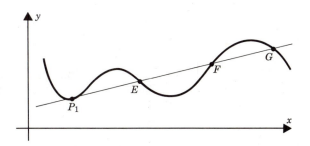

This tangent line touches the curve at four points, P_1, E, F, and G.

---

[1]If this limit is ∞, then $x = x_1$ is the equation of the tangent line to C at P_1.

Example 1

Purpose   To find the slope and the equation of the tangent line to a curve at a point on the curve:

Problem   Find the slope of the tangent line and the equation of the tangent line to $y = x^2$ at the point $(1,1)$.

Solution   The slope of the tangent line to $y = x^2$ at $x = 1$ is

$$\lim_{\Delta x \to 0} \frac{f(1 + \Delta x) - f(1)}{\Delta x}$$

$$= \lim_{\Delta x \to 0} \frac{(1 + \Delta x)^2 - (1)^2}{\Delta x}$$

$$= \lim_{\Delta x \to 0} \frac{1 + 2\,\Delta x + (\Delta x)^2 - 1}{\Delta x}$$

$$= \lim_{\Delta x \to 0} \frac{\Delta x(2 + \Delta x)}{\Delta x}$$

$$= \lim_{\Delta x \to 0} (2 + \Delta x) = 2.$$

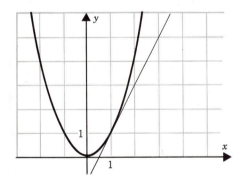

Therefore, the tangent line has a slope, $m = 2$. The equation of the tangent line is found by using the point-slope form $y - y_1 = m(x - x_1)$ where (x_1, y_1) is $(1,1)$ and $m = 2$. Therefore, the equation is $y - 1 = 2(x - 1)$ or $y = 2x - 1$. The tangent line and its slope are different "animals." The tangent line is $y = 2x - 1$ and the slope of the tangent line is 2. At any other point of $y = x^2$ the tangent line will have a different slope. You find the slope of the tangent line and the equation of the tangent line to $y = x^2$ at $(3,9)$. Your answer should be a slope of 6 and an equation, $y - 9 = 6(x - 3)$ or $y = 6x - 9$.

Example 2

Purpose   To find the slope of the tangent line to a curve at any point of the curve:

Problem   Find a general expression for the slope of the tangent line to $y = x^2 + 2x - 1$.

Solution Let us find the slope of the tangent line to $y = f(x) = x^2 + 2x - 1$ at $(x_1, f(x_1))$, a point on the graph of f when $x = x_1$. Let m denote the desired slope.

$$m = \lim_{\Delta x \to 0} \frac{f(x_1 + \Delta x) - f(x_1)}{\Delta x}$$

$$= \lim_{\Delta x \to 0} \frac{[(x_1 + \Delta x)^2 + 2(x_1 + \Delta x) - 1] - [x_1^2 + 2x_1 - 1]}{\Delta x}$$

$$= \lim_{\Delta x \to 0} \frac{x_1^2 + 2x_1 \Delta x + (\Delta x)^2 + 2x_1 + 2 \Delta x - 1 - x_1^2 - 2x_1 + 1}{\Delta x}$$

$$= \lim_{\Delta x \to 0} \frac{2x_1 \Delta x + (\Delta x)^2 + 2 \Delta x}{\Delta x} = \lim_{\Delta x \to 0} \frac{\Delta x(2x_1 + \Delta x + 2)}{\Delta x}$$

$$= \lim_{\Delta x \to 0} (2x_1 + \Delta x + 2) = 2x_1 + 2$$

The slope of the tangent line to the curve $y = x^2 + 2x - 1$ at any point on the curve is: $m = 2x + 2$. (Notice that we drop the subscript 1—it has served its purpose.)

Now, we could use this expression to obtain the slope at particular points on the curve. For example:

At $(-1, -2)$, the slope of the tangent line to $y = x^2 + 2x - 1$ is $m = 2(-1) + 2 = 0$.

At $(2, 7)$, the slope of the tangent line is $m = 2(2) + 2 = 6$.

At $(-3, 2)$, the slope is $m = 2(-3) + 2 = -4$.

Graphically:

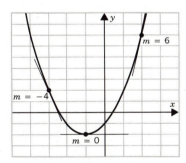

The Derivative of a Function

Once again we note the extreme stroke of good fortune that has occurred. Assume that the function f is represented by $y = f(x)$. Then, not only is the $\lim_{\Delta x \to 0} \dfrac{f(x_1 + \Delta x) - f(x_1)}{\Delta x}$ useful in finding the slope of the tangent line to a curve but it has many other applications in a great variety of different fields. Indeed, in many of these applications, tangent lines and slopes are never mentioned and, moreover, have nothing whatsoever to do with the problem being solved. For this reason, a special name is given to the $\lim\limits_{\Delta x \to 0} \dfrac{f(x_1 + \Delta x) - f(x_1)}{\Delta x}$.

It is called the *derivative* of f at $x = x_1$ and is often denoted by $f'(x_1)$, read "f prime of x_1," or $\dfrac{dy}{dx}\Big]_{x=x_1}$.

The importance of the derivative concept leads us to a slight generalization of the above definition of $f'(x_1)$.

DEFINITION

Derivative of a Function. If f is a function defined by $y = f(x)$ and x is any number in the domain of f, then the *derivative of f* is the function f' given by

$$f'(x) = \lim_{\Delta x \to 0} \frac{\Delta y}{\Delta x} = \lim_{\Delta x \to 0} \frac{f(x + \Delta x) - f(x)}{\Delta x},$$

providing this limit exists.

Other symbols for $f'(x)$ include $D_x y$, y', and $\dfrac{dy}{dx}$.

Now let us develop the derivative given a function f, represented by $y = f(x)$.

1. $y = f(x)$
2. $y + \Delta y = f(x + \Delta x)$
3. $\Delta y = f(x + \Delta x) - f(x)$
4. $\dfrac{\Delta y}{\Delta x} = \dfrac{f(x + \Delta x) - f(x)}{\Delta x}$ is a difference quotient which represents an average rate of change.
5. Exact rate of change y with respect to x is

$$\lim_{\Delta x \to 0} \frac{\Delta y}{\Delta x} = \lim_{\Delta x \to 0} \frac{f(x + \Delta x) - f(x)}{\Delta x}$$

or

6. Derivative of f is given by

$$\frac{dy}{dx} = f'(x) = y' = \lim_{\Delta x \to 0} \frac{\Delta y}{\Delta x} = \lim_{\Delta x \to 0} \frac{f(x + \Delta x) - f(x)}{\Delta x}$$

(providing this limit exists).

This procedure for finding the derivative is often called the delta-process.

Caution. Once again we mention: the derivative of f, called f', does have one application in that it gives us the slope of the tangent line to the curve. However, this is just one of the many applications of the derivative. Now, we should think of derivatives as defined above and not necessarily related to tangent lines.

Example 3

Purpose   To find the derivative by use of the definition:

Problem A   Find $f'(x)$ if $f(x) = \dfrac{1}{x}$.

Solution

$$y = f(x) = \frac{1}{x}$$

$$y + \Delta y = f(x + \Delta x) = \frac{1}{x + \Delta x}$$

$$\Delta y = f(x + \Delta x) - f(x) = \frac{1}{x + \Delta x} - \frac{1}{x}$$

$$= \frac{x - (x + \Delta x)}{x(x + \Delta x)} = \frac{-\Delta x}{x(x + \Delta x)}$$

$$\frac{\Delta y}{\Delta x} = \frac{\dfrac{-\Delta x}{x(x + \Delta x)}}{\Delta x} = \frac{-1}{x(x + \Delta x)}$$

$$\lim_{\Delta x \to 0} \frac{\Delta y}{\Delta x} = \lim_{\Delta x \to 0} \frac{-1}{x(x + \Delta x)} = -\frac{1}{x^2}$$

$$f'(x) = -\frac{1}{x^2}$$

Problem B   In economics the term marginal cost indicates the added cost of producing one extra unit of a commodity. If $C(x)$ represents a cost function, the derivative, $C'(x)$, is the rate of change of cost per one unit change in x. Thus, the marginal cost is given by $C'(x)$. To produce x patio sets ($x \geq 3$) the Wicker Company found its total cost to be $C(x) = 2x^2 - 12x + 45$. Find the marginal cost, $C'(x)$.

Solution

$$C(x) = 2x^2 - 12x + 45$$

$$C(x + \Delta x) = 2(x + \Delta x)^2 - 12(x + \Delta x) + 45$$

$$C(x + \Delta x) - C(x) = [2x^2 + 4x\,\Delta x + 2(\Delta x)^2 - 12x - 12\,\Delta x + 45]$$
$$- (2x^2 - 12x + 45)$$

$$= 4x\,\Delta x + 2(\Delta x)^2 - 12\,\Delta x$$

$$\frac{C(x + \Delta x) - C(x)}{\Delta x} = \frac{4x\,\Delta x + 2(\Delta x)^2 - 12\,\Delta x}{\Delta x} = 4x + 2\,\Delta x - 12$$

$$\lim_{\Delta x \to 0} \frac{C(x + \Delta x) - C(x)}{\Delta x} = \lim_{\Delta x \to 0} (4x + 2\,\Delta x - 12) = 4x - 12$$

$$C'(x) = 4x - 12$$

Problem C A culture plate of bacteria is being observed and a bacteria count is taken every hour. It is inferred from this information that the equation $N = 50\sqrt{t} + 300$ accurately estimates the number of bacteria after t hours. What is the general equation for the instantaneous rate of growth of the bacteria culture?

Solution
$$\frac{dN}{dt} = \lim_{\Delta t \to 0} \frac{\Delta N}{\Delta t} = \lim_{\Delta t \to 0} \frac{50\sqrt{t + \Delta t} + 300 - (50\sqrt{t} + 300)}{\Delta t}$$

$$= \lim_{\Delta t \to 0} 50 \left(\frac{\sqrt{t + \Delta t} - \sqrt{t}}{\Delta t} \right)$$

Since the limit cannot be taken in this form, it is necessary to apply a special type of simplification that involves multiplying the numerator and denominator by $(\sqrt{t + \Delta t} + \sqrt{t})$. Therefore,

$$\lim_{\Delta t \to 0} 50 \left(\frac{\sqrt{t + \Delta t} - \sqrt{t}}{\Delta t} \cdot \frac{\sqrt{t + \Delta t} + \sqrt{t}}{\sqrt{t + \Delta t} + \sqrt{t}} \right)$$

$$= \lim_{\Delta t \to 0} 50 \left(\frac{t + \Delta t - t}{\Delta t(\sqrt{t + \Delta t} + \sqrt{t})} \right)$$

$$= \lim_{\Delta t \to 0} 50 \left(\frac{\Delta t}{\Delta t(\sqrt{t + \Delta t} + \sqrt{t})} \right)$$

$$= \lim_{\Delta t \to 0} \frac{50}{\sqrt{t + \Delta t} + \sqrt{t}} = \frac{50}{2\sqrt{t}} = \frac{25}{\sqrt{t}}$$

Problem D Given the identity function f defined by $f(x) = x$, find the derivative of f.

Solution $\lim_{\Delta x \to 0} \frac{f(x + \Delta x) - f(x)}{\Delta x} = \lim_{\Delta x \to 0} \frac{(x + \Delta x) - (x)}{\Delta x} = \lim_{\Delta x \to 0} 1 = 1$

We have proved that if $f(x) = x$, the derivative is $f'(x) = 1$.

Problem E Find $f'(x)$ if the constant function f is defined by $f(x) = c$, c any constant.

Solution $\lim_{\Delta x \to 0} \frac{\Delta y}{\Delta x} = \lim_{\Delta x \to 0} \frac{f(x + \Delta x) - f(x)}{\Delta x} = \lim_{\Delta x \to 0} \frac{c - c}{\Delta x} = \lim_{\Delta x \to 0} \frac{0}{\Delta x}$

$$= \lim_{\Delta x \to 0} 0 = 0$$

If $c = 1$, the function could also be given by $f(x) = 1 = x^0$ and $f'(x)$ would be 0. Therefore, $\frac{d}{dx}(x^0) = 0$.

We have proved that the derivative of a constant function is zero. This we would expect, since the rate of change of a constant is zero.

Note. In Problem A, $f'(x)$ is negative for all $x \ne 0$, in Problem C, $f'(t)$ is positive for all $t > 0$, and in Problem E, $f'(x)$ is zero. As is shown by these examples the first derivative may be positive, negative, or zero.

In summary, this chapter has three main ideas. First, we discuss the intuitive concept of a limit. Then the limit is used to define the derivative, one of the two most important concepts of the calculus. The third important topic we discuss is that of a continuous function. Indeed, there exists a relationship between a function having a derivative at a point and the function being continuous at that point.

   If a function has a derivative at a point, then the function is said to be differentiable at that point. It can be shown (see Problem 39, Exercise 2.6) that if a function is differentiable at a point, then the function must be continuous at that point.

Exercise 2.6

1. Given the function f defined by $f(x) = x^2 - 2x - 3$.
    (a) Find $f'(x)$.
    (b) Find $f'(3)$ using the result from part (a).
    (c) What is the slope of the tangent line to the curve at $x = 3$?
    (d) Find the equation of the tangent line to the curve at $x = 3$.
    (e) Find $f'(-\frac{1}{2})$ and the equation of the tangent line to the curve at $x = -\frac{1}{2}$.
    (f) Find the equation of the normal line to the curve at $x = -\frac{1}{2}$. (The normal line is a line perpendicular to the tangent line at the point of tangency.)
    (g) Find $f'(1)$.
    (h) What special property does the tangent line have at $x = 1$?

2. Given the function f defined by $f(x) = 2x^2 + x - 10$.
    (a) Find $f'(x)$.
    (b) Find $f'(1)$ using the result from part (a).
    (c) What is the slope of the tangent line to the curve at $x = 1$?
    (d) Find the equation of the tangent line to the curve at $x = 1$.
    (e) Find $f'(0)$ and the equation of the tangent line to the curve at $x = 0$.
    (f) Find $f'(-1)$ and the equation of the tangent line to the curve at $x = -1$.
    (g) Find the equation of the normal line to the curve at $x = -1$. (The normal line is a line perpendicular to the tangent line at the point of tangency.)
    (h) Find $f'(-\frac{1}{4})$ and the equation of the tangent line to the curve at $x = -\frac{1}{4}$. What special property does this tangent line have?

3. Given the identity function $y = f(x) = x$, find $f'(x)$. What is the value of $f'(-4), f'(0), f'(6)$, and $f'(c)$ (c, any constant)? How does this relate to the slope of the tangent line at a point on the graph of this function?

4. Given the constant function $y = f(x) = c$, find $f'(x)$. What is the value of $f'(-3), f'(0), f'(7)$? How does this relate to the slope of the tangent at a point on the graph of this function?

5. A company determines its revenue, $R(x)$, as a function of the number of units produced, x, to be $R(x) = 3x + 4$. Find the marginal revenue at $x = 2$ by using the definition of the derivative.

6. Let $P(t)$ equal the expected price of one share of stock in a business company at any time t. A broker determines that $P(t) = \dfrac{3}{10,000} t^2$. Find the rate of change of $P(t)$ at $t = 3$ by using the definition of the derivative.

In Problems 7 to 18, find the derivative by forming a difference quotient and taking the limit.

7. $y = f(x) = \frac{1}{2}x + 1$

8. $y = f(x) = \frac{1}{3}x + 4$

9. $y = f(x) = 5 - 4x$

10. $y = f(x) = x^2 - 7x + 16$

11. $y = f(x) = x^2 + 6x + 5$

12. $y = f(x) = 3 + x - x^2$

13. $y = f(x) = \dfrac{3}{x}$

14. $y = f(x) = \dfrac{5}{x + 2}$

15. $y = f(x) = \dfrac{1}{x + 4}$

16. $y = f(x) = \sqrt{3 - 2x}$

17. $y = f(x) = \sqrt{2x + 5}$

18. $y = f(x) = \sqrt{1 - x}$

(*Note.* Refer to Exercise 2.2, Problems 7 to 18.)

In Problems 19 to 26, find the derivative.

19. $y = 7x + 13$

20. $y = 3(x - 2)$

21. $y = 5 - \frac{1}{4}x$

22. $y = 3x^2 + 7x - 9$

23. $y = 4 - x^2$

24. $y = \dfrac{1}{x^2}$

25. $y = \sqrt{x}$

26. $y = x^3 + 1$

27. The curve $y = 4 - 3x^2$ has a tangent line with slope -2. Find the equation of this tangent line.

28. The curve $y = \sqrt{2x - 1}$ has a tangent line whose slope is $\frac{1}{3}$. Find the equation of this tangent line.

29. A car is traveling on a road whose curve is the equation $R(x) = 4 - x^2$ from the point $(-2,0)$ to the point $(2,0)$. In which direction is the car heading at the instant it gets to the point $(1,3)$? (*Hint.* Instantaneous direction is given by the tangent line at the given point.)

30. A particle of blood moves along an artery according to the equation $s = 8t$. Find the exact or instantaneous velocity of the particle at any time t.

31. If the sugar level of the blood is increased 50 units above the normal

fasting level, the amount of insulin secreted by the pancreas is $y = 25t$. What is the instantaneous change in insulin level with respect to time?

32. Usually, from the sale of x units of a commodity the total revenue rises as the sales increase, reaches a peak or maximum, and then declines. This results from the fact that more sales occur by reducing the price. A company finds its total revenue may be determined by $R(x) = \$[90,000 - (x - 300)^2]$. Find the rate of change in total revenue with respect to x when $x = x_1$, 200, 300, 400. (*Hint.* Simplify the function by removing the parenthesis and combining like terms; then take the derivative.)

33. A specialty company has determined the total daily cost for producing x units of their commodity to be $\$(\frac{1}{2}x^2 + x + 2)$. Find the marginal cost. The marginal cost is the derivative of the cost function. (*Note.* Exercise 2.2, Problem 20.)

34. The weekly cost function for a manufacturer is $\$(x^2 + 3x + 400)$. Find the marginal cost.

In Problems 35 to 38, find the velocity, $v = \lim\limits_{\Delta t \to 0} \dfrac{\Delta s}{\Delta t}$.

35. $s = f(t) = 32t + 6$ 37. $s = f(t) = t^2 + 2$

36. $s = f(t) = 16t^2$ 38. $s = f(t) = \frac{1}{2}at^2 + v_0t + s_0$
        (a, v_0, s_0 are constants)

39. Prove, if the function f is differentiable at $x = x_1$, then f must be continuous at $x = x_1$, by verifying each of the following steps.

GIVEN. f is differentiable at $x = x_1$ or $f'(x_1)$ exists.

PROVE. f is continuous at $x = x_1$.

Proof

1. $f'(x_1) = \lim\limits_{\Delta x \to 0} \dfrac{f(x_1 + \Delta x) - f(x_1)}{\Delta x}$ exists.

2. $f(x_1 + \Delta x) = \dfrac{f(x_1 + \Delta x) - f(x_1)}{\Delta x} \cdot \Delta x + f(x_1)$

3. $\lim\limits_{\Delta x \to 0} f(x_1 + \Delta x) = \lim\limits_{\Delta x \to 0} \left[\dfrac{f(x_1 + \Delta x) - f(x_1)}{\Delta x} \cdot \Delta x + f(x_1) \right]$

4. $\lim\limits_{\Delta x \to 0} f(x_1 + \Delta x) = \lim\limits_{\Delta x \to 0} \dfrac{f(x_1 + \Delta x) - f(x_1)}{\Delta x} \cdot \lim\limits_{\Delta x \to 0} \Delta x + \lim\limits_{\Delta x \to 0} f(x_1)$

5. $\lim\limits_{\Delta x \to 0} f(x_1 + \Delta x) = f'(x_1) \cdot 0 + f(x_1)$

6. $\lim\limits_{\Delta x \to 0} f(x_1 + \Delta x) = f(x_1)$

Therefore, f is continuous at $x = x_1$ by the alternate definition for continuity.

40. Problem 39 proves that if function f is differentiable at $x = x_1$, then f must be continuous at $x = x_1$.

(a) Form the converse of this theorem. Show that this converse is "false" by following steps (b), (c), and (d) below.

(b) Show that $y = f(x) = |x|$ is continuous at $x = 0$ by verifying that $f(0) = \lim_{x \to 0} f(x)$.

(c) Set up the difference quotient $\dfrac{\Delta y}{\Delta x} = \dfrac{|0 + \Delta x| - |0|}{\Delta x}$ to find the derivative of $y = f(x) = |x|$ at $x = 0$.

(d) Show that $\lim_{\Delta x \to 0} \dfrac{|\Delta x|}{\Delta x}$ does not exist by noting that the right-hand and left-hand limits are not equal.

■ 2.7 CHAPTER REVIEW

Important Ideas

change in x	limit of a quotient
change in y	∞ and $-\infty$
secant line	tangent line to a curve
average rate of change	slope of a tangent line to a curve
difference quotient	derivative
limit of a function	if $f(x) = x$, then $f'(x) = 1$
right-hand limit	if $f(x) = c$, then $f'(x) = 0$
left-hand limit	continuous function
limit of a sum	discontinuous function
limit of a product	differentiable function

■ REVIEW EXERCISE (optional)

1. Explain what is meant by the limit of a function.

In Problems 2 to 15, evaluate the limit.

2. $\lim_{x \to 4} \dfrac{3 + x - x^2}{x + 7}$

3. $\lim_{x \to 2} \dfrac{x^2 - 2x}{x - 2}$

4. $\lim_{x \to -2^-} \dfrac{2}{(x + 2)^4}$

5. $\lim_{x \to 3} \dfrac{2x^2 - 5x - 3}{x(x - 3)}$

6. $\lim_{x \to 4/3} \dfrac{4 - 3x}{2x - 7}$

7. $\lim_{x \to \infty} \dfrac{4 - 3x}{2x - 7}$

8. $\lim_{x \to 2} \dfrac{x^3 + 7}{2 + 5x}$

9. $\lim_{x \to 0} \dfrac{3x^2}{x^2 - 3}$

10. $\lim\limits_{x\to\infty} \dfrac{3x^2}{x^2 - 3}$

11. $\lim\limits_{x\to\infty} \dfrac{3x}{x^2 - 3}$

12. $\lim\limits_{x\to\infty} \dfrac{3x^2}{x - 3}$

13. $\lim\limits_{\Delta x\to 0} (4x^2 - 7x + 9\,\Delta x)$

14. $\lim\limits_{\Delta x\to 0} \dfrac{12}{\sqrt{2 + \Delta x} + \sqrt{2}}$

15. $\lim\limits_{\Delta x\to 0} \dfrac{3}{x(x + \Delta x)}$

16. If $f(x) = \begin{cases} 3 - x, & x \neq 5 \\ 4, & x = 5 \end{cases}$, find $\lim\limits_{x\to 5} f(x)$.

17. If $f(x) = \dfrac{9 - x^2}{3 - x}$, show that $f(3)$ does not exist but the $\lim\limits_{x\to 3} f(x)$ does exist.

18. Consider the following graphs of the functions. Find the limit, if it exists.

(a) $\lim\limits_{x\to 2} f(x) =$

(b) $\lim\limits_{x\to 4} f(x) =$

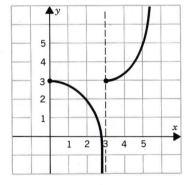

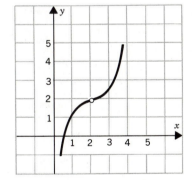

(c) $\lim\limits_{x\to 3} f(x) =$

(d) $\lim\limits_{x\to 2} f(x) =$

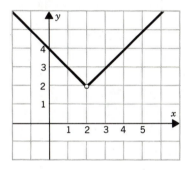

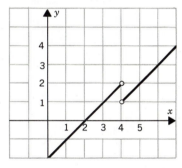

19. Given $g(x) = \begin{cases} -2, & x < 0 \\ 0, & x = 0 \\ 2, & x > 0 \end{cases}$

(a) Sketch the function.

(b) Find $\lim\limits_{x\to 0^-} g(x)$.

(c) Find $\lim\limits_{x\to 0^+} g(x)$.

(d) Does $\lim\limits_{x\to 0} g(x)$ exist?

20.

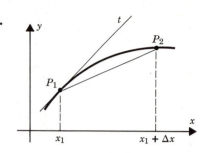

Given the function f defined by $y = f(x)$ and the points $P_1(x_1, f(x_1))$ and $P_2(x_1 + \Delta x, f(x_1 + \Delta x))$. Find:
(a) Δy.
(b) Slope of the secant line $P_1 P_2$.
(c) Slope of the tangent line t.
(d) What is meant by the tangent line to the curve at P_1?

21. What is meant by the slope of the tangent line at a point on the curve?

22. Discuss:
(a) The slope of the tangent line to the curve at $x = a$ exists.

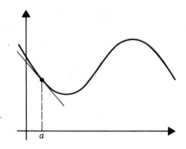

(b) The slope of the tangent line to the curve at $x = a$ does not exist.

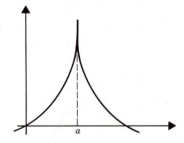

23. Define the derivative of a function.

24. The function f is defined by $f(x) = 3x^2 - 10x - 8$.
(a) Find $f'(x)$.
(b) Find $f'(0)$.
(c) What is the slope of the tangent line to the curve at $x = 0$?
(d) Find the equation of the tangent line to the curve at $x = 0$.
(e) Find $f'(4)$ and the equation of the tangent line to the curve at $x = 4$.
(f) Find the equation of the normal line to the curve at $x = 4$. (The nor-

mal line is a line perpendicular to the tangent line at the point of tangency.)

(g) Find $f'(\frac{5}{3})$.

(h) What special property does the tangent line have at $x = \frac{5}{3}$?

25. Find the equations of the tangent lines to the curve $y = \dfrac{3}{x - 2}$ whose slopes are -3.

26. A car is traveling on a road whose curve fits the equation $R(x) = 4 - x^2$ from the point $(-2,0)$ to the point $(2,0)$. When it gets to the point $(-1,3)$, a tire comes off and rolls away. Assuming the land is flat, in which direction would the tire roll? (*Hint.* The tire would roll in the direction of the tangent line at the given point.)

27. The supply function for a certain item is given by $y = \frac{3}{4}x + 4$ where y is the selling price and x is the number of items supplied. Determine the marginal supply. The marginal supply is the derivative of the supply function.

28. The daily demand function for a certain item is given by $y = -\frac{3}{5}x + 5$ where y is the selling price and x is the number of items in demand. Determine the marginal demand. The marginal demand is the derivative of the demand function.

In Problems 29 to 34 find the derivative by the definition, $\displaystyle\lim_{\Delta x \to 0} \dfrac{f(x + \Delta x) - f(x)}{\Delta x}$.

29. $y = f(x) = 16 - 5x$

30. $y = f(x) = -3$

31. $y = f(x) = \dfrac{-4}{x + 2}$

32. $y = f(x) = 2x^2 - 7x + 12$

33. $y = f(x) = 11 + 9x - x^2$

34. $y = f(x) = \sqrt{3x - 1}$

35. A tire manufacturer has determined his daily cost for producing x radial tires by the cost function $C(x) = \$(.2x^2 + 4x + 100)$. Find the marginal cost, $C'(x)$.

36. A Greyhound bus starting at New York City travels along a straight turnpike. The distance traveled is given by the function $s = f(t) = 5t^2 + 35t$, (s miles, t hours). Find the velocity, $v = f'(t)$.

37. For what values of x are the following functions continuous?

(a) $f(x) = x(x - 5)^3$

(b) $f(x) = \dfrac{1}{x^2 - 4}$

(c) $f(x) = \dfrac{x^2 - 2x + 1}{x - 1}$

(d) $f(x) = \sqrt{x + 3}$

38. Is $f(x) = \dfrac{x^2 - 1}{x - 1}$ continuous at $x = 0$? At $x = -1$? At $x = 1$? Discuss the reason for your answer.

39. A doctor uses the following table to determine the correct dosage of a drug:

Mass of patient (kg)	Less than 20	20 to 50	Greater than 50
Dosage (cc)	0	1	2

(a) Draw a graph of the function relating the correct dosage, y, to the mass of the patient, x.

(b) For what values is the graph discontinuous?

40. (a) Give an example of a function that is continuous for all values of x in the domain of the function.

(b) Give an example of a function that is not continuous for some value of x in the domain of the function.

(c) Give an example of a function that is continuous but does not have a derivative for some x in the domain of the function. (See Problem 40, Exercise 2.6.)

41. Given $g(x) = \begin{cases} 5 - 2x & \text{if } x \neq 1 \\ 6 & \text{if } x = 1 \end{cases}$

(a) Show $\lim_{\Delta x \to 0} g(1 + \Delta x) \neq g(1)$.

(b) What conclusion do you reach about the continuity of g at $x = 1$?

(c) What must be true about the function g at x if $\lim_{\Delta x \to 0} g(x + \Delta x) = g(x)$?

THREE

Differentiation Techniques

3.1 TECHNIQUES OF DIFFERENTIATION

In the preceding chapter, it became apparent that when we find derivatives by the use of the definition,

$$f'(x) = \lim_{\Delta x \to 0} \frac{f(x + \Delta x) - f(x)}{\Delta x},$$

we often become involved in a long and tedious process. In particular, this is true if the function f is complicated, for example, $f(x) = (2x^2 + \frac{1}{2}x)^{5/2}$.

We began our techniques of differentiation in the previous chapter by finding that $\frac{d}{dx}(c) = 0$, where c is any constant, and $\frac{d}{dx}(x) = 1$. With these two short-cuts, we can immediately write certain derivatives, such as $\frac{d}{dx}(10) = 0$, $\frac{d}{dx}(\pi) = 0$, and $\frac{d}{dx}(x) = 1$. We would like to develop other techniques that will enable us to find the derivative quickly.

By using the above definition for $f'(x)$, it can be established that

$$\frac{d}{dx}(x) = 1, \qquad \frac{d}{dx}(x^2) = 2x, \qquad \frac{d}{dx}(x^3) = 3x^2,$$

$$\frac{d}{dx}(x^{1/2}) = \frac{1}{2}x^{-1/2}, \qquad \frac{d}{dx}(x^{-3}) = -3x^{-4}$$

This leads to the conjecture that $\frac{d}{dx}(x^n) = nx^{n-1}$ when n is a real number.

This can be proved, and the proof for the special case when n is a positive integer may be found in Appendix III, page 394. We will state this result as Theorem 3.1a.

THEOREM 3.1α

Power Rule. If n is a real number, then $\frac{d}{dx}(x^n) = nx^{n-1}$.

Therefore, the first new shortcut of this chapter is $\frac{d}{dx}(x^n) = nx^{n-1}$, when n is any real number. In words, the derivative of x^n is n times x to a power diminished (reduced) by 1. Here and in subsequent theorems we assume that the derivative does exist for the values of x under consideration.

Example 1

Purpose   To use the above theorem to find certain derivatives quickly:

Problems   Find $\dfrac{d}{dx}(x^{100})$, $\dfrac{d}{dx}(x^{\sqrt{2}})$, $\dfrac{d}{dx}(x^{5/2})$, $\dfrac{d}{dx}(x^{-2})$.

Solutions $\dfrac{d}{dx}(x^{100}) = 100x^{99}$

$\dfrac{d}{dx}(x^{\sqrt{2}}) = \sqrt{2}x^{\sqrt{2}-1}$

$\dfrac{d}{dx}(x^{5/2}) = \tfrac{5}{2}x^{(5/2)-1} = \tfrac{5}{2}x^{3/2}$

$\dfrac{d}{dx}(x^{-2}) = -2x^{-2-1} = -2x^{-3}$

Another theorem for consideration is the theorem that tells us that the derivative of the sum of two (or more) differentiable functions is equal to the sum of the derivatives of each function.

Example 2

Purpose   To illustrate the quick approach of finding the derivative of a sum of two functions:

Problem   Find $\dfrac{d}{dx}(x^{10} + x^5)$.

Solution   Assuming that the derivative of the sum of two functions is equal to the sum of the respective derivatives:

$$\frac{d}{dx}(x^{10} + x^5) = \frac{d}{dx}(x^{10}) + \frac{d}{dx}(x^5) = 10x^9 + 5x^4$$

We now state the theorem.

THEOREM 3.1b

Derivative of a Sum.   For any value of x where the functions g and h are differentiable, the function $[g + h]$ is also differentiable with its derivative given by

$$\frac{d}{dx}[g(x) + h(x)] = g'(x) + h'(x).$$

GIVEN. Let x be any value where the functions g and h are differentiable, that is, $g'(x)$ and $h'(x)$ exist.

TO PROVE. $\dfrac{d}{dx}[g(x) + h(x)] = g'(x) + h'(x)$.

Proof

1. $y = f(x) = g(x) + h(x)$
2. $y + \Delta y = f(x + \Delta x) = g(x + \Delta x) + h(x + \Delta x)$
3. $\Delta y = f(x + \Delta x) - f(x) = g(x + \Delta x) + h(x + \Delta x) - g(x) - h(x)$
4.   Regroup terms.
     $\Delta y = [g(x + \Delta x) - g(x)] + [h(x + \Delta x) - h(x)]$
5.   Divide both sides of the equation by Δx:
     $$\dfrac{\Delta y}{\Delta x} = \dfrac{g(x + \Delta x) - g(x)}{\Delta x} + \dfrac{h(x + \Delta x) - h(x)}{\Delta x}$$
6.   Take the limit of both sides of the equation as $\Delta x \to 0$:
     $$\lim_{\Delta x \to 0} \dfrac{\Delta y}{\Delta x} = \lim_{\Delta x \to 0} \dfrac{g(x + \Delta x) - g(x)}{\Delta x} + \lim_{\Delta x \to 0} \dfrac{h(x + \Delta x) - h(x)}{\Delta x}$$
7. $\dfrac{dy}{dx} = g'(x) + h'(x)$

     or, $\dfrac{d}{dx}[g(x) + h(x)] = g'(x) + h'(x)$.

As a corollary to this theorem, for any value of x where the functions g and h are differentiable, the function $[g - h]$ is also differentiable with its derivative given by

$$\dfrac{d}{dx}[g(x) - h(x)] = g'(x) - h'(x).$$

The proof of the corollary is left as an exercise for the student.

Example 3

Purpose   To illustrate an application of the theorem and corollary for the derivative of a sum and the derivative of a difference:

Problem   Find $\dfrac{d}{dx}(x^5 - x^2 + 5)$.

Solution $\dfrac{d}{dx}(x^5 - x^2 + 5) = \dfrac{d}{dx}(x^5) - \dfrac{d}{dx}(x^2) + \dfrac{d}{dx}(5)$

$\qquad\qquad = 5x^4 - 2x + 0 = 5x^4 - 2x$

We now turn our attention to the problem of finding, by a shortcut method, the derivative of a product. *Caution.* Even though:

1. The *derivative* of the sum of two or more functions is equal to the sum of the derivatives of the separate functions, and

2. The *limit* of the product of two or more functions is equal to the product of the limits of the separate functions, we will find that:

3. *The derivative of the product of two functions does **not** equal the product of the separate derivatives.*

This may be seen by noting that the correct answer for $\dfrac{d}{dx}(4x)$ is 4, but $\dfrac{d}{dx}(4) \cdot \dfrac{d}{dx}(x) = 0 \cdot 1 = 0$ would yield a different result, which is not $\dfrac{d}{dx}(4x)$.

The correct formula for finding the derivative of a product is given in Theorem 3.1c. (The proof is found in Appendix III, page 395.)

THEOREM 3.1c

Product Rule. For any value of x where the functions g and h are differentiable, the function $[g \cdot h]$ is also differentiable with its derivative given by

$$\frac{d}{dx}[g(x) \cdot h(x)] = g(x) \cdot h'(x) + h(x) \cdot g'(x).$$

Example 4

Purpose To illustrate the quick method of finding the derivative of the product of two functions:

Problem A Find $\dfrac{d}{dx}(x^5 \cdot \sqrt{2x + 5})$.

Solution The derivative of the product of two functions is equal to the first function times the derivative of the second function plus the second function times the derivative of the first function. Also in Exercise 2.6, Problem 17, page 116, you found

$$\frac{d}{dx}(\sqrt{2x + 5}) = \frac{1}{\sqrt{2x + 5}}.$$

$$\frac{d}{dx}(x^5 \cdot \sqrt{2x + 5}) = x^5 \cdot \frac{d}{dx}(\sqrt{2x + 5}) + \sqrt{2x + 5} \cdot \frac{d}{dx}(x^5)$$

$$= x^5 \cdot \frac{1}{\sqrt{2x + 5}} + \sqrt{2x + 5}\,(5x^4)$$

$$= \frac{x^5}{\sqrt{2x + 5}} + 5x^4\sqrt{2x + 5}$$

Problem B Find $\dfrac{d}{dx}(x^3 + 7)(x^2 - x + 4)$.

Solution $\dfrac{d}{dx}[(x^3 + 7)(x^2 - x + 4)] = (x^3 + 7) \cdot \dfrac{d}{dx}(x^2 - x + 4)$

$$+ (x^2 - x + 4) \cdot \dfrac{d}{dx}(x^3 + 7)$$

$$= (x^3 + 7)(2x - 1) + (x^2 - x + 4)(3x^2)$$

$$= 5x^4 - 4x^3 + 12x^2 + 14x - 7$$

A valuable corollary to Theorem 3.1c verifies that $\dfrac{d}{dx}[c \cdot h(x)] = c \cdot h'(x)$.

COROLLARY

For any value of x where the function h is differentiable and c is a constant, the function $[c \cdot h]$ is also differentiable with its derivative given by

$$\dfrac{d}{dx}[c \cdot h(x)] = c \cdot h'(x).$$

GIVEN. $h'(x)$ exists and c is a constant.

TO PROVE. $\dfrac{d}{dx}[c \cdot h(x)] = c \cdot h'(x)$.

Proof

From the product rule, Theorem 3.1c.

$$y = c \cdot h(x)$$

$$\dfrac{dy}{dx} = c \cdot \dfrac{d}{dx}[h(x)] + h(x) \cdot \dfrac{d}{dx}(c)$$

$$= c \cdot h'(x) + h(x) \cdot 0$$

$$\dfrac{d}{dx}[c \cdot h(x)] = c \cdot h'(x)$$

Example 5

Purpose To illustrate finding the derivative of the product of a constant times a function of x:

Problem A $y = 3\sqrt{x}$ or $y = 3x^{1/2}$. Find $\dfrac{dy}{dx}$.

Solution $\dfrac{dy}{dx} = 3 \cdot \dfrac{d}{dx}(x^{1/2})$

$\qquad\qquad = 3 \cdot \dfrac{1}{2}x^{-(1/2)}$

$\qquad \dfrac{dy}{dx} = \dfrac{3}{2\sqrt{x}}$

Problem B $y = 3\sqrt{x} \cdot (4x - 17)$. Find $\dfrac{dy}{dx}$.

Solution $\dfrac{dy}{dx} = 3\sqrt{x} \cdot \dfrac{d}{dx}(4x - 17) + (4x - 17) \cdot 3\dfrac{d}{dx}(x^{1/2})$.

$\qquad\qquad = 3\sqrt{x} \cdot 4 + (4x - 17) \cdot 3\left(\dfrac{1}{2}x^{-(1/2)}\right)$

$\qquad \dfrac{dy}{dx} = 12\sqrt{x} + \dfrac{3(4x - 17)}{2\sqrt{x}}$

Problem C If the function f is defined by $s = f(t)$, (s, feet; t, seconds), and the motion of an object is on a straight line, the exact rate of change, $\dfrac{ds}{dt} = \lim\limits_{\Delta t \to 0} \dfrac{\Delta s}{\Delta t}$, is the exact or instantaneous velocity, v. Find v, given $s = 16t^2 - 3\sqrt[3]{t} + 9$.

Solution $s = 16t^2 - 3t^{1/3} + 9$

$\qquad v = 16(2t) - 3\left(\dfrac{1}{3}t^{-(2/3)}\right) + 0$

$\qquad v = 32t - \dfrac{1}{\sqrt[3]{t^2}}$

Problem D The total cost per day, in dollars, for The Electron Company to manufacture x transformers is given by $C(x) = 600 + 40x - 10x^2 + x^3$. Find the marginal cost, $C'(x)$.

Solution $C(x) = 600 + 40x - 10x^2 + x^3$
$\qquad C'(x) = 0 + 40 - 20x + 3x^2$
$\qquad C'(x) = 40 - 20x + 3x^2$

In addition to finding the derivative of the product of two functions, we would also like a quick method for finding the derivative of the quotient of two functions. As the derivative of a product does not equal the product of the derivatives, also *the derivative of a quotient does **not** equal the quotient of the derivatives*. Theorem 3.1d states the correct formula for finding the derivative of a quotient. (The proof is found in Appendix III, page 397.)

THEOREM 3.1d

Quotient Rule. For any value of x where the functions g and h are differentiable and $h(x) \neq 0$, the function $\left[\dfrac{g}{h}\right]$ is also differentiable with its derivative given by:

$$\frac{d}{dx}\left[\frac{g(x)}{h(x)}\right] = \frac{h(x) \cdot g'(x) - g(x) \cdot h'(x)}{[h(x)]^2}.$$

Example 6

Purpose   To illustrate the use of the quotient rule to find the derivative:

Problem   Find y' if $y = \dfrac{3x^3}{x - 7}$, $(x - 7) \neq 0$.

Solution   The derivative of the quotient of two functions is equal to the denominator times the derivative of the numerator minus the numerator times the derivative of the denominator all divided by the denominator squared.

$$y' = \frac{(x - 7)\dfrac{d}{dx}(3x^3) - (3x^3)\dfrac{d}{dx}(x - 7)}{(x - 7)^2}$$

$$y' = \frac{(x - 7)(9x^2) - 3x^3(1)}{(x - 7)^2} = \frac{6x^3 - 63x^2}{(x - 7)^2} = \frac{3x^2(2x - 21)}{(x - 7)^2}$$

Example 7

Purpose   To use the shortcut techniques of this section to find derivatives:

Problem A   Given $y = (3 + 6x - 2x^2)(\sqrt{x} - 4)$, find y',

Solution $y' = (3 + 6x - 2x^2) \cdot \dfrac{1}{2}x^{-(1/2)} + (\sqrt{x} - 4)(6 - 4x)$

$$y' = \frac{3 + 6x - 2x^2}{2\sqrt{x}} + (\sqrt{x} - 4)(6 - 4x)$$

Problem B   Given $y = \dfrac{3x^2 + 16}{8 - x^3}$, find y'.

Solution $y' = \dfrac{(8 - x^3)(6x) - (3x^2 + 16)(-3x^2)}{(8 - x^3)^2}$

$$y' = \frac{3x^4 + 48x^2 + 48x}{(8 - x^3)^2} = \frac{3x(x^3 + 16x + 16)}{(8 - x^3)^2}$$

Problem C   The total cost, in dollars, for the output of x automobile batteries is given to be $C(x) = \frac{1}{8}x^2 + 10x + 206$. Find the marginal cost, $C'(x)$.

Solution $C(x) = \frac{1}{8}x^2 + 10x + 206$
$C'(x) = \frac{1}{8} \cdot 2x + 10 + 0 = \frac{1}{4}x + 10.$

Problem D   The average cost, $\overline{C}(x)$, is defined to be $\dfrac{C(x)}{x}$ and the marginal average cost is $\dfrac{d}{dx}[\overline{C}(x)]$, the derivative of the average cost with respect to x. Find the average cost and marginal average cost using the cost function, $C(x) = \frac{1}{8}x^2 + 10x + 206$.

Solution $\overline{C}(x) = \dfrac{C(x)}{x} = \dfrac{\frac{1}{8}x^2 + 10x + 206}{x} = \frac{1}{8}x + 10 + \dfrac{206}{x}$

$\dfrac{d}{dx}[\overline{C}(x)] = \dfrac{d}{dx}(\frac{1}{8}x + 10 + 206x^{-1}) = \frac{1}{8} + 0 + 206(-1)x^{-2} = \frac{1}{8} - \dfrac{206}{x^2}.$

Review of Differentiation Techniques

Let us review the previous techniques of differentiation by stating the theorem and showing another example for each case.

1. Derivative of x to a power.

$$\frac{d}{dx}(x^n) = nx^{n-1}$$

Example. $\dfrac{d}{dx}(x^6) = 6x^5$

2. Derivative of the sum or difference of two functions.

$$\frac{d}{dx}[g(x) \pm h(x)] = g'(x) \pm h'(x)$$

Example. $\dfrac{d}{dx}\left(x^2 + \sqrt{x} - \dfrac{1}{x^2}\right) = \dfrac{d}{dx}(x^2) + \dfrac{d}{dx}(x^{1/2}) - \dfrac{d}{dx}(x^{-2})$

$$= 2x + \tfrac{1}{2}x^{-1/2} + 2x^{-3}$$

$$= 2x + \dfrac{1}{2\sqrt{x}} + \dfrac{2}{x^3}$$

3. Derivative of the product of two functions.

$$\frac{d}{dx}[g(x) \cdot h(x)] = g(x) \cdot h'(x) + h(x) \cdot g'(x)$$

Example. $\dfrac{d}{dx}[(x^5 - x)(x^2 + 7)] = (x^5 - x)\dfrac{d}{dx}(x^2 + 7)$

$$+ (x^2 + 7)\dfrac{d}{dx}(x^5 - x)$$

$$= (x^5 - x)(2x) + (x^2 + 7)(5x^4 - 1)$$

$$= 7x^6 + 35x^4 - 3x^2 - 7$$

3a. Derivative of a constant times a function.

$$\frac{d}{dx}[c \cdot h(x)] = c \cdot h'(x)$$

Example. $\dfrac{d}{dx}(5x^{10}) = 5 \cdot \dfrac{d}{dx}(x^{10}) = 5 \cdot 10x^9 = 50x^9$

4. Derivative of the quotient of two functions.

$$\frac{d}{dx}\left[\frac{g(x)}{h(x)}\right] = \frac{h(x) \cdot g'(x) - g(x) \cdot h'(x)}{[h(x)]^2}$$

Example. $\dfrac{d}{dx}\left[\dfrac{x^2}{3x^3 - 7}\right] = \dfrac{(3x^3 - 7)\dfrac{d}{dx}(x^2) - (x^2)\dfrac{d}{dx}(3x^3 - 7)}{(3x^3 - 7)^2}$

$$= \frac{(3x^3 - 7)(2x) - (x^2)(9x^2)}{(3x^3 - 7)^2}$$

$$= \frac{-3x^4 - 14x}{(3x^3 - 7)^2}$$

In summary:

1. Derivative of x to a power.

$$\frac{d}{dx}(x^n) = nx^{n-1}$$

2. Derivative of the sum or difference of two functions.

$$\frac{d}{dx}[g(x) \pm h(x)] = g'(x) \pm h'(x)$$

3. Derivative of the product of two functions.

$$\frac{d}{dx}[g(x) \cdot h(x)] = g(x) \cdot h'(x) + h(x) \cdot g'(x)$$

3a. Derivative of a constant times a function.

$$\frac{d}{dx}[c \cdot h(x)] = c \cdot h'(x)$$

4. Derivative of the quotient of two functions.

$$\frac{d}{dx}\left[\frac{g(x)}{h(x)}\right] = \frac{h(x) \cdot g'(x) - g(x) \cdot h'(x)}{[h(x)]^2}$$

Exercise 3.1

In Problems 1 to 50, use the differentiation formulas to find the derivatives of the following functions.

1. $y = 3x^2 + 6$

2. $y = x^4 - 5x^3 + 2x^2 - 2x$

3. $y = \frac{1}{3}x^3 - x$

4. $y = -5x^3 + x^2$

5. $y = -\frac{1}{5}x^5 + 2x^3 - 72x - 4$

6. $s = -10t^{10}$

7. $y = \frac{1}{6}(3 + 2x - x^2)$

8. $u = \frac{1}{4}(v^4 + 2v^2 + 1)$

9. $s = \frac{3}{4}t^4 - \frac{4}{3}t^3 + \frac{1}{2}t^2 - 14$

10. $y = \dfrac{3 + 2x - x^2}{6}$

11. $y = \dfrac{x^3 - 6x^2 - 3}{3}$

12. $s = \sqrt{2}(t^2 - t)$

13. $u = \frac{1}{2}(v^2 + 7 - v^{-2})$

14. $u = 5v^3 - v + \sqrt[3]{v}$

15. $y = \dfrac{3}{x^2}$

16. $y = x^5 - 5x^2 - \dfrac{1}{x^2}$

17. $y = \dfrac{x^5}{5} - \dfrac{1}{2}x^{-2}$

18. $y = \dfrac{4}{x^3} + 3\sqrt{x}$

19. $y = \sqrt{x} + \dfrac{1}{\sqrt{x}}$

20. $y = x + \dfrac{1}{x}$

21. $y = \dfrac{x^3}{3} + \dfrac{1}{x^3}$

22. $y = 5\sqrt[3]{x^2}$

23. $w = 3\sqrt{u} + \dfrac{2}{\sqrt{u}}$

24. $u = v^4 + \dfrac{1}{v^3}$

25. $s = \dfrac{9}{t^9}$

26. $u = \dfrac{v^4}{4} + \dfrac{1}{v^4}$

27. $s = \sqrt{t} + \dfrac{16}{t^2}$

28. $y = (2x - 7)(-15x^4 + 93)$

29. $y = (2 - x^3)(x^2 + 2)$

30. $y = (5x^2 + 2)(4\sqrt{x} + 1)$

31. $y = -7x^4 \cdot \sqrt[4]{x^3}$

32. $y = (x^2 + 4)(x^2 - 4)$

33. $u = 3\sqrt[3]{v}(1 - v)$

34. $s = (16t^2 - t)(2 + \sqrt[3]{t^2})$

35. $s = (1 + 2t + 3t^2)(1 - t - 2t^2)$

36. $y = (\sqrt{x})^3 + (\sqrt[3]{x})^2$

37. $w = \dfrac{12u^2 + 3u - 8}{3u}$

38. $y = \dfrac{x^2 - x - 2}{-2x}$

39. $y = \dfrac{x + 3}{x - 3}$

40. $y = \dfrac{4 - x^3}{4 + x^3}$

41. $s = \dfrac{\sqrt{t}}{t - 1}$

42. $y = \dfrac{7 - x - x^3}{\sqrt{x}}$

43. $y = \dfrac{-1}{6x^3 - x + 12}$

44. $u = \dfrac{3}{7v^5 - 9v^2 + v}$

45. $y = \dfrac{12(3x^3 - 16)}{4x - x^4}$

48. $s = \dfrac{3(9 - t^2)}{t^4 + 2}$

46. $y = \dfrac{(x^2 - 5)(2 - x)}{x^3}$

49. $y = \dfrac{(x + 1)(1 - x^2)}{3 + x^3}$

47. $u = \dfrac{4v^2 + 5}{v^3 - 2v - 1}$

50. $y = \dfrac{(x + 2)(3x^2 - 4)}{3 - x^3}$

51. Given $y = 3x^2 + 5$.
(a) Graph the function.
(b) What is the first derivative?
(c) Graph the first derivative function on a separate set of axes.

52. Given $y = x^2 - 6x + 9$.
Find:
(a) the derivative of the function.
(b) the slope of the tangent line to the curve at $x = 0$; $x = 3$.
(c) the equation of the tangent line to the curve at $x = 0$; $x = 3$.
Sketch the curve and the tangent lines.

53. Given $y = (x^2 - 4)(x + 3)$.
Find:
(a) the derivative of the function.
(b) the slope of the tangent line to the curve at $x = -1$; $x = 3$.
(c) the equation of the tangent line to the curve at $x = -1$; $x = 3$.

54. Given $y = \dfrac{x}{x - 2}$.
Find:
(a) the derivative of the function.
(b) the slope of the tangent line to the curve at $x = 1$.
(c) the equation of the tangent line to the curve at $x = 1$.
(d) the equation of the normal line to the curve at $x = 1$.
(The normal line is a line perpendicular to the tangent line.)

55. Find the equations of the tangent line and the normal line (line $\perp$ to the tangent line) to the curve $y = 7 - 7x^2 + 2x^3$ at the point $(2, -5)$.

56. Find the equation of the tangent line with slope 3 to the curve $y = 2\sqrt{x}$.

57. $y = 7 + 5x - x^2$. Find the point on the curve when $f'(x) = 0$. Sketch the curve and the tangent line to the curve at that point.

58. The revenue function is equal to $7.50 per unit sold. Find the derivative of the revenue function. What is the meaning of this derivative?

59. The total cost function (in dollars) for the Jones Corp. is found to be $C(x) = \frac{1}{2}x^2 + 4x + 24$.
Find:
(a) the marginal cost, the rate of change (increase) in total cost with respect to x.

(b) the average cost.
(c) the marginal average cost.
(*Note*. See Example 7, Problem D.)

60. Given the total cost function, in dollars, for x units of a commodity to be $C(x) = \frac{1}{3}x^3 + 3x^2 - 7x + 16$.
Find:
(a) the marginal cost.
(b) the average cost.
(c) the marginal average cost.
(*Note*. See Example 7, Problem D.)

61. The demand function for a certain item is given by $y = \dfrac{20,000}{50 + x}$ where y is the commodity price and x is the quantity demanded. Determine the marginal demand. The marginal demand is the derivative of the demand function.

62. The demand function may be expressed as $y = D(x)$ where y is the price of the commodity and x is the quantity demanded. The total revenue from the sale of x units, $R(x)$, is $x \cdot D(x)$. If $D(x) = 2 - \frac{4}{5}x$, find the total revenue function. The marginal revenue is the derivative of the total revenue with respect to x, $R'(x)$. Find the marginal revenue. In general, show that the marginal revenue is equal to $D(x) + x \cdot D'(x)$.

63. Variables other than x and y are frequently used in business and economics books. The notation for price and quantity are P and Q, respectively. Let P equal the price of the commodity and Q equal the quantity demanded. If the total revenue, R, from the sale of Q units is $R = P \cdot Q$, show that the marginal revenue is $R' = P + Q \cdot P'$.

64. Find the total revenue, marginal revenue, and the change in price resulting from a change in quantity demanded for the demand function $P = 4 - Q^2$ where Q equals the quantity demanded. (See Problem 63.)

65. The supply function for a certain item is given by $y = 100 - \dfrac{50}{x - 1}$ where y is the selling price and x is the number of items supplied. Determine the marginal supply. The marginal supply is the derivative of the supply function.

66. The velocity of blood can be described by the graph $s = 8\sqrt{t}$ as it travels from the heart, through the aorta and other major arteries and finally through capillaries, consequently losing velocity.
(a) What is the average velocity from $t = 4$ to $t = 9$?
(b) What is the instantaneous velocity at $t = 4$? at $t = 9$?

67. The sales department of a local business firm has determined the following function that gives the sales per day of a new product as a function of the day. $S(t) = 4500 + .3t^{3/2}$. Find the rate at which the sales are increasing on the twenty-fifth day.

68. A polynomial function is defined by $f(x) = a_0x^n + a_1x^{n-1} + \cdots + a_{n-1}x + a_n$ where n is a nonnegative integer and $a_0, a_1, \ldots, a_n$ are real numbers.

(a) Find $f'(x)$.

(b) Is $f(x)$ defined for all x? Explain.

(c) Is $f'(x)$ defined for all x? Explain.

    You have now established that a polynomial function is differentiable for all x and continuous for all x. (See Exercise 2.4–2.5, Problem 55, and Problem 39 in Exercise 2.6.)

69. Prove the Corollary of Theorem 3.1b. For any value of x where the functions g and h are differentiable, the function $[g - h]$ is also differentiable with its derivative given by

$$\frac{d}{dx}[g(x) - h(x)] = g'(x) - h'(x).$$

70. Verify the Corollary to Theorem 3.1c, $\dfrac{d}{dx}[c \cdot h(x)] = c \cdot h'(x)$ using the definition of the derivative. Therefore, $\dfrac{d}{dx}[c \cdot h(x)] = \lim\limits_{\Delta x \to 0} \dfrac{c \cdot h(x + \Delta x) - c \cdot h(x)}{\Delta x}$; then show that the limit is $c \cdot h'(x)$.

71. Show that:

$$D_x[g(x) \cdot h(x) \cdot k(x)] = g(x)h(x)k'(x) + g(x)h'(x)k(x) + g'(x)h(x)k(x).$$

72. Show that:

(a) $\dfrac{d}{dx}[f(x)]^2 = 2f(x)f'(x)$. *Hint.* Consider $[f(x)]^2$ as $f(x) \cdot f(x)$.

(b) $\dfrac{d}{dx}[f(x)]^3 = 3[f(x)]^2f'(x)$. *Hint.* Consider $[f(x)]^3$ as $[f(x)]^2 \cdot f(x)$ and make use of the result from part (a).

(c) What would you expect $\dfrac{d}{dx}[f(x)]^n$ to be? This will be discussed in the next section.

3.2   THE CHAIN RULE

One of the most important uses of the chain rule is in finding the derivative of a function raised to a power. For example, we would quickly like to find $\dfrac{dy}{dx}$ if $y = (3x^2 - 4)^{100}$. To determine a short method for finding $\dfrac{dy}{dx}$, let us investigate a simpler problem.

Consider the problem of finding $\dfrac{dy}{dx}$ if $y = u^2$ and u in turn is defined by $x^3 - 4$. Now, by substitution, y may be written as a function of x.

$$y = u^2 \quad \text{and} \quad u = x^3 - 4$$
$$y = (x^3 - 4)^2$$
$$y = x^6 - 8x^3 + 16$$
$$\frac{dy}{dx} = 6x^5 - 24x^2 = 6x^2(x^3 - 4)$$

or

$$\frac{dy}{dx} = 2(x^3 - 4)(3x^2) \tag{1}$$

This algebraic substitution may become very cumbersome, and we want to avoid it. That is, we want to use only the two equations

$$\begin{cases} y = u^2 \\ u = x^3 - 4 \end{cases}$$

and, without substituting at first, be able to obtain $\dfrac{dy}{dx}$. The important theorem that enables us to do this is called the *Chain Rule*. Basically, the chain rule says that if certain conditions are met, then $\dfrac{dy}{dx} = \dfrac{dy}{du} \cdot \dfrac{du}{dx}$. That is $\dfrac{dy}{dx}$ may be found by taking the derivative of y (treating u as an independent variable) times the derivative of u (treating u as the dependent variable and x as the independent variable). From our example:

If $\begin{cases} y = u^2 \\ \\ u = x^3 - 4, \end{cases}$ then

$$\frac{dy}{du} = 2u = 2(x^3 - 4)$$

$$\frac{du}{dx} = 3x^2$$

and

$$\frac{dy}{dx} = \frac{dy}{du} \cdot \frac{du}{dx} = 2(x^3 - 4)(3x^2).$$

Compare with equation (1) and notice the same answer.

Another indication that $\dfrac{dy}{dx}$ may be found by use of $\dfrac{dy}{dx} = \dfrac{dy}{du} \cdot \dfrac{du}{dx}$, is the following argument. If $\Delta u \neq 0$, then $\dfrac{\Delta y}{\Delta x} = \dfrac{\Delta y}{\Delta u} \cdot \dfrac{\Delta u}{\Delta x}$; if u is a continuous function of x as $\Delta x \to 0$, $\Delta u \to 0$.

Therefore:

$$\lim_{\Delta x \to 0} \frac{\Delta y}{\Delta x} = \lim_{\Delta x \to 0} \left(\frac{\Delta y}{\Delta u} \cdot \frac{\Delta u}{\Delta x} \right)$$

$$= \lim_{\Delta x \to 0} \frac{\Delta y}{\Delta u} \cdot \lim_{\Delta x \to 0} \frac{\Delta u}{\Delta x}$$

$$= \lim_{\Delta u \to 0} \frac{\Delta y}{\Delta u} \cdot \lim_{\Delta x \to 0} \frac{\Delta u}{\Delta x}$$

or

$$\frac{dy}{dx} = \frac{dy}{du} \cdot \frac{du}{dx}.$$

This discussion is not a proof of the chain rule. It is included to motivate the statement of Theorem 3.2a, which follows. A complete proof of the chain rule is beyond the scope of this book. Also, in the statement of the chain rule, the functions g and h are combined to form a composite function of g and h denoted by $g(h(x))$, which is a function of x.

THEOREM 3.2a

The Chain Rule. If $y = g(u)$ and $u = h(x)$ define $y = g(h(x)) = F(x)$ and if $\dfrac{dy}{du}$ and $\dfrac{du}{dx}$ exist, then

$$\frac{dy}{dx} = \frac{dy}{du} \cdot \frac{du}{dx}.$$

This theorem is now applied in Example 1.

Example 1

Purpose   To illustrate the use of the chain rule in finding derivatives:

Problem A $y = g(u) = u^3$; $u = h(x) = 3x^2 + 9$. Find $\dfrac{dy}{dx}$.

Solution

If
$$\begin{cases} y = u^3 \\ \\ u = 3x^2 + 9, \end{cases}$$
then
$$\frac{dy}{du} = 3u^2 = 3(3x^2 + 9)^2$$
$$\frac{du}{dx} = 6x$$

and

$$\frac{dy}{dx} = \frac{dy}{du} \cdot \frac{du}{dx} = 3(3x^2 + 9)^2(6x).$$

Problem B $y = g(u) = u^2 - \sqrt{u}; \; u = h(x) = 5x - x^3$. Find $\dfrac{dy}{dx}$.

Solution

If $\begin{cases} y = u^2 - u^{1/2} \\ \\ u = 5x - x^3, \end{cases}$ then $\begin{aligned} \frac{dy}{du} &= 2u - \tfrac{1}{2}u^{-1/2} = 2(5x - x^3) - \frac{1}{2\sqrt{5x - x^3}} \\ \\ \frac{du}{dx} &= 5 - 3x^2 \end{aligned}$

and

$$\frac{dy}{dx} = \frac{dy}{du} \cdot \frac{du}{dx} = \left[2(5x - x^3) - \frac{1}{2\sqrt{5x - x^3}} \right](5 - 3x^2).$$

Problem C $y = g(u) = \dfrac{1}{u^3}; \; u = h(x) = x^2 - 7$. Find $\dfrac{dy}{dx}$.

Solution

If $\begin{cases} y = u^{-3} \\ \\ u = x^2 - 7, \end{cases}$ then $\begin{aligned} \frac{dy}{du} &= -3u^{-4} = \frac{-3}{(x^2 - 7)^4} \\ \\ \frac{du}{dx} &= 2x \end{aligned}$

and

$$\frac{dy}{dx} = \frac{dy}{du} \cdot \frac{du}{dx} = -\frac{3}{(x^2 - 7)^4}(2x).$$

Extensions of the Chain Rule to Find Derivatives

We wish to use the chain rule to obtain the formula for finding the derivatives of quantities raised to powers. Also, we will use the chain rule in another technique known as *implicit differentiation*. These two techniques will greatly increase our ability to quickly find the derivatives of many functions.

The Chain Rule Used to Find Derivatives of Quantities Raised to Powers

Let us take $y = (3x^2 - 4)^{100}$. The problem we are considering is that of finding $\dfrac{dy}{dx}$.

Now, it is true that we could expand $(3x^2 - 4)^{100}$ by multiplication or, preferably, by the binomial theorem obtaining an expansion that contains 101 terms. Then, we could differentiate each term and note that the derivative of $y = (3x^2 - 4)^{100}$ is the sum of the derivatives of the 101 terms. What an unwieldly problem!

Instead, let us substitute $u = 3x^2 - 4$, then $y = (3x^2 - 4)^{100}$ can be expressed as the chain, $y = u^{100}$ and $u = 3x^2 - 4$. Now, by the chain rule,

$$\text{If} \begin{cases} y = u^{100} \\ \\ u = 3x^2 - 4, \end{cases} \text{then} \quad \begin{aligned} \frac{dy}{du} &= 100u^{99} = 100(3x^2 - 4)^{99} \\ \\ \frac{du}{dx} &= 6x \end{aligned}$$

$$\frac{dy}{dx} = \frac{dy}{du} \cdot \frac{du}{dx} = 100(3x^2 - 4)^{99}(6x).$$

or

$$\frac{d}{dx}(3x^2 - 4)^{100} = 100(3x^2 - 4)^{99}(6x)$$
$$= 600x(3x^2 - 4)^{99}.$$

The results of this example may be generalized by the following theorem.

THEOREM 3.2b

If the function f is differentiable and n is any real number, then $[f]^n$ is also differentiable with its derivative given by:

$$\frac{dy}{dx} = n[f(x)]^{n-1} \cdot f'(x),$$

for all x where $[f(x)]^{n-1}$ is a real number. That is, the derivative of a quantity to the nth power is equal to the product of n, the quantity to the power $(n - 1)$, and the derivative of the quantity.

GIVEN. The function f is differentiable and $y = [f(x)]^n$, n is a real number, and $[f(x)]^{n-1}$ is a real number.

TO PROVE. $\dfrac{dy}{dx} = n[f(x)]^{n-1} \cdot f'(x)$.

Proof

1. Let $f(x) = u$, then $y = [f(x)]^n$ is given by $y = u^n$ and $u = f(x)$.

2. By the chain rule, $\dfrac{dy}{dx} = \dfrac{dy}{du} \cdot \dfrac{du}{dx}$

3. If $\begin{cases} y = u^n \\ \\ u = f(x), \end{cases}$ then $\begin{aligned} \dfrac{dy}{du} &= nu^{n-1} = n[f(x)]^{n-1} \\ \\ \dfrac{du}{dx} &= \dfrac{d}{dx}[f(x)] = f'(x) \end{aligned}$

4. Therefore, $\dfrac{dy}{dx} = \dfrac{dy}{du} \cdot \dfrac{du}{dx} = n[f(x)]^{n-1}f'(x)$.

Example 2

Purpose To find the derivative of a quantity raised to a power:

Problem If $y = (3x^5 - 7x^2 - 4)^{10}$, find $\dfrac{dy}{dx}$.

Solution Employ the previous theorem and note that in actual practice the u substitution is not made.

$$\frac{dy}{dx} = 10(3x^5 - 7x^2 - 4)^9 \cdot \frac{d}{dx}(3x^5 - 7x^2 - 4)$$

$$\frac{dy}{dx} = 10(3x^5 - 7x^2 - 4)^9(15x^4 - 14x).$$

Example 3

Purpose To use Theorem 3.2b in conjunction with other basic theorems:

Problem A $y = (4x^3 - x)^7\sqrt{3x^2 + 5x}$. Find $\dfrac{dy}{dx}$.

Solution Since the function is a product, apply the product rule first.

$$y = (4x^3 - x)^7(3x^2 + 5x)^{1/2}$$

$$\frac{dy}{dx} = (4x^3 - x)^7 \cdot \frac{d}{dx}(3x^2 + 5x)^{1/2} + (3x^2 + 5x)^{1/2} \cdot \frac{d}{dx}(4x^3 - x)^7$$

$$= (4x^3 - x)^7 \cdot \tfrac{1}{2}(3x^2 + 5x)^{-1/2} \cdot \frac{d}{dx}(3x^2 + 5x)$$

$$+ (3x^2 + 5x)^{1/2} \cdot 7(4x^3 - x)^6 \cdot \frac{d}{dx}(4x^3 - x).$$

$$\frac{dy}{dx} = (4x^3 - x)^7 \cdot \tfrac{1}{2}(3x^2 + 5x)^{-1/2}(6x + 5)$$

$$+ (3x^2 + 5x)^{1/2} \cdot 7(4x^3 - x)^6(12x^2 - 1).$$

(Simplification is left for the student.)

Problem B $y = \dfrac{3}{(x^2 - 7x + 2)^4}$. Find y'.

Solution Method I. Since the function is a quotient apply the quotient rule first.

$$y = \frac{3}{(x^2 - 7x + 2)^4}$$

$$y' = \frac{(x^2 - 7x + 2)^4 \cdot \dfrac{d}{dx}(3) - 3 \cdot \dfrac{d}{dx}(x^2 - 7x + 2)^4}{[(x^2 - 7x + 2)^4]^2}$$

$$= \frac{(x^2 - 7x + 2)^4 \cdot (0) - 3 \cdot (4)(x^2 - 7x + 2)^3(2x - 7)}{(x^2 - 7x + 2)^8}$$

$$= \frac{-12(x^2 - 7x + 2)^3(2x - 7)}{(x^2 - 7x + 2)^8}$$

$$y' = \frac{-12(2x - 7)}{(x^2 - 7x + 2)^5}.$$

Method II

$$y = \frac{3}{(x^2 - 7x + 2)^4} = 3(x^2 - 7x + 2)^{-4}$$

$$y' = 3(-4)(x^2 - 7x + 2)^{-5}(2x - 7)$$

$$y' = \frac{-12(2x - 7)}{(x^2 - 7x + 2)^5}.$$

Method II is usually preferred as a more expedient procedure when the numerator is a constant.

There is another technique that could be used to find y' in Example 3, Problem A. This technique, called logarithmic differentiation, is introduced in Chapter 6.

Exercise 3.2

In Problems 1 to 4, find $\dfrac{dy}{dx}$.

1. $y = g(u) = u^{10}$ and $u = h(x) = x^2 + x$

2. $y = g(u) = \sqrt{u}$ and $u = h(x) = 7 - x^3$

3. $y = g(u) = u^3 + 2u + 1$ and $u = h(x) = \dfrac{1}{x^2 - 4}$

4. $y = g(u) = u^2 + \dfrac{1}{u^2}$ and $u = h(x) = x^2 - 5x$

In Problems 5 to 22, find the derivative.

5. $y = (5x + 2)^4$

6. $y = (4 - 3x^2)^7$

7. $s = (3t^3 - 4t + 1)^3$

8. $f(w) = (w^2 + w)^{-1}$

9. $g(v) = (1 - v)(v^2 + 3)^2$

10. $h(x) = 3x^3 + (3x)^3$

11. $y = \dfrac{1}{3x^2 - 7}$

12. $y = \sqrt{2x}$

13. $y = \sqrt[3]{3 - x^2}$

14. $g(x) = \dfrac{-16}{(x^3 - 1)^2}$

15. $y = x^{4/3}(3 + 4x)^3$

16. $f(x) = \dfrac{4x}{\sqrt[3]{6x + 1}}$

17. $y = 3x + \sqrt{\dfrac{3}{x}}$

18. $f(x) = \left(x - \dfrac{1}{x}\right)^2$

19. $y = \sqrt{7 - 2x} + \dfrac{1}{\sqrt{7 - 2x}}$

20. $s = \dfrac{(t^2 + 2t)^2}{t^3 - 3}$

21. $y = \left(\dfrac{x - 4}{2x - 3}\right)^2$

22. $y = \sqrt{x + \sqrt{x}}$

23. The function f is defined by $f(x) = (3x - 5)^2$.
(a) Find the first derivative.
(b) Find the equation of the tangent line to the curve at $x = 2$.
(c) Is there a horizontal tangent line [i.e., can $f'(x) = 0$]?

24. Given $y = (x - 3)(x + 3)^2$.
(a) Find the first derivative of the function.
(b) For what value(s) of x is the first derivative equal to zero? Determine the tangent line to the curve at these value(s) of x.
(c) For what value(s) of x is the first derivative positive (i.e., greater than zero)?
(d) For what value(s) of x is the first derivative negative (i.e., less than zero)?

25. Given $y = \sqrt{x^2 + 1}$.
Find:
(a) the slope of the tangent line to the curve at any point $(x = x_1)$.
(b) the slope of the tangent line to the curve at $x = 0$. Identify this line.

26. Find the coordinates of the point(s) on the graph of $y = (2x - 3)^3$ where the slope of the tangent line to the curve is 0; where the slope of the tangent line to the curve is 6.

27. Given the curve $y = (x^2 + 4)^{3/2}$.
(a) What is the slope of the tangent line at $x = 0$?
(b) For what value of x is the slope of the tangent equal to $\sqrt{45}$?

28. The daily demand function for a certain product is $y = \sqrt{\frac{1}{9}x^2 - 4x + 36}$ where y is the price of the product and x is the quantity demanded. Marginal demand is the derivative of the demand function. Find the marginal demand.

29. The revenue function for a company is $R = 10\sqrt{2Q + 5} + \dfrac{10}{Q}$ where Q is the quantity demanded and R is the revenue in dollars. Find the marginal revenue. Marginal revenue is the derivative of the revenue function. What is the marginal revenue when Q is 10 units?

30. The monthly production cost for a firm, cost in hundreds of dollars, is given by $C(x) = 175(x - \sqrt{2x} - 90)$ where x is the number of units produced, $45 \le x \le 143$. Find the marginal cost at a production output of:
(a) 63 units per month.   (b) 117 units per month.

31. A person's learning level, L, can be represented as the number of years, t, of the person's education by $L = 3(\frac{1}{2}t - 2)^2$ for $4 \le t \le 20$. How fast is the learning level changing when $t = 12$? When $t = 16$?

32. If the function f is defined by $s = f(t) = (4t^2 - 3t + 1)^2$, find:
(a) the velocity at $t = t_0$.   (b) the velocity at $t = 1$.
$\left(\text{Recall: velocity, } v = \dfrac{ds}{dt}. \right)$

33. The distance a particle travels in t seconds can be determined by the equation $s = \sqrt{1 + t^2}$. What is the velocity of the particle for $t = 0$? $t = 2$? $t = 4$?

34. The radius of a spherical balloon is increasing at a rate of $\frac{1}{2}$ inch per minute $\left(\dfrac{dr}{dt} = \frac{1}{2} \right)$. The formula for computing the volume of a sphere with a known radius is $V = \frac{4}{3}\pi r^3$. By using this formula and the chain rule $\left(\dfrac{dV}{dt} = \dfrac{dV}{dr} \cdot \dfrac{dr}{dt} \right)$, determine the change in volume with respect to time, when the radius is 5 inches.

3.3 HIGHER ORDER DERIVATIVES

There are many instances in the calculus and statistics where we need to deal with, and apply, so-called higher order derivatives. A higher order derivative is the derivative of another derivative. For example, the second derivative of a function is the derivative of the first derivative of the function. This will be made clearer by considering the following examples.

Example 1

Purpose To illustrate finding the second-, third-, and fourth-order derivatives:

Problem Given $y = x^3 + 2x^2 - 7x + 5$. Find the first-, second-, third-, and fourth-order derivatives for this function.

Solution	*Derivative*	*Symbols for*	*The Derivative Is:*

Solution

First $\quad \dfrac{dy}{dx}; \; y'; \; D_x y \qquad\qquad \dfrac{dy}{dx} = 3x^2 + 4x - 7$

Second $\quad \dfrac{d^2y}{dx^2}; \; y''; \; D_x^2 y \qquad \dfrac{d^2y}{dx^2} = \dfrac{d}{dx}\left(\dfrac{d}{dx}\right) = \dfrac{d}{dx}(3x^2 + 4x - 7)$

$$= 6x + 4$$

Third $\quad \dfrac{d^3y}{dx^3}; \; y'''; \; D_x^3 y \qquad \dfrac{d^3y}{dx^3} = \dfrac{d}{dx}\left(\dfrac{d^2y}{dx^2}\right) = \dfrac{d}{dx}(6x + 4) = 6$

Fourth $\quad \dfrac{d^4y}{dx^4}; \; y^{iv}; \; D_x^4 y \qquad \dfrac{d^4y}{dx^4} = \dfrac{d}{dx}\left(\dfrac{d^3y}{dx^3}\right) = \dfrac{d}{dx}(6) = 0$

In fact, for this function all higher order derivatives after the fourth will also be equal to zero. Why?

Example 2

Purpose To illustrate finding a second-order derivative:

Problem A If $y = \sqrt[3]{x^2 - 2x + 4}$, find y''.

Solution $y'' = \dfrac{d^2y}{dx^2} = \dfrac{d}{dx}\left(\dfrac{dy}{dx}\right).$

First find y'. $\quad y = (x^2 - 2x + 4)^{1/3}$

$$y' = \frac{1}{3}(x^2 - 2x + 4)^{-(2/3)} \cdot \frac{d}{dx}(x^2 - 2x + 4)$$

$$y' = \frac{1}{3}(x^2 - 2x + 4)^{-(2/3)}(2x - 2)$$

Now $\quad y'' = \dfrac{d^2y}{dx^2} = \dfrac{d}{dx}\left(\dfrac{dy}{dx}\right) = \dfrac{d}{dx}(y')$

Therefore, $\quad y'' = \dfrac{d}{dx}\left[\dfrac{1}{3}(x^2 - 2x + 4)^{-(2/3)}(2x - 2)\right]$

$$= \frac{1}{3}\left[(x^2 - 2x + 4)^{-(2/3)} \cdot \frac{d}{dx}(2x - 2)\right.$$

$$\left. + (2x - 2) \cdot \frac{d}{dx}(x^2 - 2x + 4)^{-(2/3)}\right]$$

$$y'' = \frac{1}{3}\left[(x^2 - 2x + 4)^{-(2/3)} \cdot 2\right.$$

$$\left. + (2x - 2) \cdot \left(-\frac{2}{3}\right)(x^2 - 2x + 4)^{-(5/3)}(2x - 2)\right]$$

From this point it is merely a "cleanup" operation to put this second derivative of y with respect to x into its most concise form.

Problem B   Given $y = x^3 - 9x^2 - 21x - 5$.
(a) Find y' and determine the values of x for which $y' = 0$ and $y' > 0$.
(b) Find y'' and determine the values of x for which $y'' = 0$ and $y'' > 0$.

Solution   (a) $y = x^3 - 9x^2 - 21x - 5$

$$y' = 3x^2 - 18x - 21$$
$$y' = 3(x^2 - 6x - 7) = 3(x + 1)(x - 7)$$
$$y' = 0 \quad \text{when} \quad x = -1 \quad \text{and} \quad x = 7.$$
$$y' > 0 \quad \text{when} \quad x > 7 \quad \text{and} \quad x < -1.$$

(b) $y' = 3x^2 - 18x - 21$

$$y'' = 6x - 18 = 6(x - 3)$$
$$y'' = 0 \quad \text{when} \quad x = 3.$$
$$y'' > 0 \quad \text{when} \quad x > 3.$$

(See Section 3.5, When May the Values of a Function Change Sign?)

In the next chapter, "Applications of the Derivative," you will find that valuable information is attained by examining the first and second derivatives and their signs be they positive, negative, or zero. Also, one of the immediate uses of higher order derivatives is in Taylor's theorem, which has great importance in mathematics and statistics. However, the lack of time compels us to place Taylor's theorem in Appendix VI. Classes with the time available may wish to consider this topic.

In Exercise 3.3–3.4, Problems 1–14 will give you practice in solving problems requiring the use of higher order derivatives.

3.4   IMPLICIT DIFFERENTIATION

Consider the two equations $y = x^2 + 2$ and $xy + 2y = 4$. The first equation, $y = x^2 + 2$, is in the form $y = f(x)$. This form is called *explicit form* for a function. However, the second equation, $xy + 2y - 4 = 0$ is not in the form $y = f(x)$. It is in the form $f(x,y) = 0$. True, we could rewrite $xy + 2y - 4 = 0$ as $y = \dfrac{4}{x + 2}$, and now the function would be in explicit form, $y = f(x)$. However, in its original form, $xy + 2y - 4 = 0$, we will say that the function is specified in *implicit form*. To summarize:

Example	Basic Form	Form is Called
$y = x^2 + 2$	$y = f(x)$	Explicit form
$xy + 2y - 4 = 0$	$f(x,y) = 0$	Implicit form
$x^2 + y^3 + 2y - 4 = 0$	$f(x,y) = 0$	Implicit form

DEFINITIONS

Explicit Form, Implcit Form. If a function is given by an equation of the type $y = f(x)$, the function is said to be in *explicit form*. However, a function defined by an equation of the form $f(x,y) = 0$ is said to be expressed in *implicit form*.

Example 1

Purpose   To classify functions as being expressed in implicit or explicit forms:

Problem   Assuming that the following do represent functions, classify their forms as being implicit or explicit.

(A) $y = x^2 - 2x - 4$ (C) $x + y = 5$
(B) $y = \sqrt{x}$ (D) $x^2y^2 + y = 6$

Solution   (A) and (B) are explicit. The form is $y = f(x)$.

(C) and (D) are implicit. $f(x,y) = x + y - 5 = 0.$
$$f(x,y) = x^2y^2 + y - 6 = 0.$$

The purpose of making a distinction between the two forms of specifying a function is that at present we know how to take the derivative of a function only if it is in explicit form, $y = f(x)$. What, then, do we do about the derivatives of functions when the functions are expressed in implicit form?

One obvious answer would be to switch the form from implicit to explicit and then take the derivative. From one of our examples cited previously, this approach would have worked well.

For, $xy + 2y - 4 = 0$, find $\dfrac{dy}{dx}$.

$$xy + 2y - 4 = 0 \leftrightarrow y = \frac{4}{x + 2} = 4(x + 2)^{-1}$$

$$\frac{dy}{dx} = 4 \cdot (-1)(x + 2)^{-2} \cdot \frac{d}{dx}(x + 2)$$

$$\frac{dy}{dx} = -4(x + 2)^{-2} \cdot 1 = -4(x + 2)^{-2} = \frac{-4}{(x + 2)^2}.$$

However, in many implicit forms, it is difficult and sometimes impossible to solve for y in terms of x. For example, to solve $xy^5 - x^2y^4 + y^3 - 1 = 0$ for y in terms of x is just too much trouble and probably is impossible. Yet, we would like to find $\dfrac{dy}{dx}$. To find the derivative of y with respect to x when the function is expressed implicitly, we resort to a technique known as *implicit differentiation*. We illustrate this technique with several examples.

Example 2

Purpose   To illustrate the technique of implicit differentiation:

Problem   If $y^5 - y^3 - x^2 + 3x + 4 = 0$, find $\dfrac{dy}{dx}$.

Solution   Assume that there is at least one differentiable function defined by $y = f(x)$, implied by this equation.

Remember that x is the independent variable and y is the dependent variable. This is important, since we know that for the independent variable x,

$$\frac{d}{dx}(x^2) = 2x \text{ and } \frac{d}{dx}(3x) = 3.$$

However, y is not an independent variable. y is, in turn, a function of x, $y = f(x)$. Therefore, we must take derivatives of y^n or $[f(x)]^n$ with respect to x by the chain rule. For example:

$$\frac{d}{dx}([f(x)]^5) = 5[f(x)]^4 \cdot f'(x)$$

or

$$\frac{d}{dx}(y^5) = \frac{d}{dy}(y^5) \cdot \frac{dy}{dx} = 5y^4 \cdot y'$$

and

$$\frac{d}{dx}([f(x)]^3) = 3[f(x)]^2 \cdot f'(x)$$

or

$$\frac{d}{dx}(y^3) = \frac{d}{dy}(y^3) \cdot \frac{dy}{dx} = 3y^2 \cdot y'.$$

Returning to our example:

$$y^5 - y^3 - x^2 + 3x + 4 = 0$$

$$\frac{d}{dx}(y^5) - \frac{d}{dx}(y^3) - \frac{d}{dx}(x^2) + \frac{d}{dx}(3x) + \frac{d}{dx}(4) = \frac{d}{dx}(0)$$

$$5y^4y' - 3y^2y' - 2x + 3 + 0 = 0$$

Solve for y':

$$y'(5y^4 - 3y^2) = 2x - 3$$

or

$$y' = \frac{dy}{dx} = \frac{2x - 3}{5y^4 - 3y^2}.$$

Thus, we have found $\dfrac{dy}{dx}$ without ever solving the original equation for $y = f(x)$. True, the derivative does contain y's. However, this is no disadvantage, since in most problems we are evaluating the derivative at a particular point. For example, at $(4,1)$,

$$\frac{dy}{dx}\bigg]_{(4,1)} = \frac{2x-3}{5y^4 - 3y^2}\bigg]_{(4,1)} = \frac{2 \cdot 4 - 3}{5 \cdot 1^4 - 3 \cdot 1^2} = \frac{5}{2}.$$

Example 3

Purpose To illustrate the technique of implicit differentiation in conjunction with a previously learned technique, the product rule:

Problem If $xy - x^2y^3 + 3x = 4$, find y'.

Solution Remember, by the chain rule,

$$\frac{d}{dx}(y) = \frac{d}{dy}(y) \cdot \frac{dy}{dx} = 1 \cdot y' = y'$$

and

$$\frac{d}{dx}(y^3) = \frac{d}{dy}(y^3) \cdot \frac{dy}{dx} = 3y^2 \cdot y'.$$

Applying the product rule to xy and x^2y^3 and differentiating both sides of the equation with respect to x:

$$x \cdot \frac{d}{dx}(y) + y \cdot \frac{d}{dx}(x) - \left[x^2 \cdot \frac{d}{dx}(y^3) + y^3 \cdot \frac{d}{dx}(x^2)\right] + \frac{d}{dx}(3x) = \frac{d}{dx}(4)$$

$$x \cdot y' + y \cdot 1 - x^2 \cdot 3y^2y' - y^3 \cdot 2x + 3 = 0.$$

Solve for y':

$$y'(x - 3x^2y^2) + y - 2xy^3 + 3 = 0$$

$$y'(x - 3x^2y^2) = -y + 2xy^3 - 3$$

$$y' = \frac{-y + 2xy^3 - 3}{x - 3x^2y^2}.$$

Example 4

Purpose Given $f(x,y) = 0$. We assume that there is at least one differentiable function of the form $y = f(x)$ that we can construct. If there are several constructable functions, which of the several functions has its derivative given by the technique of implicit differentiation?

Problem $x^2 + y^2 = 4$. Find y' by implicit differentiation. Which of the two differentiable functions constructable from $x^2 + y^2 = 4$ has its derivative given by the technique of implicit differentiation?

Solution   Remember, by the chain rule,

$$\frac{d}{dx}(y^2) = \frac{d}{dy}(y^2) \cdot \frac{dy}{dx} = 2yy'.$$

Therefore, differentiating both sides of the equation with respect to x:

$$x^2 + y^2 = 4$$

$$2x + 2yy' = 0$$

$$2yy' = -2x$$

$$y' = \frac{-2x}{2y} = -\frac{x}{y}$$

Now, from $x^2 + y^2 = 4$, we can construct two differentiable functions in explicit form:

Function 1: $y = \sqrt{4 - x^2}$, $|x| \leqslant 2$ *Function 2: $y = -\sqrt{4 - x^2}$, $|x| \leqslant 2$*

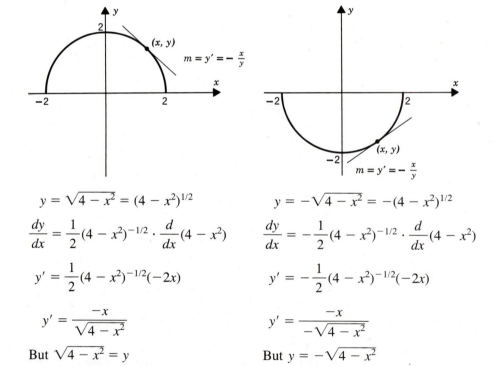

$$y = \sqrt{4 - x^2} = (4 - x^2)^{1/2} \qquad\qquad y = -\sqrt{4 - x^2} = -(4 - x^2)^{1/2}$$

$$\frac{dy}{dx} = \frac{1}{2}(4 - x^2)^{-1/2} \cdot \frac{d}{dx}(4 - x^2) \qquad \frac{dy}{dx} = -\frac{1}{2}(4 - x^2)^{-1/2} \cdot \frac{d}{dx}(4 - x^2)$$

$$y' = \frac{1}{2}(4 - x^2)^{-1/2}(-2x) \qquad\qquad y' = -\frac{1}{2}(4 - x^2)^{-1/2}(-2x)$$

$$y' = \frac{-x}{\sqrt{4 - x^2}} \qquad\qquad\qquad y' = \frac{-x}{-\sqrt{4 - x^2}}$$

But $\sqrt{4 - x^2} = y$ But $y = -\sqrt{4 - x^2}$

Therefore, $y' = -\dfrac{x}{y}$ Therefore, $y' = -\dfrac{x}{y}$

By implicit differentiation, we previously found $y' = -\dfrac{x}{y}$.

Therefore, the technique of implicit differentiation gives us $\dfrac{dy}{dx}$ for the two differentiable functions we have constructed. That is, the derivative obtained by implicit differentiation is the derivative of each of the differentiable functions constructed from $f(x,y) = 0$.

Exercise 3.3–3.4

In Problems 1 to 10, find the second derivative.

1. $y = x^4 - 3x^2 + 2x - 9$

6. $y = \dfrac{1}{x} + x$

2. $s = 16t^2 - 16t + 5$

7. $y = \sqrt{2x}$

3. $y = ax^2 + bx + c$

8. $y = \sqrt{x^2 - 1}$

4. $s = t(4 - 3t)^2$

9. $y = \dfrac{\sqrt[3]{1 - x}}{5}$

5. $P = (Q^2 + 3)(2 - Q)$

10. $y = \dfrac{x^3}{x^3 - 1}$

11. Given $y = (x + 2)(x - 2)^2$
  (a) Find the first derivative of the function.
  (b) For what value(s) of x is the first derivative equal to zero? Describe the tangent line to the curve at these value(s) of x.
  (c) For what value(s) of x is the first derivative positive (i.e., greater than zero)?
  (d) For what value(s) of x is the first derivative negative (i.e., less than zero)?
  (e) Find the second derivative of the function.
  (f) For what value(s) of x is the second derivative equal to zero?
  (g) For what value(s) of x is the second derivative positive?
  (h) For what value(s) of x is the second derivative negative?
  (i) Sketch the function.

12. If $s = f(t)$ defines the motion of an object on a straight line, find the velocity and acceleration for the following functions at the specified times. Velocity is the rate of change of distance with respect to time, $v = f'(t) = \dfrac{ds}{dt}$ and acceleration is the rate of change of velocity with respect to time, $a = f''(t) = \dfrac{dv}{dt}$.
  (a) $s = f(t) = t^3 - 5t + 4$; $t = t_0$, $t = 3$
  (b) $s = f(t) = t^2 - \dfrac{1}{t}$; $t = t_0$, $t = 2$
  (c) $s = f(t) = \sqrt{t^2 + 9}$; $t = t_0$, $t = 4$

13. Given the distance equation $s = (t - 4)^2$.
    (a) Graph this equation.
    (b) Determine and graph the velocity function.
    (c) Determine and graph the acceleration function. (Refer to Problem 12.)

14. (a) Given $y = f(x) = x^4$. Find y', y'', y''', y^{iv}.
    (b) If $y = f(x) = x^n$ (n is a positive integer), what might you conclude to be the nth derivative of the function? Therefore, $D_x^n x^n = ?$

15. Assuming that the following represent functions, classify their forms as being explicit or implicit.

    (a) $x^2 + 2y^2 + y^3 = 7$ (c) $2y - \dfrac{x}{y} = y$

    (b) $y = \dfrac{1}{x} + x$ (d) $3\sqrt{x} + \sqrt{xy} = 4$

In Problems 16 to 25, use implicit differentiation to find $\dfrac{dy}{dx}$.

16. $x^2 + 2x + y^2 = 5$ 21. $x^2 y^2 + 3xy - 4y = 10$
17. $3x^2 - 2y + y^2 = 3$ 22. $x^2 + 2xy^2 - 4x - 3y^2 = 0$
18. $3x^2 + xy - y^2 = 75$ 23. $\sqrt{x} + \sqrt{y} = 9$
19. $x^2 - 2xy + 3y^2 = 16$ 24. $\sqrt{xy} = 4$

20. $(x + y)^2 + (x - y)^2 = a^2$ 25. $\dfrac{1}{x} + \dfrac{1}{y} = 1$

26. Find the equations of the tangent lines to the curve $x^2 + y^2 = 25$ at the points where the x coordinate is 3.

27. The quality of an item, y, is dependent on the number x, which is to be mass produced. The quality for a given number can be found by the equation $x^{-2} - y^3 = 0$. What is $\dfrac{dy}{dx}$ when $x = 1$?

28. In stating the power rule, $\dfrac{d}{dx}(x^n) = nx^{n-1}$, n is any real number. If n is a rational number of the form $\dfrac{p}{q}$ where p and q are integers and $q \neq 0$, show by implicit differentiation that $\dfrac{d}{dx}(x^{p/q}) = \dfrac{p}{q} x^{p/q-1}$. *Hint.* Let $y = x^{p/q}$ or $y^q = x^p$ and differentiate implicitly where $\dfrac{d}{dx}(y^q) = q \cdot y^{q-1} \cdot y'$.

3.5 WHEN MAY THE VALUES OF A FUNCTION CHANGE SIGN?

As a prelude to Chapter 4, "Applications of the Derivative," we often need to determine the values of x at which the sign of the derivative changes from $+$ to $-$ or from $-$ to $+$. Since the derivative is a function, this is part of the broader question: When May the Values of a Function Change Sign? To answer this question, examine the two graphs that follow.

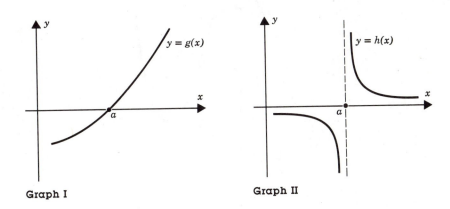

Graph I   Graph II

We see in Graph I that $y = g(x)$ changes sign at $x = a$ from $y = g(x) < 0$ to $y = g(x) > 0$. Also, in Graph I, at $x = a$, $y = g(a) = 0$. In Graph II, $y = h(x)$ changes sign at $x = a$ from $y = h(x) < 0$ to $y = h(x) > 0$. But $y = h(x)$ is discontinuous at $x = a$. Therefore, the conclusion is:

If $y = f(x)$ changes sign ($+$ to $-$, or $-$ to $+$) at $x = a$, then (1) $y = f(a) = 0$ or (2) $y = f(x)$ is discontinuous at $x = a$; that is:

The only place that the values of a function may change sign is where the function has a value of zero or is discontinuous.

This conclusion is extremely important and is a recurring theme throughout the next chapter. Notice that it does not say the values of a function *must* change sign at $(a, f(a))$ if $f(a) = 0$ or f is discontinuous at $x = a$. The conclusion is used to generate values of x where $f(x)$ *may* change sign, and then these values of x are tested further to see if $f(x)$ does actually change sign.

Example 1

Purpose To discover values of x where the values of a function may change sign:

Problem A Investigate $y = x^3 - 3x^2 + 4$ to find values of x where the values of the function may change sign. Verify by a graph when the values of the function do change sign.

Solution $y = x^3 - 3x^2 + 4 = (x - 2)^2(x + 1)$. $y = 0$ when $x = 2$, $x = -1$. The values of x where the function may possibly change sign are at 2 and -1. In this example the function changes sign only at $x = -1$.

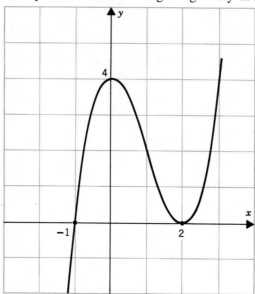

Problem B Investigate $y = \dfrac{1}{x - 3}$ to find values of x where the values of the function may change sign. Verify by a graph when the values of the function do change sign.

Solution $y = \dfrac{1}{x - 3}$ is discontinuous at $x = 3$. There is no value of x for which the function has a value of zero. Therefore, the only possible place that the values of this function may change sign is at $x = 3$. The following graph verifies that the values of the function do, indeed, change sign at $x = 3$.

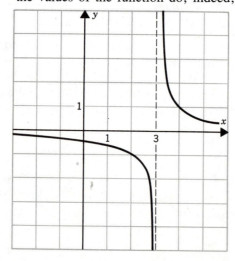

Notice that to determine the sign of a function on a certain interval, provided the interval does not contain points where the function is zero or undefined, it is sufficient to find the sign of the function at any value of x in that interval. Then the function will have the same sign for all x in the interval. This procedure is illustrated in the following example.

Example 2

Purpose   To find values of x where the first derivative function is positive, negative, and zero:

Problem   The function f is defined by $f(x) = 4x^3 - 3x^2 - 60x + 14$. Find the values of x where $f'(x)$ is positive, negative, and zero.

Solution

$$f(x) = 4x^3 - 3x^2 - 60x + 14$$
$$f'(x) = 12x^2 - 6x - 60$$
$$= 6(2x^2 - x - 10)$$
$$= 6(x + 2)(2x - 5)$$
$$f'(x) = 6(x + 2)(2x - 5) = 0 \text{ if } x = -2 \text{ or } x = \tfrac{5}{2}.$$

if $x < -2$,	then	$x + 2 < 0$ $2x - 5 < 0$	and $\quad f'(x) > 0$
if $-2 < x < \tfrac{5}{2}$,	then	$x + 2 > 0$ $2x - 5 < 0$	and $\quad f'(x) < 0$
if $x > \tfrac{5}{2}$,	then	$x + 2 > 0$ $2x - 5 > 0$	and $\quad f'(x) > 0.$

Notice that any value of $x < -2$ when substituted in $f'(x)$ will show $f'(x) > 0$. Similarly, select any x, $-2 < x < \tfrac{5}{2}$, and substitute this value of x into $f'(x)$. For example, if $x = -1$, then $f'(-1) = -42 < 0$. Therefore $f'(x) < 0$ for all $-2 < x < \tfrac{5}{2}$. Also, if $x = 3$, $f'(3) = 30 > 0$. Therefore $f'(x) > 0$ for all $x > \tfrac{5}{2}$.
To summarize:

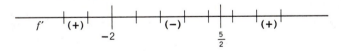

3.6 THE MEAN VALUE THEOREM FOR DERIVATIVES (OPTIONAL UNLESS THE THEOREMS OF CHAPTER 4 ARE TO BE PROVED)

This mean value theorem is of great importance in theoretical mathematics. Although we will not prove this theorem, we will show that it is geometrically plausible. This mean value theorem is an example of an existence theorem, that is, if certain conditions are met, it guarantees the existence of a certain property.

Let the function f be defined and continuous for all $a \leq x \leq b$. Also, let $f'(x)$ exist for all $a < x < b$. Connect points $P(a, f(a))$ and $Q(b, f(b))$ with secant line PQ.

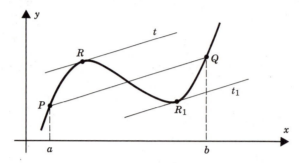

It seems plausible that there will exist at least one point $R(c, f(c))$ on the curve such that the tangent line t to $y = f(x)$ at R is parallel to PQ. In fact, in our drawing, we strongly suspect that there is another point, R_1, such that the tangent line to $y = f(x)$ at R_1 is also parallel to PQ. The important idea, however, is that there exists, at least, one point $R(c, f(c))$, $a < c < b$, such that the tangent line to $y = f(x)$ at R is parallel to secant line PQ.

Recall that the slope of secant line PQ is $\dfrac{f(b) - f(a)}{b - a}$ and that the slope of tangent line t is $\dfrac{dy}{dx}\bigg]_{x=c}$ or $f'(c)$. Therefore, we are saying that if certain conditions are met, we may always find, at least, one value of c, $a < c < b$, such that:

$$\begin{array}{c} \text{The slope of the tangent line} \\ \text{to } y = f(x) \text{ at } R(c, f(c)) \end{array} = \begin{array}{c} \text{The slope of the} \\ \text{secant line } PQ \end{array}$$

or
$$f'(c) = \frac{f(b) - f(a)}{b - a}, \text{ where } a < c < b.$$

Let us now formally state the mean value theorem for derivatives.

THEOREM 3.6

The Mean Value of Theorem for Derivatives. Let the function f be continuous for all $a \leq x \leq b$ and let $f'(x)$ exist for all $a < x < b$. Then, there exists at least one value of x, call it c, where $a < c < b$, such that

$$f'(c) = \frac{f(b) - f(a)}{b - a}.$$

Example 1

Purpose   To illustrate the use of this mean value theorem to find the value of $x = c$, which this theorem guarantees does exist:

Problem   Given $y = f(x) = x^2$ for $-1 \leq x \leq 3$. Find the value of $x = c$ guaranteed by the mean value theorem for derivatives.

Solution $f(x) = x^2$ is a polynomial so it is continuous everywhere, including $-1 \leq x \leq 3$ (see Exercise 2.4–2.5, Problem 55). Also, $f'(x) = 2x$, which exists for all $-1 < x < 3$.

$$a = -1; \; b = 3$$
$$f(b) = f(3) = 3^2 = 9$$
$$f(a) = f(-1) = (-1)^2 = 1$$
$$f'(x) = 2x; \; f'(c) = 2(c) = 2c \qquad \text{where } -1 < c < 3$$

$$\frac{f(b) - f(a)}{b - a} = \frac{9 - 1}{3 - (-1)} = \frac{8}{4} = 2$$

$$\frac{f(b) - f(a)}{b - a} = f'(c)$$

$$2 = 2c$$

$$1 = c$$

Graphically:

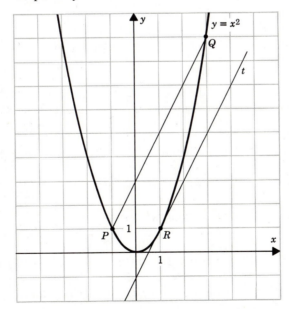

$$\text{Slope of } PQ = \frac{9 - 1}{3 - (-1)} = 2$$

Slope of tangent line to $y = x^2$ at $x = 1$ is $f'(1) = 2 \cdot 1 = 2$. PQ has slope 2; and t has slope 2.

One of the chief uses of this mean value theorem is in the proof of other theorems. This theorem will be used in the proof of certain theorems in subsequent parts of this book.

Exercise 3.5–3.6

In Problems 1 to 6, determine the values of x for which the functions are positive and negative.

1. $f(x) = x^2 - 6x - 7$

2. $f(x) = 7 + 6x - x^2$

3. $f(x) = -x^2 + 5x - 6$

4. $f(x) = (x - 5)^2(x + 3)$

5. $f(x) = \dfrac{1}{x + 2}$

6. $f(x) = \dfrac{x^3}{2 - x}$

In Problems 7 to 12, determine the values of x for which the values of the first derivative of the function are positive and negative.

7. $f(x) = 4 - 3x$

8. $f(x) = 2x^2 + 4x + 6$

9. $f(x) = x^3 - 9x^2 + 27x - 10$

10. $f(x) = \frac{1}{3}x^3 - 4x + 12$

11. $f(x) = x^3 - 7x^2 + 8x + 3$

12. $f(x) = \dfrac{x}{x - 1}$

In Problems 13 to 18, determine the values of x for which the values of the second derivative of the function are positive and negative.

13. $f(x) = x^2 - 10x + 21$

14. $f(x) = x^3 - 7x^2 + 8x + 3$

15. $f(x) = x^3 - 9x^2 + 27x - 10$

16. $f(x) = x^4 - 2x^3 - 36x^2 + 3x$

17. $f(x) = x^4 + 4x^3 + 6x^2 + 7$

18. $f(x) = \dfrac{x}{x + 1}$

19. The function f is defined by $f(x) = x + \dfrac{1}{x}$.

(a) Find $f'(x)$.
(b) Find the values of x for which $f'(x) = 0$. Describe the tangent lines to the curve at these values of x.
(c) For what values of x is $f'(x) > 0$? $f'(x) < 0$?
(d) Find $f''(x)$.
(e) For what values of x is $f''(x) > 0$? $f''(x) < 0$?

In Problems 20 to 25, find the value c guaranteed by the mean value theorem for the function f defined by $f(x)$.

20. $f(x) = 2x^2 - 8x + 5,\ 1 \leqslant x \leqslant 4$

21. $f(x) = x^2 - 2x + 3,\ -1 \leqslant x \leqslant 2$

22. $f(x) = x^2 - 2x - 3,\ -1 \leqslant x \leqslant 3$

23. $f(x) = x^3 + 1,\ 1 \leqslant x \leqslant 4$

24. $f(x) = \sqrt{x + 3},\ 1 \leqslant x \leqslant 6$

25. $f(x) = \dfrac{1}{x},\ \frac{1}{2} \leqslant x \leqslant 2$

26. A biologist has determined the following function that gives the bacteria population in an infected person as a function of time: $B(t) = t^4 + 1000$. Find the value of $x = c$ guaranteed by the mean value theorem for derivatives on the interval $1 \leqslant x \leqslant 3$.

■ 3.7 CHAPTER REVIEW

Important Ideas

$$\frac{d}{dx}(x^n) = n \cdot x^{n-1}$$

$$\frac{d}{dx}[g(x) + h(x)] = g'(x) + h'(x)$$

$$\frac{d}{dx}[g(x) - h(x)] = g'(x) - h'(x)$$

$$\frac{d}{dx}[g(x) \cdot h(x)] = g(x) \cdot h'(x) + h(x) \cdot g'(x)$$

$$\frac{d}{dx}[c \cdot h(x)] = c \cdot h'(x)$$

$$\frac{d}{dx}\left[\frac{g(x)}{h(x)}\right] = \frac{h(x)g'(x) - g(x)h'(x)}{[h(x)]^2}$$

chain rule

$$\frac{d}{dx}[f(x)]^n = n \cdot [f(x)]^{n-1} \cdot f'(x)$$

higher order derivatives
implicit differentiation
when may the values of a function change sign?
(the mean value theroem)

■ **REVIEW EXERCISE (optional)**

In Problems 1 to 50 use the differentiation formulas to find the first derivatives.

1. $y = 6$

2. $f(x) = 2 - 7x$

3. $g(x) = x^2 + 4x - 10$

4. $s = (t + 1)^2$

5. $y = ax^2 + bx + c$

6. $u = v^4 - 12$

7. $f(x) = 5x^6 - 2x^3 + x$

8. $y = \frac{1}{2}(x^4 + 8x^3 - 6x)$

9. $s = 6(6 + t^3)^3$

10. $u = \dfrac{v^2 - \sqrt{v}}{2}$

11. $y = (x + 7)(2 - x^2)$

12. $y = x(x^3 - 4x + 1)$

13. $s = 2t^2 + (2t)^2$

14. $g(x) = \sqrt{3}(x - x^2)^4$

15. $g(x) = x(2x + 3)^4$

16. $s = \sqrt{16t}$

17. $f(t) = 4 + \dfrac{1}{t}$

18. $g(x) = x^{1/2} - x^{-1/2}$

19. $y = \dfrac{13}{x^2}$

20. $u = \dfrac{2}{3\sqrt{v}}$

21. $y = \dfrac{\sqrt{9 - x^2}}{5}$

22. $f(x) = \dfrac{15}{\sqrt[3]{3x^2}}$

23. $y = (x + 1)(x^{-1/2})$

24. $f(x) = (16 - 3x^2)^5$

25. $y = \dfrac{4}{\sqrt{x^3 + 3}}$

26. $y = \dfrac{1}{2x} + \sqrt{2x}$

27. $y = \sqrt{2x} + \sqrt{\dfrac{2}{x}}$

28. $y = \dfrac{x^2 - 5x - 14}{x}$

29. $y = \dfrac{10 - 7x}{x^2}$

30. $f(x) = \dfrac{6}{(1 - 2x^2)^3}$

31. $y = \dfrac{2}{x}(x^4 + x - 7)$

32. $f(x) = \dfrac{\sqrt{2x - 1}}{x^3}$

33. $y = \dfrac{2 - x}{x - 2}$

34. $y = \dfrac{x + 1}{\sqrt{x}}$

35. $y = \dfrac{\sqrt{x}}{x + 1}$

36. $y = \dfrac{2x - 1}{x^2 + 2}$

37. $u = \dfrac{1 + x^2}{x^2 - 1}$

38. $h(x) = \dfrac{2 + x^2}{3 + 4x + x^2}$

39. $f(x) = \dfrac{x^2 - 2x + 1}{x^2 + 2x + 1}$

40. $s = \dfrac{(t^2 + 1)^2}{2 + t^2}$

41. $y = \left(\dfrac{7 + x}{2x - 3}\right)^2$

42. $f(t) = \left(\dfrac{3t^2 + 1}{1 - 5t}\right)^2$

43. $y = \dfrac{2x + 3}{x + 4}(7x + 1)$

44. $y = \dfrac{4 - x}{(x + 3)^3(5 + x^2)}$

45. $x^2 + y^2 = 49$

46. $x^2 + y^2 - y^3 = 0$

47. $xy + 7x - 10 = 0$

48. $xy^2 + y^3 = 5$

49. $x^2y + 16y = 4$

50. $xy + y^2 = 11$

51. Find the equation of the tangent line to the curve $y = \dfrac{6}{x}$ at $(2,3)$.

52. Evaluate $\lim\limits_{\Delta x \to 0} \dfrac{[(x + \Delta x)^2 + 4] - [x^2 + 4]}{\Delta x}$. This is the derivative of what function?

53. Evaluate $\lim\limits_{\Delta x \to 0} \dfrac{[3(x + \Delta x) - 2]^2 - [3x - 2]^2}{\Delta x}$. Of what function is this the derivative?

54. Given $y = 7 + 8x^2 - x^4$.
   (a) Find the first derivative.
   (b) Find the equation of the tangent line to the curve at $x = 1$.
   (c) Find the point(s) on the curve where $y' = 0$.

55. The function f is defined by $f(x) = \sqrt{4x + 13}$.
   (a) Find $f'(x)$.
   (b) Find the equation of the tangent line to the curve at $x = 3$.
   (c) Is there a horizontal tangent line? [i.e., can $f'(x) = 0$?]
   (d) Find $f''(x)$.
   (e) Can $f''(x) > 0$? Explain.

56. Given $y = x^3 + 3x^2 - 9x + 12$.
   (a) Find the first derivative of the function.
   (b) For what value(s) of x is the first derivative equal to zero? Describe the tangent line to the curve at these value(s) of x.
   (c) For what value(s) of x is the first derivative positive?
   (d) For what value(s) of x is the first derivative negative?
   (e) Find the second derivative of the function.
   (f) For what value(s) of x is the second derivative equal to zero?
   (g) For what value(s) of x is the second derivative positive?
   (h) For what value(s) of x is the second derivative negative?

57. The population of sparrows nesting in a given area is recorded each spring. It is found that the population grows each year according to the equation $N(t) = 300t + .3t^2$. How many sparrows are nesting in the fifth spring? What is the rate of growth in population for the fifth year?

58. Given the demand function $D(x) = -\frac{3}{4}x^2 + 9x + 9$.
   (a) Find the change in price necessary to bring about a one-unit change in the quantity demanded, that is, $D'(x)$.
   (b) Find the total revenue.
   (c) Find the marginal revenue.

59. Given the cost function, $C(x) = \dfrac{3x^2}{x + 5} + 400$; find the marginal cost.

60. If $s = f(t)$ defines the motion of an object on a straight line, then the velocity, $v = \dfrac{ds}{dt}$ and acceleration, $a = \dfrac{dv}{dt}$. If $s = f(t) = \sqrt{t^2 + 5}$,

(a) find the velocity at $t = 2$, and

(b) show that the acceleration at $t = 2$ is $\frac{5}{27}$.

61. An object is falling such that its distance above the ground after t seconds can be found by the equation $s = 155 - 16t^2$.

(a) What is the velocity of the falling object at the instant when $t = t_1$ seconds?

(b) What is the instantaneous velocity for t equal to 1 second? 2 seconds? 3 seconds?

62. Find the equations of tangent lines to the curve $x^2 - y^2 = 5$ at $x = 3$.

63. Find the equation of the tangent line to the curve $\sqrt{x} + \sqrt{y} = 4$ at $x = 1$.

64. Find the value c guaranteed by the mean value theorem for the function f defined by $f(x)$.

(a) $f(x) = x^2 + 3x - 3$, $0 \leqslant x \leqslant 4$

(b) $f(x) = \sqrt{x + 1}$, $0 \leqslant x \leqslant 8$

■ IN RETROSPECT

At the beginning of this chapter, our only method for finding a derivative was the definition:

$$f'(x) = \lim_{\Delta x \to 0} \frac{f(x + \Delta x) - f(x)}{\Delta x}.$$

This chapter has been concerned with the methods for arriving at the correct answer for $f'(x)$ without actually using this definition.

The reason for sidestepping the use of the definition for finding $f'(x)$ is that finding derivatives by definition is often a lengthy and time-consuming process. Therefore, we have now examined techniques that give us the same answer, $f'(x)$, as the definition, but we arrive at this answer much sooner. Do you wonder why you had to learn the definition of the derivative in the first place? To answer your question, the definition was used to prove the convenient techniques you have acquired. Also, the definition of the derivative involving the limit is needed in more advanced work including certain applications.

The next chapter will consider some of the many applications that are made of derivatives. We begin to see the derivative's power. The one concept, derivative, has applications to a multitude of problems from many different academic disciplines.

FOUR

Applications of
the Derivative

4.1 OVERVIEW OF THIS CHAPTER

This chapter has as its focal point two applications of the derivative. The first application is in the graphical analysis of functions. This application is augmented by the use of the derivative in extremum theory, that is, the use of the derivative to establish the maximum or minimum values that a function may attain.

In econometric theory, particularly, the ability to obtain the maximum value of a function is a fundamental tool. Indeed, the economist's model, used to make predictions about economic phenomena, is built on the assumption of maximizing behavior. Also, in statistics the least-squares method is based on the requirement that a certain error be minimized.

These are but two special applications of the theory in this chapter. It should be evident that graphing and extremum theory are important to you. Almost all disciplines are concerned with some aspect of the theory of maxima and minima and with acquiring the ability to readily graph functions.

Before the theory of maxima and minima can be developed, we must take a brief look at the behavior of functions. This work is a foundation for the theory of maxima and minima and for graphing techniques. The chapter culminates in applied problems that can be stated as maximum or minimum problems.

Our goal is to gain an understanding of the theory and to be able to solve applied problems like the following:

1. An automobile manufacturer can sell x cars per hour at $\$(7070 - .4x^2)$ per car. It costs the manufacturer $\$(2000x + 5200)$ to produce x cars per hour. To maximize profit, how many cars should be sold?

2. If a piece of food becomes lodged in the windpipe, the windpipe contracts to increase the velocity of airflow in order to dislodge and carry along the food. If the velocity of airflow in the windpipe is given by $V = C(r_0 - r)r^2$ where r_0 is the at rest radius of the windpipe and r is the radius of the windpipe when it contracts, find the value of r that will maximize the velocity V.

4.2 INCREASING AND DECREASING FUNCTIONS

This section is a start in "classifying" the graph of a function. We want to make mathematically precise the concepts: "the graph is rising," and "the graph is falling." Also, the theorem is proved that enables us to classify a function as *increasing* or *decreasing*.

DEFINITION

Increasing Function. A function f is said to be *increasing* on the interval $a < x < b$ if $f(x_1) < f(x_2)$ whenever x_1 and x_2 belong to the interval and $x_1 < x_2$ (see Figure 4.1).

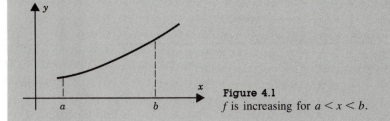

Figure 4.1
f is increasing for $a < x < b$.

In a similar manner the definition of a decreasing function is given.

DEFINITION

Decreasing Function. A function f is said to be *decreasing* on the interval $a < x < b$ if $f(x_1) > f(x_2)$ whenever x_1 and x_2 belong to the interval and $x_1 < x_2$ (see Figure 4.2).

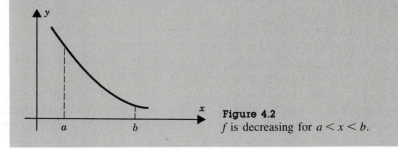

Figure 4.2
f is decreasing for $a < x < b$.

Now examine Figures 4.3 and 4.4. In Figure 4.3 the function is increasing and the slope of the tangent line at any point on the curve is positive. In Figure 4.4 the function is decreasing and the slope of the tangent line at any point on the curve is negative.

It is very dangerous to jump to a conclusion concerning slopes and increasing and decreasing functions on the basis of two figures. However, we may form a conjecture and then try to prove the conjecture. Since the derivative gives the slope of the tangent line, we "guess" that a positive slope of the tangent line or

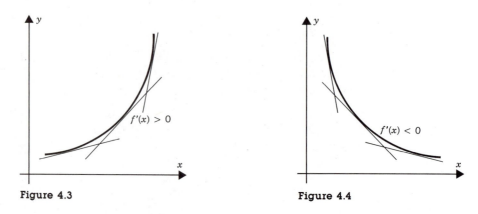

Figure 4.3   **Figure 4.4**

a positive derivative implies an increasing function and that a negative derivative implies a decreasing function.

Test for Increasing and Decreasing Functions

THEOREM 4.2a

If $f'(x) > 0$ on the interval $a < x < b$, then f is an increasing function on this interval.

GIVEN. $f'(x) > 0$ for $a < x < b$.

TO PROVE. f is an increasing function on this interval.

Proof

1. Let x_1 and x_2 be two points in $a < x < b$ with $x_1 < x_2$.

2. f is continuous for $x_1 \leq x \leq x_2$ and differentiable for $x_1 < x < x_2$. Therefore, the hypotheses for the mean value theorem are satisfied. By the mean value theorem, there exists a value c such that:

$$\frac{f(x_2) - f(x_1)}{x_2 - x_1} = f'(c), \; x_1 < c < x_2$$

    or

$$f(x_2) - f(x_1) = f'(c) \cdot (x_2 - x_1)$$

3. Since $f'(x) > 0$ for $a < x < b$ it follows that $f'(c) > 0$, and $x_1 < x_2$ implies $x_2 - x_1 > 0$.

4. $\therefore f(x_2) - f(x_1) = f'(c) \cdot (x_2 - x_1) > 0$.

5. $f(x_2) > f(x_1)$ or f is an increasing function for $a < x < b$.

> **THEOREM 4.2b**
>
> If $f'(x) < 0$ on the interval $a < x < b$, then f is a decreasing function on this interval. (The proof is left to the student.)

In conclusion, the sign of $f'(x)$ gives us information about the graph of the original function f.

Information About f Obtained from the Sign of f'(x)

$f'(x)$		f
$f'(x) > 0$	implies	f is increasing abbreviated $\uparrow$
$f'(x) < 0$	implies	f is decreasing abbreviated $\downarrow$

Also recall that if $f'(x_1) = 0$, the curve has a horizontal tangent at $x = x_1$, since the slope of the tangent line to the curve is 0.

Example 1

Purpose   To determine intervals for which a function is increasing and decreasing; also, to find values of x where the curve has horizontal tangents:

Problem A   Find the intervals for which the function $f(x) = x^3 + x^2 - 8x - 1$ is increasing and decreasing and the values of x where the curve has horizontal tangents.

Solution   We need to remember:

$$\text{If } f'(x) > 0, \quad \text{then} \quad f \text{ is increasing, } \uparrow.$$
$$\text{If } f'(x) < 0, \quad \text{then} \quad f \text{ is decreasing, } \downarrow.$$

The only place that the values of a function may change sign is where the function has a value of zero or is discontinuous. Since f' is a function, the only place the values of f' may change sign is where $f'(x) = 0$ or f' is discontinuous.

$$f(x) = x^3 + x^2 - 8x - 1$$
$$f'(x) = 3x^2 + 2x - 8 \qquad \text{Factor}$$
$$f'(x) = (x + 2)(3x - 4)$$

We must determine when $f'(x)$ is positive, negative, zero, or discontinuous. f' is never discontinuous (every polynomial is continuous for all x, see Exercise 2.4–2.5, Problem 55).

$$f'(x) = (x + 2)(3x - 4) = 0 \text{ if } x = -2 \text{ or } x = \tfrac{4}{3}.$$

$$\text{if } x < -2, \qquad \text{then} \qquad \begin{array}{l} x + 2 < 0 \\ 3x - 4 < 0 \end{array} \qquad \text{and} \qquad f'(x) > 0$$

$$\text{if } -2 < x < \tfrac{4}{3}, \qquad \text{then} \qquad \begin{array}{l} x + 2 > 0 \\ 3x - 4 < 0 \end{array} \qquad \text{and} \qquad f'(x) < 0$$

$$\text{if } x > \tfrac{4}{3}, \qquad \text{then} \qquad \begin{array}{l} x + 2 > 0 \\ 3x - 4 > 0 \end{array} \qquad \text{and} \qquad f'(x) > 0.$$

Therefore:

$x < -2$	$-2 < x < \tfrac{4}{3}$	$x > \tfrac{4}{3}$
$f'(x) > 0$, $(+)$	$f'(x) < 0$, $(-)$	$f'(x) > 0$, $(+)$
f is ↑	f is ↓	f is ↑

To summarize:

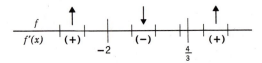

In conclusion, by Theorem 4.2a and 4.2b, the function is increasing on $x < -2$ and $x > \tfrac{4}{3}$. The function is decreasing on $-2 < x < \tfrac{4}{3}$. When x is equal to -2 and $\tfrac{4}{3}$ the curve has horizontal tangents.

Graphically:

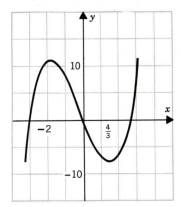

Problem B   Find the intervals for which the function $f(x) = \dfrac{x}{x + 1}$ is increasing and decreasing. Are there any horizontal tangents?

Solution
$$f(x) = \frac{x}{x + 1}$$

$$f'(x) = \frac{1}{(x + 1)^2} \qquad \text{Verify this step.}$$

f' is discontinuous at $x = -1$.

When $x > -1$, $f'(x) > 0$ which implies f is ↑.
When $x < -1$, $f'(x) > 0$ which implies f is ↑.

To summarize:

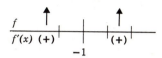

Therefore, the function is increasing for $x < -1$ and increasing for $x > -1$. Also, there are no horizontal tangents, since $f'(x)$ is never equal to zero.

Graphically:

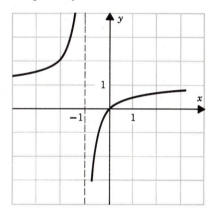

Example 2

Purpose   To illustrate the concept of an increasing or decreasing function in relation to a problem of economics:

Problem   A company producing x units of a commodity has determined that it can describe the total costs by $C(x) = \$(50 + 30x)$, the demand for its product by $D(x) = \$(90 - x)$, the total revenue by $R(x) = x \cdot D(x)$, and the profit by $P(x) = R(x) - C(x)$.

(A)   Describe the marginal cost, $C'(x)$.
(B)   Describe the marginal revenue, $R'(x)$.
(C)   When is the profit increasing? Decreasing?
(D)   Graph the profit function.

Solution   (A) $C(x) = 50 + 30x$
$C'(x) = 30$

The marginal cost is a constant $30. $C'(x) > 0$ for all x, which implies the cost function is always increasing.

(B) $R(x) = x(90 - x) = 90x - x^2$, $0 \leqslant x \leqslant 90$
$R'(x) = 90 - 2x = -2(x - 45)$
When $0 \leqslant x < 45$, R ↑.
When $45 < x \leqslant 90$, R ↓.

(C) $P(x) = (90x - x^2) - (50 + 30x) = -50 + 60x - x^2, \; x \geq 0$
   $P'(x) = 60 - 2x$
   When $0 \leq x < 30$, $P \uparrow$.
   When $x > 30$, $P \downarrow$.

(D)   Graphically:

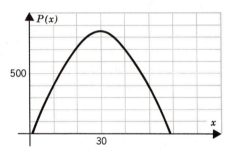

Exercise 4.2

1. From the function illustrated by the graph below, state the values of x for which the function is increasing and decreasing.

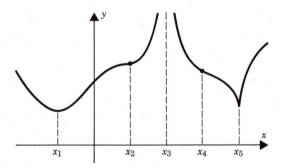

2. A first derivative function, f', is graphed below. State the values of x for which the original function f is increasing and decreasing.

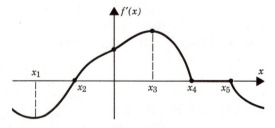

In Problems 3 to 17, determine the x values for which the functions are increasing, decreasing, or have horizontal tangents.

3. $f(x) = 3x + 4$ 4. $f(x) = 4 - 3x$

5. $f(x) = x^2 - 6x - 7$

6. $f(x) = 7 + 6x - x^2$

7. $f(x) = x^2 - 10x + 21$

8. $f(x) = 4x^2 - 7x - 15$

9. $f(x) = (x - 4)(x + 3)$

10. $f(x) = x^3 - 3x^2 - 9x + 15$

11. $f(x) = 2x^3 - 7x^2 + 4x - 2$

12. $f(x) = 2x^3 - 12x^2 + 24x$

13. $f(x) = x^3 - 3x + 3$

14. $f(x) = \frac{1}{2}x^4 - x^2$

15. $f(x) = \frac{1}{4}(x^4 - 6x^2 + 8x + 12)$

16. $f(x) = \dfrac{x^2}{x^2 + 1}$

17. $f(x) = \dfrac{x + 2}{x - 2}$

18. Graph Problems 4, 5, 6, 13, and 17.

19. A function f is defined by $f(x) = mx + b$. When is the function:
    (a) Increasing? (b) Decreasing? (c) What happens when $m = 0$?

20. A function f is defined by $f(x) = ax^2 + bx + c$, where $a > 0$. When is the function: (a) Increasing? (b) Decreasing? (c) What happens when
    $$x = -\frac{b}{2a}?$$

21. A company finds a product demand function to be $f(x) = 150 - 3x$. Is this an increasing or decreasing function? If the price, y, the company wants for each item increases, what happens to the quantity of items sold, x? If the price is decreased, what happens to the number of items sold?

22. A company finds its product supply function to be $f(x) = 2x + 5$. Is this an increasing or decreasing function?

23. The learning of a research animal at a given time, t, $(t \geq 0)$ is given by $L = 12t^{1/2} - 5$. When is the rate of learning increasing? Decreasing?

24. To test the effectiveness of a pesticide, a population of fruit flies was given the pesticide in its food. The growth in population is determined by the function $P = -10t^2 + 20t + 100$, where time is measured in weeks. $(0 \leq t \leq 4)$
    (a) Is the fruit fly population increasing or decreasing in the first week?
    (b) Determine the time intervals in which the population is increasing and decreasing.

25. A body in space is moving according to the law $s = t^{3/2} - 6t$.
    (a) Determine the time interval in which the distance traveled is in a positive direction. In a negative direction. (i.e., $s > 0$, $s < 0$.)
    (b) Determine the time interval in which the distance traveled is increasing. Decreasing. (i.e., $v > 0$, $v < 0$.)

26. A mobile home company finds that its total revenue may be determined by $R(x) = \$[160,000 - (x - 400)^2]$. When is the revenue function increasing? Decreasing?

27. An automobile manufacturer can sell x cars per hour at $\$(7070 - .4x^2)$ per

car. It costs the manufacturer $\$(2000x + 5200)$ to produce x cars per hour. Find:

(a) The marginal cost.

(b) The average cost, $\dfrac{C(x)}{x}$ or $\overline{C}(x)$.

(c) The marginal average cost, the change in average cost per unit change in output.

(d) Show that the marginal average cost $= \dfrac{(x)(\text{marginal cost}) - C(x)}{x^2}$.

(e) Is the average cost function increasing or decreasing?

(f) Verify: if the average cost function is decreasing then it is more than the marginal cost. [*Hint.* Use part (d).]

28. The daily production cost for a factory to manufacture x deluxe contour chairs is given to be $\$(500 + 14x + \frac{1}{2}x^2)$. The demand function is $\$(150 - \frac{3}{2}x)$. Find:

(a) The marginal cost.

(b) The average cost.

(c) The marginal average cost, the change in average cost per unit change in output.

(d) Show that the marginal average cost $= \dfrac{(x)(\text{marginal cost}) - C(x)}{x^2}$.

(e) For what values of x is the average cost function decreasing?

(f) When is the demand function increasing or decreasing?

29. The following changes in oxygen pressure (P_{O_2}) have been recorded.

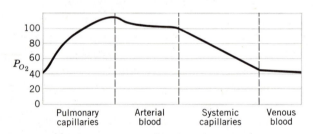

Is P_{O_2} increasing or decreasing in the

(a) Pulmonary capillaries?

(b) Arterial blood?

(c) Systemic capillaries?

(d) Venous blood?

30. Prove Theorem 4.2b. If $f'(x) < 0$ for all $a < x < b$, then f is a decreasing function on this interval.

4.3 RELATIVE EXTREMA—RELATIVE MAXIMUM AND RELATIVE MINIMUM

We now consider one of the most important applications of the derivative. Here, we use the first derivative as an aid in determining the ''high points'' or the ''low points'' on a curve. To do this precisely, we follow the mathematical procedure of defining a relative maximum and relative minimum.

DEFINITION

Relative Maximum. A function f has or attains a *relative maximum* at $x = x_1$, if there exists an interval $a < x < b$, containing x_1, such that $f(x) \leq f(x_1)$ for all x belonging to this interval (see Figure 4.5).

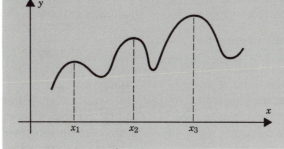

Figure 4.5
f has a relative maximum at $x = x_1$, x_2, and x_3.

DEFINITION

Relative Minimum. A function f has or attains a *relative minimum* at $x = x_1$ if there exists an interval $a < x < b$ containing x_1, such that $f(x) \geq f(x_1)$ for all x belonging to this interval (see Figure 4.6).

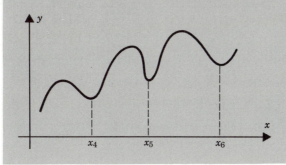

Figure 4.6
f has a relative minimum at $x = x_4$, x_5, and x_6.

First Test for Relative Maximum and Relative Minimum

We now investigate Figure 4.7 and attempt to formulate a test for a relative maximum occurring at $x = x_1$.

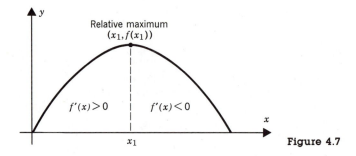

Figure 4.7

For a function to have a relative maximum at $x = x_1$:

1. f must be increasing immediately to the left of x_1, and
2. f must be decreasing immediately to the right of x_1.

Thus, f' must change sign at x_1 from $+$ to $-$.

In a similar manner investigate Figure 4.8 to formulate a test for a relative minimum at $x = x_1$.

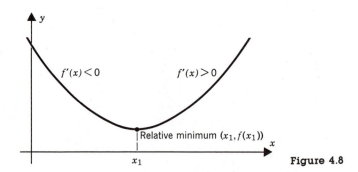

Figure 4.8

For a function to have a relative minimum at $x = x_1$:

1. f must be decreasing immediately to the left of x_1, and
2. f must be increasing immediately to the right of x_1.

Thus, f' must change sign at x_1 from $-$ to $+$.

For emphasis, these criteria are restated.

Criteria for Relative Extrema

Condition	Criteria
Relative maximum at $x = x_1$	f changes at $x = x_1$ from an increasing to a decreasing function or $f'(x)$ changes sign at $x = x_1$ from $+$ to $-$
Relative minimum at $x = x_1$	f changes at $x = x_1$ from a decreasing to an increasing function or $f'(x)$ changes sign at $x = x_1$ from $-$ to $+$

We notice from these criteria that for relative extrema to exist at $x = x_1$, $f'(x)$ must change sign at $x = x_1$. Recall that the only place a function may change sign is at a value of x where the function either has a value of zero or is discontinuous. This motivates stating the following theorem whose formal proof will not be given.

THEOREM 4.3

If the function f is continuous in the interval about x_1, and has a relative extremum (relative maximum or relative minimum) at $x = x_1$, then $f'(x_1) = 0$ or f' is discontinuous at $x = x_1$.

Theorem 4.3 gives us a means for generating candidates for values of x where f may attain a relative extremum. Such a candidate is called a critical number of f.

DEFINITION

Critical Number. Let x_1 be a number in the domain of the function f. If $f'(x_1) = 0$ or f' is discontinuous at $x = x_1$, then x_1 is called a *critical number* of f.

Our method for obtaining relative extrema will be as follows.

First Derivative Test

Given $y = f(x)$.

1. Find $f'(x)$.

2. See where $f'(x) = 0$ and where f' is discontinuous. The most common

place for f' to be discontinuous is at those values of x that cause the denominator of $f'(x)$ to be equal to zero. These values of x that cause $f'(x)$ to be zero and f' to be discontinuous generate the set of critical numbers of f. Find these critical numbers, which are candidates for the values of x where f attains a relative extrema.

3. Test each critical number to see if $f'(x)$ does change sign at that critical number.

If $f'(x)$ changes sign at a critical number from
(a) $+$ to $-$, then f has a relative maximum at $x = x_1$ (see Figure 4.7),
(b) $-$ to $+$, then f has a relative minimum at $x = x_1$ (see Figure 4.8).

Example 1

Purpose   To illustrate the technique of finding relative extrema for given functions:

Problem A   The function f is defined by $f(x) = x^2 - 4x - 5$. Find relative extrema for this function.

Solution   1.   Find $f'(x)$.
$$f(x) = x^2 - 4x - 5$$
$$f'(x) = 2x - 4$$

2.   See where $f'(x) = 0$ and where f' is discontinuous. Find critical numbers.
$f'(x) = 0$ when $x = 2$.
$x = 2$ is the critical number.

3.   Test each critical number for relative extrema.
When $x < 2$, $f'(x) < 0$.
When $x > 2$, $f'(x) > 0$.
Therefore, by the first derivative test, a relative minimum exists at $x = 2$.
The value of the relative minimum is $y = -9$.

To summarize:

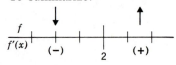

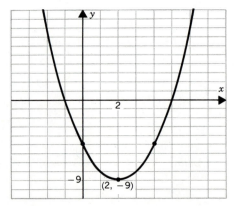

Problem B Find relative extrema for the function f defined by $f(x) = x^3 + x^2 - 8x - 1$.

Solution 1. $f(x) = x^3 + x^2 - 8x - 1$

$f'(x) = 3x^2 + 2x - 8 = (3x - 4)(x + 2)$

2. $f'(x) = 0$ when $x = -2$ and $x = \frac{4}{3}$.

$x = -2$ and $x = \frac{4}{3}$ are critical numbers.

3. In an interval about $x = \frac{4}{3}$,

when $-2 < x < \frac{4}{3}$, $f'(x) < 0$; when $x > \frac{4}{3}$, $f'(x) > 0$.

Therefore, by the first derivative test, a relative minimum will exist at $x = \frac{4}{3}$. The value of the relative minimum is $y = \frac{-203}{27}$.

In an interval about $x = -2$,

when $x < -2$, $f'(x) > 0$; when $-2 < x < \frac{4}{3}$, $f'(x) < 0$.

Therefore, by the first derivative test, a relative maximum will exist at $x = -2$. The value of the relative maximum is $y = 11$.

To summarize:

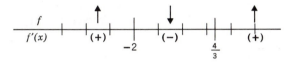

The graph of f follows.

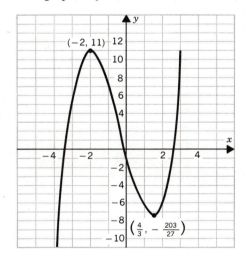

Problem C Find relative extrema, if possible, for the function f defined by $f(x) = \sqrt[3]{x}$.

Solution $f(x) = \sqrt[3]{x} = x^{1/3}$

$$f'(x) = \tfrac{1}{3}x^{-2/3} = \frac{1}{3\sqrt[3]{x^2}}$$

There are no values of x for which $f'(x) = 0$. However, if $x = 0$, then $f'(x)$ is discontinuous. Although $x = 0$ is a critical number, it is neither a relative maximum nor a relative minimum. $f'(x)$ does not change sign; in fact, $f'(x)$ is positive for all x except $x = 0$, and the function is increasing.

Graphically:

Problem D   Find relative extrema for the function f defined by $f(x) = \dfrac{1}{x}$.

Solution   1. $f(x) = \dfrac{1}{x}$.

$$f'(x) = -\frac{1}{x^2}.$$

2. f' is discontinuous at $x = 0$.

3.   When $x < 0$, $f'(x) < 0$.
When $x > 0$, $f'(x) < 0$.
There is no relative extrema, since the requirements for the first derivative test are not met. The function f is decreasing for $x < 0$ and decreasing for $x > 0$. The graph of f follows.

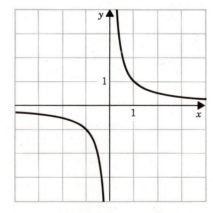

Problem E The profit for a company is determined by the profit function P defined by $P(x) = \$[(90x - x^2) - (50 + 30x)]$. Find x, the number of units that would yield a maximum profit.

Solution $P(x) = [(90x - x^2) - (50 + 30x)]$

$P(x) = -50 + 60x - x^2$

$P'(x) = 60 - 2x, \ x \geq 0$

$P'(x) = 0$ when $x = 30$

When $0 \leq x < 30$, $P'(x) > 0$.

When $x > 30$, $P'(x) < 0$.

Therefore, 30 units of the commodity will yield a relative maximum profit of $850.

Section 4.8 will give an extensive treatment of applied extrema problems.

4.4 ABSOLUTE EXTREMA—ABSOLUTE MAXIMUM AND ABSOLUTE MINIMUM

Supplementing the ideas of relative or local extrema are the ideas of global or absolute extrema. They are quite easily grasped, so we move directly to the definitions.

DEFINITION

Absolute Maximum. A function f has or attains an *absolute maximum* at $x = x_1$ if $f(x_1) \geq f(x)$ for all x belonging to the domain of f.

Note. A function may have its absolute maximum at a place of relative maximum or at an end point of its domain (see Figure 4.9).

DEFINITION

Absolute Minimum. A function f has or attains an *absolute minimum* at $x = x_1$ if $f(x_1) \leq f(x)$ for all x belonging to the domain of f.

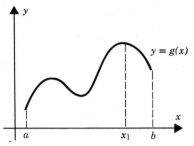

$y = g(x)$, $a \leq x \leq b$, has an absolute maximum at $x = x_1$, a value of x where $y = g(x)$ also has a relative maximum.

Figure 4.9

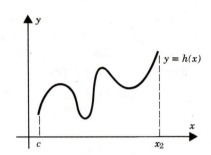

$y = h(x)$, $c \leq x \leq x_2$, has an absolute maximum at $x = x_2$, a value of x where $y = h(x)$ does not have a relative maximum. $x = x_2$ is an end point of the domain of $y = h(x)$.

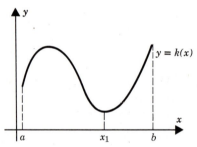

$y = k(x)$, $a \leq x \leq b$, has an absolute minimum at $x = x_1$, a value of x where $y = k(x)$ also has a relative minimum.

Figure 4.10

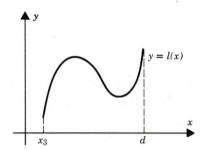

$y = l(x)$, $x_3 \leq x \leq d$, has an absolute minimum at $x = x_3$, a value of x where $y = l(x)$ does not have a relative minimum. $x = x_3$ is an end point of the domain of $y = l(x)$.

Note. A function may have its absolute minimum at a place of relative minimum or at an end point of its domain (see Figure 4.10).

This leads us to conclude that absolute extrema may occur at a place of relative extrema or at an end point of the domain of a function. If an absolute extremum occurs at an end point of the domain of a function, it is called an *end point extremum*.

Unfortunately, not all functions have absolute extrema. (See the Review Exercise, Problems 22 and 23.) However, there is a theorem in advanced calculus that guarantees that if a function f is continuous on the closed interval $a \leq x \leq b$, then f has an absolute maximum and an absolute minimum value on $a \leq x \leq b$. Assuming that f does have absolute extrema, we have the following method for obtaining them.

Method for Obtaining Absolute Extrema

Given $y = f(x)$.

1. Find $f'(x)$.
2. Find the critical numbers of f. This is accomplished by finding those values of x for which $f'(x) = 0$ or f' is discontinuous.
3. Determine the end points of the domain of f. Let a and b be the end points of the domain of f.
4. Find the corresponding values of $y = f(x)$ at each critical number and at $x = a$ and $x = b$. Select from these values of y the largest y value, if an absolute maximum is desired, or the smallest y value, if an absolute minimum is of interest.

Example 1

Purpose   To illustrate the technique of finding absolute extrema:

Problem A   Given $f(x) = x^2 - 4x - 5$, $-1 \leqslant x \leqslant 3$. Find the absolute extrema of this function.

Solution $f(x) = x^2 - 4x - 5$; $-1 \leqslant x \leqslant 3$

1. Find $f'(x)$.
    $f'(x) = 2x - 4$
2. Find critical numbers.
    $x_1 = 2$
    $f(x_1) = f(2) = -9$
3. Determine values of $f(x)$ at end points of domain.
    $a = -1$; $f(a) = f(-1) = 0$
    $b = 3$; $f(b) = f(3)\quad = -8$
4. To find absolute maximum, select the largest y value from $f(a), f(b)$, and the values of $f(x)$ at the critical numbers.
    $f(a) = f(-1) = 0$
    $f(b) = f(3) = -8$
    $f(x_1) = f(2) = -9$
    Absolute maximum value is $y = 0$.
    To find absolute minimum, select the smallest y value from $f(a), f(b)$, and the values of $f(x)$ at the critical numbers.
    $f(a) = f(-1) = 0$
    $f(b) = f(3) = -8$
    $f(x_1) = f(2) = -9$
    Absolute minimum value is $y = -9$.

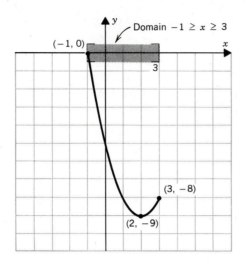

Problem B Given $f(x) = x^3 + x^2 - 8x - 1$ with domain, $-4 \leq x \leq 2$. Find the absolute extrema of this function.

Solution $f(x) = x^3 + x^2 - 8x - 1$.

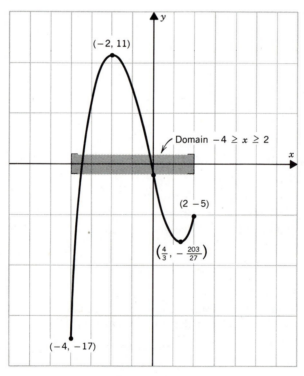

1. $f'(x) = 3x^2 + 2x - 8$
$= (3x - 4)(x + 2)$

2. $x_1 = -2$ $x_2 = \frac{4}{3}$
      $f(x_1) = f(-2) = 11$ $f(x_2) = f(\frac{4}{3}) = -\frac{203}{27}$

3. $a = -4$ $b = 2$
      $f(a) = f(-4) = -17$ $f(b) = f(2) = -5$

4. $f(a) = f(-4) = -17$
      $f(b) = f(2) = -5$
      $f(x_1) = f(-2) = 11$
      $f(x_2) = f(\frac{4}{3}) = \frac{-203}{27}$
      Absolute maximum value is $y = 11$.
      Absolute minimum value is $y = -17$.

Exercise 4.3–4.4

In Problems 1 to 20, find the relative extrema (relative maxima or relative minima) of the functions, and indicate the interval(s) on which the functions are increasing or decreasing.

1. $f(x) = 3x + 4$
2. $f(x) = x^2 - 6x - 7$
3. $f(x) = 7 + 6x - x^2$
4. $f(x) = 4x^2 - 7x - 15$
5. $f(x) = -x^2 + 10x - 21$
6. $f(x) = (x + 4)^2$
7. $f(x) = 8 - x^2$
8. $f(x) = x^3 - 3x^2 - 9x + 15$
9. $f(x) = 2x^3 - 7x^2 + 4x - 2$
10. $f(x) = 2x^3 - 12x^2 + 24x$
11. $f(x) = x^3 - 3x + 3$
12. $f(x) = \frac{1}{3}x^3 - x^2 + 3$

13. $f(x) = \frac{1}{2}x^4 - x^2$
14. $f(x) = x^4 - 2x^3$
15. $f(x) = \frac{1}{4}(x^4 - 6x^2 + 8x + 12)$
16. $f(x) = \dfrac{x^2}{x^2 + 1}$
17. $f(x) = \dfrac{x + 2}{x - 2}$
18. $f(x) = \dfrac{1}{x^2}$
19. $f(x) = 3 + x^{2/3}$
20. $f(x) = 3x^5 - 20x^3$

21. Graph Problems 2, 5, 11, 12, 13, and 18.

22. In economics, the cost functions are sometimes specified in cubic equations with the level of output designated by Q. In general, the total cost function is given by $C(Q) = aQ^3 + bQ^2 + cQ + d$ where $a, c, d > 0$ and $b < 0$. (Furthermore, the condition for the marginal cost being greater than zero at all levels of output, Q, is $b^2 < 3ac$.) From the cost function, terms commonly used are:

$$\text{Total fixed cost} = d$$
$$\text{Total variable cost} = aQ^3 + bQ^2 + cQ$$
$$\text{Marginal cost} = \frac{dC}{dQ}$$

$$\text{Average fixed cost} = \frac{d}{Q}$$

$$\text{Average variable cost} = \frac{aQ^3 + bQ^2 + cQ}{Q}$$

$$\text{Average total cost} = \frac{C}{Q}$$

If the total cost function is $C(Q) = Q^3 - 12Q^2 + 60Q + 200$,
(a) Find the average variable cost function, AC.
(b) What is the level of output, Q, that will minimize the AC function?
(c) Find the marginal cost function, MC.
(d) Graph both the marginal cost function and the average variable cost function on the same coordinate axes. Where do these graphs intersect?
(e) From the graph, further note that the marginal cost curve lies *above* the average cost curve when the average cost is increasing. What condition prevails when the average cost is decreasing?

23. Given the total cost function, $C(Q) = Q^3 - 10Q^2 + 40Q + 300$,
(a) Find the average variable cost function, AC.
(b) What is the level of output that will minimize the AC function?
(c) Find the marginal cost function, MC.
(d) Graph both the AC and MC functions on the same coordinate axes. Where do these graphs intersect? (See Problem 22.)

24. Consider the general total cost function, $C(Q) = aQ^3 + bQ^2 + cQ + d$. Find the level of output, Q, that will minimize the average variable cost function. Show this Q value is the Q coordinate for the point of intersection of the AC and MC functions. (See Problem 22.)

25. Before a big game team members are told to ''get psyched.'' A player's performance level is related to the psyche or arousal level. Up to a certain point arousal will increase performance. After this point any more arousal will cause a decrease in performance since the player will become too tense and unable to concentrate. A coach finds the relationship to be $P = -\frac{1}{50}A^2 + 2A + 22$ where P is the performance level and A is the arousal level, both on a scale from 1 to 100. For what values of the arousal level is performance increasing? When is performance decreasing? Find the relative maximum performance.

26. A psychologist gives a rat an electric shock to stimulate the rat's movement through the maze. The rat's performance time, T minutes, for completing the maze is related to the voltage of the shock, V, by the equation $T = 26 - V + .01V^2$. For what values of V is the rat's performance time decreasing? For what values of V does the rat's performance time increase? For what voltage does the rat's performance time reach a minimum value?

In Problems 27 to 32, determine the absolute extrema (absolute maximum and absolute minimum) of the functions over their respective domains. Graph each of the functions over their respective domains.

27. $f(x) = x^2 - 6x - 7; \ 1 \leqslant x \leqslant 4$

28. $f(x) = -x^2 + 10x - 21; \ 2 \leqslant x \leqslant 6$

29. $f(x) = x^3 - 3x + 3; \ -3 \leqslant x \leqslant \frac{3}{2}$

30. $f(x) = \frac{1}{3}x^3 - x^2 + 3; \ -3 \leqslant x \leqslant 2$

31. $f(x) = 3 + x^{2/3}; \ -1 \leqslant x \leqslant 8$

32. $f(x) = 2 + 4x - x^4; \ 0 \leqslant x \leqslant 2$

33. Indicate from the graph the values of x where the function has or attains a relative extrema (relative maximum and relative minimum).

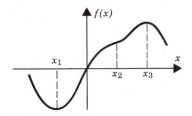
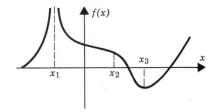

34. Indicate from the graph the values of x where the function has or attains absolute extrema (absolute maximum and absolute minimum).

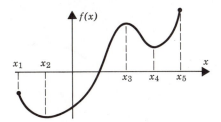
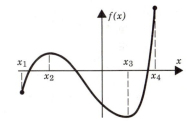

4.5 INFORMATION FROM THE SECOND DERIVATIVE—CONCAVITY AND POINT OF INFLECTION

The sign of the second derivative furnishes us useful information concerning the shape of a graph. We need to examine definitions for concave upward, concave downward, and point of inflection. Also, the theory of the second derivative gives us a second test for relative maxima and relative minima. Many people prefer this second test for relative extrema and resort to the first derivative test only if the second test fails to yield a conclusion, or whenever the task of finding the second derivative becomes prohibitive.

DEFINITION

Concave Upward. A curve is said to be *concave upward* on an interval $a < x < b$ if the curve lies above its tangent lines for all x belonging to the interval. Concave upward is abbreviated by $\cup$.

DEFINITION

Concave Downward. A curve is said to be *concave downward* on an interval $a < x < b$ if the curve lies below its tangent lines for all x belonging to the interval. Concave downward is abbreviated by $\cap$.

If the function f has a graph that is concave upward on the interval $a < x < b$, then f is said to be concave upward on the interval. A similar convention is adopted for concave downward.

Test for Concavity

We now investigate Figure 4.11 and attempt to formulate a test for concavity.

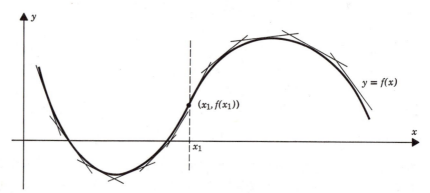

Curve lies above its tangent lines.
Curve is concave upward, $\cup$.
As x increases, the slope of the tangent line increases.
f' is an increasing function.

Curve lies below its tangent lines.
Curve is concave downward, $\cap$.
As x increases, the slope of the tangent line decreases.
f' is a decreasing function.

Figure 4.11

Thus, we see that if,

1. f is concave upward then f' is an increasing function. Therefore, the deriva-

tive of this increasing function must be positive. That is, f'' must be positive when f is concave upward.

2. f is concave downward then f' is a decreasing function. Therefore, the derivative of this decreasing function must be negative. That is, f'' must be negative when f is concave downward.

These results are summarized below.

In conclusion, the sign of $f''(x)$ gives us information about the graph of the function f.

Information About f Obtained from the Sign of f''(x)

$f''(x)$		f
$f''(x) > 0$	implies	f is concave upward abbreviated $\cup$
$f''(x) < 0$	implies	f is concave downward abbreviated $\cap$

Also of interest to us is the point or points of the curve where the tangent line crosses the curve. A point such as $(x_1, f(x_1))$ in Figure 4.11 is a point where the tangent line crosses the curve, and the sense of concavity changes, in this case, from upward to downward. This particular type of point is an example of a point of inflection.

DEFINITION

Point of Inflection. A point, $(x_1, f(x_1))$, on a curve, $y = f(x)$, where the sense of concavity of the curve changes, is called a *point of inflection.*

Since at a point of inflection, the definition requires that the sense of concavity must change, it follows that $f''(x)$ must change sign at $x = x_1$. Again, f'' is a function, and if $f''(x)$ changes sign at $x = x_1$, then either $f''(x_1) = 0$ or f'' is a discontinuous function at $x = x_1$. This yields Theorem 4.5.

THEOREM 4.5

If the function f has a point of inflection at $x = x_1$, then either $f''(x_1) = 0$ or f'' is discontinuous at $x = x_1$.

Theorem 4.5 gives us a means of testing candidates for values of x where f may have a point of inflection. Such a candidate is called a hypercritical number of f.

DEFINITION

Hypercritical Number. Let x_1 be a number in the domain of the function f. If $f''(x_1) = 0$ or f'' is discontinuous at $x = x_1$, then x_1 is called a *hypercritical number of f*.

The systematic procedure for finding points of inflection is as follows:

Method for Obtaining Points of Inflection

Given $y = f(x)$.

1. Find $f''(x)$.

2. Determine those values of x for which $f''(x) = 0$ or f'' is discontinuous. These values of x are known as hypercritical numbers.

3. Test to see if $f''(x)$ changes signs at a hypercritical number. If $f''(x)$ changes signs at hypercritical number x_1, then f has a point of inflection at $(x_1, f(x_1))$, providing x_1 is a number in the domain of f.

Example 1

Purpose   To find the interval(s) in which the given function is concave upward and concave downward and to find points of inflection for a given function:

Problem A   Find the intervals for which the function $f(x) = x^3 + x^2 - 8x - 1$ is concave upward and concave downward and find the points of inflection.

Solution
1. Find $f''(x)$.
$$f(x) = x^3 + x^2 - 8x - 1$$
$$f'(x) = 3x^2 + 2x - 8$$
$$f''(x) = 6x + 2$$

2. Determine the values for x where $f''(x) = 0$ or where f'' is discontinuous to find hypercritical numbers.
$f''(x) = 0$ when $x = -\frac{1}{3}$.
$x = -\frac{1}{3}$ is the hypercritical number.

3. Does $f''(x)$ change signs at the hypercritical number? If $f''(x)$ changes sign at a hypercritical number, then this number is the x coordinate of a point of inflection.
$f''(x) = 6x + 2$
In an interval about $x = -\frac{1}{3}$.
when $x > -\frac{1}{3}$, $f''(x) > 0$;
when $x < -\frac{1}{3}$, $f''(x) < 0$.

Therefore, when $x > -\frac{1}{3}$, f is $\cup$; when $x < -\frac{1}{3}$, f is $\cap$ and $(-\frac{1}{3}, \frac{47}{27})$ is a point of inflection.

To summarize:

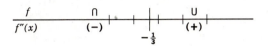

Graph of f:

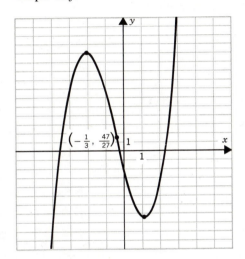

Problem B   The function f is defined by $f(x) = \dfrac{1}{x}$. Is the curve concave upward for all allowable values of x?

Solution   1. $f(x) = \dfrac{1}{x}$

$f'(x) = -\dfrac{1}{x^2}$

$f''(x) = \dfrac{2}{x^3}$

2. $f''(x) = 0$ for no values of x.
$f''(x)$ does not exist when $x = 0$.

3. $f''(x) = \dfrac{2}{x^3}$

When $x > 0$, $f''(x) > 0$.
When $x < 0$, $f''(x) < 0$.
No, the curve is not concave upward for all allowable values of x, it is

concave upward when $x > 0$ and concave downward when $x < 0$. $f(0)$ does not exist so a point of inflection does not exist.

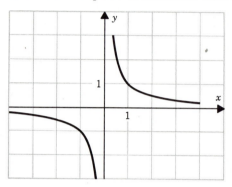

4.6 SECOND TEST FOR RELATIVE MAXIMA AND RELATIVE MINIMA

Examine Figures 4.12a and 4.12b.

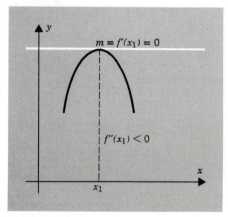

Figure 4.12a
Relative maximum at $x = x_1$.

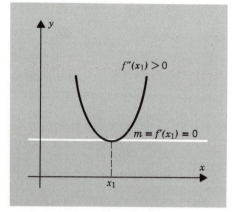

Figure 4.12b
Relative minimum at $x = x_1$.

These two drawings lead us to an alternate test for relative maxima and relative minima.

Second Derivative Test

Let f' and f'' exist at every point of an interval $a < x < b$ that contains x_1 and let $f'(x_1) = 0$.

  I.   If $f''(x_1) < 0$, f has a relative maximum at $x = x_1$.

  II. If $f''(x_1) > 0$, f has a relative minimum at $x = x_1$.

 III. If $f''(x_1) = 0$, this test fails.

That the test fails if $f''(x_1) = 0$ is easily seen by investigating these three graphs.

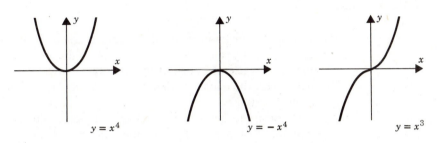

$$y = x^4 \qquad y = -x^4 \qquad y = x^3$$

Consider the functions defined by $y = x^4$, $y = -x^4$, and $y = x^3$. Each of these three functions has its second derivative equal to zero at $x = 0$. However, $y = x^4$ has a relative minimum at $x = 0$, $y = -x^4$ has a relative maximum at $x = 0$, and $y = x^3$ has neither a relative minimum nor a relative maximum at $x = 0$. Therefore, if $f''(x_1) = 0$, this test gives no conclusion and the first derivative test must be used to establish relative maxima or relative minima.

Example 1

Purpose To use the second test for relative maxima or relative minima:

Problem A Refer to Section 4.3, Example 1, Problem A, page 176. Find the relative extrema for the function defined by $f(x) = x^2 - 4x - 5$.

Solution $f(x) = x^2 - 4x - 5$

$f'(x) = 2x - 4$ when $x = 2$; $f'(x) = 0$

$f''(x) = 2$

$f''(2) = 2$

Therefore, since $f''(2) > 0$ and $f'(2) = 0$, there exists a relative minimum when $x = 2$. The value of the relative minimum is -9.

Problem B Refer to Section 4.3, Example 1, Problem B, page 177. Find the relative extrema for the function defined by $f(x) = x^3 + x^2 - 8x - 1$.

Solution
$$f(x) = x^3 + x^2 - 8x - 1$$
$$f'(x) = 3x^2 + 2x - 8 = (3x - 4)(x + 2)$$

When $x = \frac{4}{3}$, and -2; $\quad f'(x) = 0$

$$f''(x) = 6x + 2$$

$f'(-2) = 0$	$f'(\frac{4}{3}) = 0$
$f''(-2) < 0$	$f''(\frac{4}{3}) > 0$

Therefore, a relative maximum exists when $x = -2$. The value of the relative maximum is $y = 11$.

Therefore, a relative minimum exists when $x = \frac{4}{3}$. The value of the relative minimum is $y = \frac{-203}{27}$

4.7 THE USE OF MAXIMUM, MINIMUM, AND POINTS OF INFLECTION IN GRAPHING

We now apply our knowledge of increasing and decreasing functions, relative extrema, concavity, and points of inflection to the analysis of graphs. This does not rule out our use of points of intersection, knowledge of symmetry, and the like. However, the one thing that we will avoid is the laborious, tedious task of constructing extensive tables of x and y values. There will be certain times when we must construct these tables, but let us agree to do so only as a last resort.

A summary of our analysis to the first and second derivative, with their applications to graphs, follows. Assume that the function, f, is defined in the given interval.

Derivative	*Nature of the Function*	*Graph of the Function*

Increasing, decreasing, and concavity

$f'(x) > 0$ $f''(x) > 0$	f is $\uparrow$ f is $\cup$	
$f'(x) > 0$ $f''(x) < 0$	f is $\uparrow$ f is $\cap$	
$f'(x) < 0$ $f''(x) > 0$	f is $\downarrow$ f is $\cup$	
$f'(x) < 0$ $f''(x) < 0$	f is $\downarrow$ f is $\cap$	

Relative extrema

$f'(x_1) = 0$
$f'(x) > 0,\ x < x_1$
$f'(x) < 0,\ x > x_1$
or
$f'(x_1) = 0$
$f''(x_1) < 0$

Relative maximum
at $x = x_1$

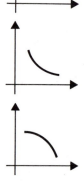

Derivative	*Nature of the Function*	*Graph of the Function*
$f'(x_1) = 0$ $f'(x) < 0, \; x < x_1$ $f'(x) > 0, \; x > x_1$ or $f'(x_1) = 0$ $f''(x_1) > 0$	Relative minimum at $x = x_1$	

Example 1

Purpose To illustrate "sophisticated" techniques of graphing:

Problem A Graph $f(x) = \frac{1}{4}x^4 - \frac{3}{2}x^2 + 2x + 1$.

Solution 1. Find $f'(x)$. Determine critical numbers which may yield relative extrema and determine the intervals where f is increasing, ↑, and decreasing, ↓.

$$f'(x) = x^3 - 3x + 2$$
$$f'(x) = (x - 1)^2(x + 2)$$
$$f'(x) = 0 \text{ when } x = 1 \text{ and } -2.$$

Therefore, 1 and −2 are critical numbers.

When $x < -2$, $f'(x) < 0$ and f is ↓
   $-2 < x < 1$, $f'(x) > 0$ and f is ↑
     $x > 1$, $f'(x) > 0$ and f is ↑

To summarize:

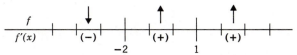

$(-2, f(-2)) = (-2, -5)$ is a relative minimum.

2. Find $f''(x)$. Determine hypercritical numbers that may yield points of inflection and determine the intervals where f is concave upward, ∪, and concave downward, ∩.

$$f''(x) = 3x^2 - 3$$
$$f''(x) = 3(x + 1)(x - 1)$$
$$f''(x) = 0 \text{ when } x = 1 \text{ and } x = -1$$

Therefore, 1 and −1 are hypercritical numbers.

When $x < -1$, $f''(x) > 0$ and f is ∪
   $-1 < x < 1$, $f''(x) < 0$ and f is ∩
     $x > 1$, $f''(x) > 0$ and f is ∪

To summarize:

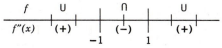

$(1, f(1)) = (1, \frac{7}{4})$ and $(-1, f(-1)) = (-1, -\frac{9}{4})$ are points of inflection.

3. Graph $f(x) = \frac{1}{4}x^4 - \frac{3}{2}x^2 + 2x + 1$

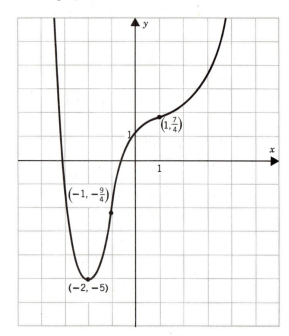

Note: It is not necessary to plot additional points.

Problem B Graph $f(x) = x(6 - x)^2$.

Solution 1. Find $f'(x)$, determine the critical numbers, and intervals where f is $\uparrow$ and $\downarrow$.

$$f'(x) = 3(x - 6)(x - 2)$$

The student should check this result!
To summarize:

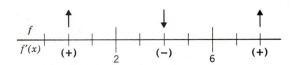

2. Find $f''(x)$, determine hypercritical numbers, and intervals where f is $\cup$ and $\cap$.

$$f''(x) = 6(x - 4)$$

To summarize:

f		$\cap$		$\cup$	
$f''(x)$		$(-)$	4	$(+)$	

3. Graph $f(x) = x(6 - x)^2$

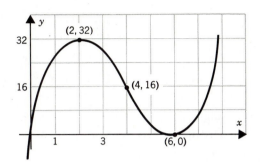

Note: (1) $f(x) = 0$ when $x = 0$ and $x = 6$

(2) extrema at $(2,32)$ and $(6,0)$

(3) point of inflection at $(4,16)$

Problem C Make a sketch of a function f defined in the interval about $x = 2$ where $f'(2) = 0$, $f'(x) > 0$ for $x \neq 2$; $f''(x) < 0$ for $x < 2$, and $f''(x) > 0$ for $x > 2$.

Solution To summarize the conditions given by the first and second derivatives

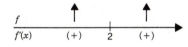   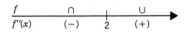

Sketch of f:

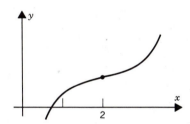

Exercise 4.5–4.7

In Problems 1 to 8, find $f''(x)$, the point(s) of inflection and determine the intervals in which the function is concave upward and concave downward.

1. $f(x) = x^2 + x + 21$

2. $f(x) = 16 - x^2$

3. $f(x) = x^3 - 3x^2 + 4$

4. $f(x) = 2x^3 + 3x^2 - 12x - 9$

5. $f(x) = 2x^3 + 4x - 2$

6. $f(x) = \frac{1}{2}x^4 - x^2$

7. $f(x) = \frac{1}{12}(x^4 - 8x^3 + 18x^2 + 9)$

8. $f(x) = 3x^5 - 20x^3$

In Problems 9 to 18, find the relative extrema (relative maxima or relative minima) of the function using the second derivative test for relative maxima and relative minima.

9. $f(x) = x^2 - 6x - 7$

10. $f(x) = 7 + 6x - x^2$

11. $f(x) = (x + 4)^2$

12. $f(x) = x^3 - 3x^2 + 4$

13. $f(x) = 2x^3 - 12x^2 + 24x$

14. $f(x) = \frac{1}{3}x^3 - x^2 + 3$

15. $f(x) = 2x^3 - 9x^2 + 12x + 1$

16. $f(x) = x^3 - 3x + 3$

17. $f(x) = \frac{1}{4}(x^4 - 8x^3 + 18x^2 + 9)$

18. $f(x) = x^4 - 2x^3$

In Problems 19 to 45, find $f'(x)$ and $f''(x)$. Locate extrema (state whether a relative maximum or minimum) and point(s) of inflection. Indicate the intervals in which the function is increasing, decreasing, concave upward or concave downward. Graph each function.

19. $f(x) = x^2 - 8x + 12$

20. $f(x) = 15 + 2x - x^2$

21. $f(x) = (x - 3)^2$

22. $f(x) = -(x - 3)^2$

23. $f(x) = 9 - x^2$

24. $f(x) = x^3 - 3x^2 - 9x + 15$

25. $f(x) = 2x^3 - 9x^2 + 12x + 1$

26. $f(x) = x^3 - 5x^2 + 3x + 9$

27. $f(x) = x^3 - 12x + 1$

28. $f(x) = \frac{1}{3}x^3 + x^2 + x - 1$

29. $f(x) = \frac{1}{3}x^3 - x^2 + 3$

30. $f(x) = 2x^3 - 12x^2 + 24x$

31. $f(x) = x^3 - 9x^2 + 15x$

32. $f(x) = 3 + 3x - x^3$

33. $f(x) = \frac{1}{4}(x^4 - 6x^2 + 8x + 12)$

34. $f(x) = \frac{1}{2}x^4 - x^2$

35. $f(x) = x^4 - 2x^3$

36. $f(x) = \frac{1}{12}(x^4 - 8x^3 + 18x^2 + 9)$

37. $f(x) = (x - 1)^2(x + 1)^2$

38. $f(x) = x^4 + 4x^3 - 16x + 2$

39. $f(x) = 3x^5 - 20x^3$

40. $f(x) = 3 + x^{2/3}$

41. $f(x) = (1 + x)^{1/3}$

42. $f(x) = x + \dfrac{1}{x}$

43. $f(x) = x - \dfrac{1}{x}$

44. $f(x) = x\sqrt{x + 1}$

45. $f(x) = \sqrt{2x - x^2}$

46. Sketch the function f in the interval about $x = 3$ where $f'(3) = 0, f'(x) < 0$ for $x < 3$, $f'(x) > 0$ for $x > 3$, and $f''(x) > 0$.

47. Sketch the function f in the interval about $x = a$ where $f'(x) > 0$ for $x < a$, $f'(x) < 0$ for $x > a$, and $f''(x) > 0$ for $x \neq a$.

48. Sketch the function f in the interval about $x = x_1$ where $f'(x) > 0$ for $x < x_1, f'(x) < 0$ for $x > x_1, f''(x) > 0$ for $x < x_1$ and $f''(x) < 0$ for $x > x_1$.

4.8 APPLIED EXTREMA PROBLEMS

This type of problem occurs in every branch of knowledge. In this section you will see how the calculus relates to your field of interest. However, the problems are selected primarily to show the concept of maximizing or minimizing a quantity and, when appropriate, the problems are related to a particular academic discipline.

Basic Extremum Problem. The problem contains a quantity that is to be maximized or minimized. Let us call this quantity Q.

Solution to a Basic Extremum Problem

1. First you must write an equation containing Q, the quantity to be maximized or minimized. Q should be a function of a single independent variable, that is, $Q = f(x)$.

   Warning! If you are not able to write this equation for Q, the quantity to be maximized or minimized, don't waste your time churning your wheels by taking derivatives. You may take derivatives all over your paper but with no equation for Q, there will never be any solution to the problem.

2. Find the absolute extrema for Q by following the previously described technique for obtaining absolute extrema.

   To help you gain a feeling for these applied extrema problems, let us look at a variety of examples.

Example 1

Purpose To apply this technique to the solution of applied extrema problems:

Problem A What number exceeds its square by a maximum amount?

Solution Let x be the desired number.

$$M = x - x^2$$

Maximize M, $$M' = 1 - 2x$$

$$M' = 0 \text{ when } x = \tfrac{1}{2}.$$

Test $x = \tfrac{1}{2}$ by the first or second derivative test. Using the second derivative test, $M'' = -2$. Since $M'' < 0$, $x = \tfrac{1}{2}$ is the number that exceeds its square by a maximum amount.

Problem B If a piece of food becomes lodged in the windpipe, the windpipe contracts to increase the velocity of airflow in order to dislodge and carry along the food. If the velocity of airflow in the windpipe is given by $V = C(r_0 - r)r^2$ where r_0 is

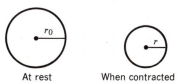

At rest   When contracted

the at rest radius of the windpipe and r is the radius of the windpipe when it contracts, find the value of r that will maximize the velocity V.

Solution   To find the maximum velocity, take the derivative of $V(r) = C(r_0 - r)r^2$ and find the extrema by setting $V'(r)$ equal to zero.

$$V(r) = C(r_0 - r)r^2$$
$$= C(r_0 r^2 - r^3)$$
$$V'(r) = C(2r_0 r - 3r^2)$$
$$= Cr(2r_0 - 3r)$$
$$V'(r) = 0$$
$$0 = Cr(2r_0 - 3r)$$

$$Cr = 0 \quad \text{or} \quad 2r_0 - 3r = 0$$
$$r = 0 \qquad\qquad 3r = 2r_0$$
$$r = \tfrac{2}{3}r_0$$

Since the radius of the windpipe cannot be zero, the maximum velocity must be produced when the windpipe is $\tfrac{2}{3}r_0$. X rays taken during coughing have shown that the human body contracts its windpipe to $\tfrac{2}{3}$ of its rest radius, thus producing maximum velocity. See Problem 47 in Exercise 4.8 to verify that $r = \tfrac{2}{3}r_0$ is an absolute maximum.

Graphically:

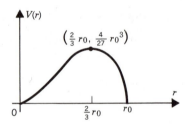

Problem C   A manufacturing company has found that by ordering goods at precise times they can make a larger profit on their product. This is often called inventory control. This particular company has determined that the following function describes profit as a function of inventory ordering time: $P(t) = 12t - .25t^2$, where t is the time in days after the inventory drops below 2000 parts. Determine the time when parts should be reordered to maximize the profit.

Solution
$$P(t) = 12t - .25t^2$$
$$P'(t) = 12 - .5t$$
$$P'(t) = 0$$
$$12 - .5t = 0$$
$$t = \frac{12}{.5} = 24 \text{ days}$$

$P''(t) = -.5$. By the second derivative test, the maximum occurs at $t = 24$.

Therefore, the company should reorder 24 days after the inventory drops below 2000 parts.

Problem D A rectangular piece of cardboard is 5 inches by 8 inches. Identical small squares are cut from each corner and the remaining cardboard section is folded to form a box without a cover. Find the dimensions of the box that will yield the largest volume. What is the value of this maximum volume?

Solution The quantity to be maximized, in this instance the volume of a box, must be written as an equation.

Since $V = l \cdot w \cdot h$, for our particular problem,
$$l = 8 - 2x, \quad w = 5 - 2x, \quad h = x$$
and $V = f(x) = (8 - 2x)(5 - 2x)x$

where the domain of f is $0 \leqslant x \leqslant \frac{5}{2}$.

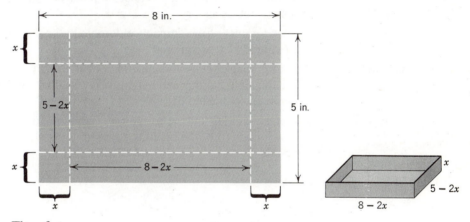

Therefore

$$V = f(x) = (8 - 2x)(5 - 2x)x$$
$$V = f(x) = 2(2x^3 - 13x^2 + 20x)$$
$$V' = f'(x) = 2(6x^2 - 26x + 20)$$
$$V' = f'(x) = 4(3x - 10)(x - 1)$$

$f'(x) = 0$ when $x = 1$, $x = \frac{10}{3}$. Since $x = \frac{10}{3}$ is not in the domain of f it is not considered.

The candidates for absolute maximum are $x = 0$, 1, and $\frac{5}{2}$ (the values of x which cause $f'(x) = 0$ and the end points of the domain).

$$f(0) = 0 \text{ cubic inches}$$
$$f(1) = 18 \text{ cubic inches}$$
$$f(\tfrac{5}{2}) = 0 \text{ cubic inches}$$

Therefore, when x is 1 the dimensions of the largest box will be 6 inches by 3 inches by 1 inch and the volume will be 18 cubic inches.

Graphically:

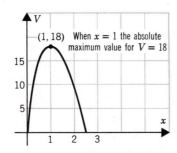

(1, 18)   When $x = 1$ the absolute maximum value for $V = 18$

Note   (1)   The graph in this example is included to further illustrate our concept of extrema. In working the exercises a pictorial idea is usually helpful, the graph is not necessary.

(2)   The end points of the domain seldom yield the desired extrema; however, they should not be ignored.

Problem E   A 125π cubic foot circular therapeutic pool is to be built for a nursing home. Find the dimensions, radius and height, of a pool that can be constructed with a minimal amount of material.

Solution   The building material will be a minimum amount for a minimum amount of surface area. Therefore, the quantity to be minimized is the surface area of the circular pool.

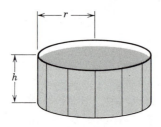

The circular base has an area of πr^2. The lateral surface area is $2\pi rh$. The total surface area, $A = \pi r^2 + 2\pi rh$. To write the area, A, as a function of a single variable, reread the problem and notice that the volume is given.

$$V = 125\pi \qquad \text{and} \qquad V = \pi r^2 h.$$
$$125\pi = \pi r^2 h$$
$$\frac{125}{r^2} = h$$

Now the total surface area, $\pi r^2 + 2\pi rh$, can be rewritten as a function of r, or

$$A = f(r) = \pi r^2 + 2\pi r\left(\frac{125}{r^2}\right)$$

$$= \pi r^2 + \frac{250\pi}{r}$$

$$A' = f'(r) = 2\pi r - \frac{250\pi}{r^2}$$

$$A' = f'(r) = 0, \quad \text{so } 2\pi r - \frac{250\pi}{r^2} = 0$$

$$2\pi r^3 = 250\pi$$

$$r^3 = 125$$

$$r = 5$$

$$\text{When } r = 5, \quad h = \frac{125}{r^2} = \frac{125}{25} = 5$$

Therefore, the pool's dimensions should be radius 5 feet and height 5 feet.

Problem F An automobile manufacturer can increase sales by decreasing the price. The manufacturer can sell x cars per hour at $\$(7070 - .4x^2)$ per car and it costs $\$(2000x + 5200)$ to produce x cars per hours. To maximize profit how many cars should be sold?

Solution Recall that $P(x) = R(x) - C(x)$ and that $R(x) = x \cdot D(x)$.

$$D(x) = \$(7070 - .4x^2)$$

$$R(x) = \$x(7070 - .4x^2)$$

$$C(x) = \$(2000x + 5200)$$

$$P(x) = \$[x(7070 - .4x^2) - (2000x + 5200)]$$

$$P(x) = 7070x - .4x^3 - 2000x - 5200$$

$$P'(x) = 7070 - 1.2x^2 - 2000$$

$$P'(x) = 0$$

$$0 = 7070 - 2000 - 1.2x^2$$

$$1.2x^2 = 5070 \qquad\qquad (1)$$

$$x^2 = \frac{5070}{1.2}$$

$$x^2 = 4225$$

$$x = 65$$

Therefore, the manufacturer should sell 65 cars per hour to maximize his profits.

An alternate way to determine the x value that will yield the maximum profit is the following.

Since $P(x) = R(x) - C(x)$ and $P'(x) = R'(x) - C'(x)$, you may solve for x by setting $R'(x) - C'(x) = 0$ or $R'(x) = C'(x)$. In words we are saying that the maximum profit will occur when the marginal revenue and marginal cost are equal. To solve the same problem:

$$R(x) = 7070x - .4x^3 \quad \text{and} \quad R'(x) = 7070 - .12x^2$$

$$C(x) = 2000x + 5200 \quad \text{and} \quad C'(x) = 2000$$

$$R'(x) = C'(x)$$

$$7070 - .12x^2 = 2000$$

$$5070 = .12x^2 \quad \text{[See (1) above]}$$

$$65 = x$$

Problem G   A charter airline advertisement reads, "Ski Excursion to Austria for the Holidays. Cost $222 per passenger for 100 passengers with a refund of $3 for each 10 passengers in excess of 100." Find the number of passengers that will maximize the amount of money the airline receives.

Solution   Let x be the number of passengers and $R(x)$, the revenue that is to be maximized. The revenue from each passenger is $[222 - \frac{3}{10}(x - 100)]$, for $x \geq 100$. The total revenue is $R(x) = x[222 - \frac{3}{10}(x - 100)]$.

$$R(x) = 252x - \tfrac{3}{10}x^2$$

$$R'(x) = 252 - \tfrac{3}{5}x$$

$$R'(x) = 0; \qquad 252 - \tfrac{3}{5}x = 0$$

$$252 = \tfrac{3}{5}x$$

$$420 = x$$

420 passengers will maximize the airline's revenue.

Problem H   Another type of problem whose solution requires the use of derivatives is referred to as a related rate problem. In this kind of problem, one is concerned with the rate of change of two (or more) variables, say x and y, with respect to time, t. The key to doing this problem is to use the chain rule. To illustrate: if $y = x^2$, then take the derivative of both x and y with respect to t. Therefore, $\frac{dy}{dt} = 2x\frac{dx}{dt}$. Consider this problem. In a biological experiment a person's lung capacity is examined by the rate at which he can inflate a balloon. If the volume of the balloon is increasing at a rate of 48 cm^3 per sec at what rate is the radius increasing when the radius is 5 cm?

Solution $V = \frac{4}{3}\pi r^3$. Using the chain rule, take the derivative of both V and $\frac{4}{3}\pi r^3$ treating t as the independent variable.

$$\frac{dV}{dt} = 4\pi r^2 \frac{dr}{dt}$$

Substitute the known values for $\dfrac{dV}{dt}$ and r, $\dfrac{dV}{dt} = 48 \dfrac{\text{cm}^3}{\text{sec}}$, $r = 5$ cm,

$$48 \frac{\text{cm}^3}{\text{sec}} = 4\pi(5 \text{ cm})^2 \frac{dr}{dt}$$

$$\frac{dr}{dt} = \frac{48 \dfrac{\text{cm}^3}{\text{sec}}}{4\pi(25 \text{ cm}^2)} = \frac{48 \dfrac{\text{cm}^3}{\text{sec}}}{100\pi \text{ cm}^2}$$

$$\frac{dr}{dt} = \frac{12}{25\pi} \frac{\text{cm}}{\text{sec}} \approx .15 \frac{\text{cm}}{\text{sec}}$$

Exercise 4.8

1. A psychologist determines that the capacity to learn new ideas depends on age, where $C(t) = 24 + 60t - \frac{3}{2}t^2$ (t in years). At what age is your capacity to learn a maximum?

2. A calculator manufacturer performs a series of experiments to determine how the quality of their product varies with the number of calculators produced per hour. The quality, Q, is measured on a scale of 0 to 1000, where 1000 represents a perfect product. Suppose the manufacturer determines $Q(x) = 875 + 10x - x^2$, where x is measured in 100 units per hour. Find the number of calculators per hour that will produce the maximum quality, and the quality at this point.

3. The concentration of a chemical in the bloodstream of a person t hours after the chemical has been introduced can be described by the function $K(t) = \dfrac{2t}{16 + t^3}$. At what time is the concentration a maximum?

4. A company undertakes an extensive advertising campaign and finds that, on increasing its advertising, the profit increases. The profit, in this case, is the net of total costs of goods plus advertising expenditures. $P(x) = 130 + 80x - x^2$ is the relation of profit, $P(x)$, to advertising, x, in hundreds of dollars.
    (a) What amount of advertising gives the maximum profit?
    (b) What is the maximum profit?

5. The following function gives the sales of a certain product as a function of the number of days after the advertising campaign was completed; $S(t) = 400 + 50t - 4t^2$. Find the rate of change of the sales at $t = 3$ and the time at which the sales are a maximum.

6. A man is on a diet where an additive is combined with his food. If x ounces of additive is added to every pound of his normal food, the man's total consumption as a function of x is given by $C(x) = 1.5x^2 - 6x + 7$. Find the amount of additive which will minimize the man's consumption.

7. The cost of manufacturing an item is given by the cost equation where $C(x) = .1x^2 - 10x^{3/2}$.
    (a) What is the equation for the marginal cost?
    (b) How many items need to be sold for a minimum cost?
    (c) What is the marginal cost for the number of items sold for minimum cost? Will this be true for all cost equations?

8. A manufacturer of lighting fixtures has daily production costs of $C(x) = \$\left(\dfrac{x}{3} + \dfrac{4(12)^3}{x} + 100\right)$, $x > 0$. How many fixtures, x, should be produced daily to minimize the manufacturer's cost?

9. A pharmaceutical company has determined that a realistic model for the relationship between blood pressure (systolic) and the amount of a new drug designed to reduce the systolic pressure is given by the equation $P = P_i + 2x - x^2$. P is measured in millimeters of mercury and x is the number of milligrams of the drug administered 30 minutes before the reading of the pressure P. P_i is the systolic pressure before the medicine is taken and will vary with each patient. However, P_i must fall in the range $200 \leq P_i \leq 220$ for the model to be valid.
    (a) Determine the rate of change of the systolic pressure P with respect to the amount of drug, x, administered.
    (b) Determine the minimum amount of the drug that can be administered and still be effective.
    (c) If a particular patient has an initial systolic pressure of 200 and 10 milligrams of the drug is administered, what will the pressure be 30 minutes later?

10. A psychologist gives a rat electronic shocks to stimulate the rats movement through a maze. The rat is given a score depending on how close he comes to the goal. The following function relates the rat's score to the number of shocks per minute: $S(x) = 12 + 9x - 1.5x^2$. How many shocks per minute should be given to maximize the rat's score?

11. The rhinovirus is responsible for the common cold. Once the virus enters the body it begins to multiply. However, at a certain point the virus will begin to die. The number of viral invaders present on a given day may be estimated by the function, $N(t) = -\frac{3}{4}t^4 + 3t^3 + 5$ where N is in billions. Determine the day on which there is a maximum number of viral particles.

12. A particle moves along a straight line such that $s = f(t) = t^4 - 8t^3 + 18t^2$.
    (a) Find the velocity and acceleration functions.
    (b) Find the absolute extrema (both maximum and minimum) of the velocity and acceleration over the time interval $0 \leq t \leq 4$. (Acceleration is rate of change in velocity with respect to time t.)

13. A stone is thrown upward, with an initial velocity of 48 feet per second from a second floor window 28 feet from the ground. $s = h(t) =$

$-16t^2 + 48t + 28$ gives the height of the stone from the ground after t seconds.
(a) Find the velocity of the stone after $t = t_0$ seconds.
(b) At what time is the velocity zero?
(c) What is the maximum height of the stone?
(d) What is the acceleration?
(e) When does the stone hit the ground?

14. $\$(\frac{1}{3}x^2 + 34x + 40)$ is the total cost of production per day for x stereotape players, and they are sold for $\$(62 - \frac{1}{4}x)$ per player.
    (a) Show that the profit function is $P(x) = (62 - \frac{1}{4}x)x - (\frac{1}{3}x^2 + 34x + 40)$.
    (b) What daily output will produce a maximum profit? (Assume the company can sell all the stereotape players it makes.)
    (c) As an alternate method, use the profit maximizing rule of marginal revenue is equal to marginal cost. Find the daily output to produce a maximum profit using this method. (See Example 1, Problem F).

15. The XYZ Paper Company has boxed greeting cards that cost 60 cents per box. A distributor can sell 100 boxes at $1 each, and for each cent she lowers the price, she can increase the number of boxes sold by five.
    (a) If x ($x \geqslant 100$) is the number of boxes sold, show that the total revenue, $R(x)$, is $\frac{6}{5}x - \frac{.01}{5}x^2$.
    (b) Show that the profit, $P(x)$, is $\left(\frac{6}{5}x - \frac{.01}{5}x^2\right) - .6x$.
    (c) How many boxes should she sell to maximize her profit?
    (d) What price per box will maximize her profit?
    (e) What will be her maximum profit?

16. The daily production cost for a factory to manufacture x deluxe contour chairs is given to be $\$(500 + 14x + \frac{1}{2}x^2)$. The demand function is $\$(150 - \frac{3}{2}x)$.
    (a) State the revenue function, $R(x)$.
    (b) State the profit function, $P(x)$.
    (c) How many chairs should be produced daily to maximize the profit?

17. The White Optical Company produces and sells microscopes for $250 each. The cost, in dollars, of producing x microscopes per year is $C(x) = \frac{1}{4}x^2 + 56x + 3060$.
    (a) Express the profit function.
    (b) How many microscopes should be sold to yield a maximum profit?

18. A furniture store can sell 50 easy chairs at $260 each. For every easy chair over 50 in number, the price is reduced by $2. Find the number of chairs that should be sold for maximum revenue. If the furniture store buys the easy chair from the manufacturer for $120 each, how many chairs need to be sold for maximum profit?

19. (a) A company charges $250 for a television set on orders of 60 or less sets. The charge is reduced on every set by $1 per set for each set

ordered in excess of 60. Find the largest size order the company should allow so as to receive a maximum revenue.

(b) This, in itself, is hardly a realistic problem. Although it would yield the largest gross amount, the key to a successful business is profit! The cost of manufacturing must also be considered. To make the problem feasible, find the profit the company would realize given the above conditions and the added condition that the cost of manufacturing one television set is $150.

20. A ski shop sold a ski package including skis, bindings, boots, and poles for $420. This resulted in an average of 10 sales a week for the first month. The next month the price was reduced to $320. The average number of sales per week increased to 15. If the wholesale cost for the package was $175 plus $5 shipping charge for each package, find the profit function $P(x)$. Determine the marginal profit function and the selling price that would bring in the greatest revenue per week. (Assume the demand equation is linear.)

21. The sum of two positive numbers is 16. Find the numbers if their product is to be a maximum.

22. The sum of two positive numbers is 16. Find the numbers if the product of their squares is to be a maximum.

23. The perimeter of a rectangle is 240 feet.
(a) State the length and width in terms of a single variable (if the length is x, what is the width?).
(b) Find the dimensions of the rectangle of largest area with perimeter 240 feet.

24. Find the dimensions of the rectangle of largest area whose perimeter is 100 feet.

25. Find the dimensions of the rectangle of least perimeter whose area is 196 square feet.

26. Find the area of the largest rectangle that can be inscribed in a circle of radius 2.

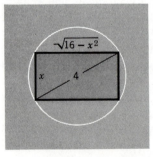

Let x represent the length of one side of the rectangle. By the Pythagorean theorem, another side will be of length $\sqrt{16 - x^2}$.

27. Find the area of the largest rectangle that can be inscribed in a circle of radius a.

28.   A rectangular field of area 5000 square feet borders a stream on one side. Find the least amount of fencing required for the other three sides.

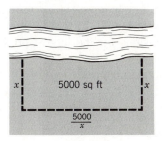

If x represents the length of one side of a rectangle, use the formula for the area of a rectangle to verify that another side is of length $\dfrac{5000}{x}$.

29.   A man wishes to plant a rectangular garden. One side of the garden will be his house, and the other three edges will be formed by a wooden fence. Find the dimensions of the largest garden the man can build using 30 feet of fencing.

30.   A manufacturer of cardboard boxes has an order for boxes that are to have a square base with no top. The volume of each box is to be 32 cubic feet.

(a) Show that the dimensions of the box are x by x by $\dfrac{32}{x^2}$ if x is the length of a square base.

(b) Show $x^2 + \dfrac{128}{x}$ expresses the surface area of the box.

(c) Find the dimensions of the box that will minimize the amount of material needed.

31.   Minimize the cost of constructing an open box whose rectangular base has a length twice as long as its width and the volume of the box is to be 72 cubic feet. The material for the base is 50 cents a square foot, and for the sides is 25 cents a square foot.

32.   A soup manufacturer needs tin cans that will hold 24 cubic inches. Find the radius of the tin can that requires the least amount of metal.

33.   Using the same conditions as Problem 32 find the minimal cost of producing a tin can if the tin for the top and bottom is 3 cents a square inch and for the side is 2 cents a square inch.

34.   A salt manufacturer wants to box his product in cardboard cylinders that would yield the lowest cost. The cardboard used for the top and bottom costs $2\frac{1}{2}$ cents per square inch, while the side is 1 cent per square inch. If the containers are to hold 40 cubic inches each, find the minimum cost.

35.   A wire, 40 inches long, is cut into two pieces, one piece shaped into a square, the other into a circle. If the sum of the areas of the square and the circle is to be a minimum, how should the wire be cut?

36.   A man has a 30-inch piece of wood. How should he cut the wood to make a picture frame that encloses the maximum area?

37. What is the radius of a cylinder of maximum volume and constant surface area?

38. A contractor is installing church windows in the shape of a rectangle surmounted by an equilateral triangle. The perimeter of each window is to be 24 feet. Find the dimensions for the largest possible window.

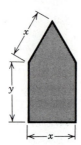

Verify the perimeter, $P = 24 = 2y + 3x$ and the area of the window, $A = xy + \left(\frac{\sqrt{3}}{4}\right)x^2$. Form A as a function of x and find the dimensions to maximize A.

39. A company is trying to decide which shaped container to use for shipping its product. The two favorable shapes are cylindrical or rectangular with a square base. Given a fixed volume, V, find the dimensions of each container that will minimize the amount of material needed (i.e., minimize the surface area). If the volume is 1 cubic foot, calculate the surface areas. Which container is more reasonably priced?

40. The B.T. Battery Company experiences a yearly demand for 40,000 of their Long Life Batteries. This demand is spread uniformly throughout the year. Experience has shown that the setup cost for a production run of Long Life Batteries is $200. However, the cost of carrying one battery in inventory is $1 per year. Also, since there is a constant rate of depletion, the average inventory size is one half of the beginning inventory. How large should each equal size production run be and how should they be scheduled to minimize total cost (setup plus inventory)?

Let x represent the size of the production run that will minimize total cost. This model will assume that:

Average inventory size per year is $\dfrac{x}{2}$.

Total inventory cost per year is $\$1 \cdot \dfrac{x}{2} = \dfrac{x}{2}$.

Number of required setups per year is $\dfrac{40,000}{x}$.

Total setup cost is $\$200 \cdot \dfrac{40,000}{x} = \dfrac{8,000,000}{x}$.

(a) Show that the total cost is $C(x) = \dfrac{x}{2} + \dfrac{8,000,000}{x}$.

(b) Find the size of the production run, x, that will minimize total cost, $C(x)$.

(c) Divide 40,000 by the value of x that will minimize $C(x)$. This will yield the number of production runs needed per year.

(d) The plant operates 260 days per year. Determine the approximate number of days between production runs for Long Life Batteries, if total cost (setup plus inventory) is to be kept at a minimum.

41. To generalize the preceding problem, consider a basic inventory control problem: to minimize the cost of producing and holding inventory. Let us develop a formula for finding the value of x, the lot size, that will minimize the total cost.

Let $x =$ the lot size in units.

$D =$ fixed demand per year (assume a uniform and fixed demand for the product).

$S =$ cost of setting up a production run.

$I =$ the annual inventory cost of carrying one unit, for average inventory.

Then $\dfrac{x}{2} =$ average annual inventory size.

$\dfrac{x}{2} \cdot I =$ total inventory cost per year.

$\dfrac{D}{x} =$ number of required setups.

$\dfrac{D}{x} \cdot S =$ total setup cost;

and the total cost function $C(x)$, is $C(x) = \dfrac{x}{2}I + \dfrac{D}{x}S$.

Find the size of the production run, x, that will minimize the total cost.

42. A manufacturer finds the yearly uniform demand for his product to be 2420 units. The cost of setting up a production run is $80 and the cost of carrying one unit in inventory is 50 cents. Find the lot size that will yield a minimal total cost. (See Problem 41.)

43. A power station is on one side of a river, which is $\frac{1}{2}$ mile wide, and a factory is 1 mile downstream on the other side of the river. It costs $3 per foot to run power lines over land and $5 per foot to run them under water. Find the most economical way to run the power lines from the power station to the factory.

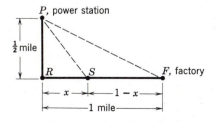

Which is the most economical way to run the power lines, $PR + RF$; PF; or $PS + SF$?

44. An electric line is to be installed from a cottage to a utility pole 3 miles west and 2 miles north of the cottage. A road running east-west passes the cottage, and a wooded area is north of the road. If the cost is $5 per mile to run the line along the road and $13 per mile to run the line through the woods, what is the most economical route to run the electric line from the cottage to the pole, and what will be this minimum cost?

45. A wholesaler supplies an Auto Supply Mart with 20 cases of high-grade oil per week. For the wholesaler to purchase an order of x cases of oil, the cost is $9.80 + .90x$ (dollars), and the average weekly cost to store the portion of an order not yet delivered to the Auto Supply Mart is x cents per week. What size order will minimize the average weekly cost of supplying the Auto Supply Mart with the cases of oil, if the wholesaler has storage space for at most 200 cases? *Hint.* Find the average weekly cost, $C(x)$, to the wholesaler by noting that $C(x)$ is the average purchase cost of an order per week plus the average storage cost per week. Thus, show that

$$C(x) = \frac{9.80 + .90x}{x/20} + \frac{1}{100}x.$$

46. The amount of money deposited in a bank is proportional to the interest rate the bank pays on this money. Furthermore, the bank can reinvest this money at 7 percent. Find the interest rate the bank should pay to maximize its profit.

47. See Example 1, Problem B.
   (a) Show that $V(r) = C(r_0 - r)r^2$ has horizontal tangent lines at $r = \frac{2}{3}r_0$ and $r = 0$.
   (b) Show that the second derivative of $V(r) = C(r_0 - r)r^2$ is less than zero when $r = \frac{2}{3}r_0$.
   (c) Show that when $r = \frac{2}{3}r_0$, an absolute maximum for $\frac{1}{2}r_0 \leqslant r \leqslant r_0$ has been obtained.

48. Gas is being pumped into a spherical balloon such that the radius is increasing at a rate of $\dfrac{3}{4\pi}$ inches per second. At what rate is the volume changing when $r = 3$?

49. A person's lung capacity is examined by the rate at which he can inflate a balloon. If the volume of the balloon is increasing at the rate of 4 inches per minute, at what rate is the radius increasing when the radius is 3 inches?

50. The correspondence between the price of gasoline per gallon, y, and the weekly supply to a station in thousands of gallons, x, is given by the equation $y = .37 + \dfrac{.45}{x}$. At what rate will the price be changing, if the weekly supply of 6000 gallons begins decreasing at a rate of 500 gallons per week?

51. Computer Problem (optional). Let $y = f(x) = \dfrac{x - 4}{x^2 - 1}$ with domain

$3 \leqslant x \leqslant 8$. Divide the interval into 50 equal parts, that is, 3, 3.1, 3.2, . . . , 7.8, 7.9, 8. Program the computer:

(a) to calculate $f(3)$, $f(3.1)$, . . . , $f(7.9)$, $f(8)$;

(b) to find the maximum value of the function in $3 \leqslant x \leqslant 8$;

(c) to print out the maximum value and where it occurs.

4.9 THE DIFFERENTIAL

The derivative was invented to find the slope of the tangent line to a curve. The differential was invented to approximate Δy, the change in y. Later in this course, we will find other uses for the differential.

We will follow the common practice of defining dx to be an independent variable that is another name for Δx, $dx = \Delta x$. A typical problem is stated: Find a method for approximating Δy.

Make these assumptions:

1. the function defined by $y = f(x)$ has a derivative in an interval containing x; and

2. over small intervals the tangent line to the curve $y = f(x)$ is a "close approximation" to the shape of the curve of $y = f(x)$.

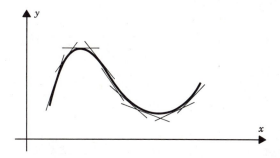

To analyze the problem and develop a solution, refer to Figure 4.13 on the following page. Let $dx = PR$. Let t, the tangent line to f at x, have slope $= f'(x)$. Approximate $\Delta y = RQ$ by assuming that line segment PT is a "close approximation" to curve PQ if dx is "small."

First, return to the derivative.

$$f'(x) = \lim_{\Delta x \to 0} \frac{f(x + \Delta x) - f(x)}{\Delta x} = \lim_{\Delta x \to 0} \frac{\Delta y}{dx}.$$

If dx is small and the limit had not been taken, we assume: $f'(x) \simeq \dfrac{\Delta y}{dx}$. That is, $\dfrac{dy}{dx}$ is only approximately equal to the difference quotient, $\dfrac{\Delta y}{dx}$, if dx is small

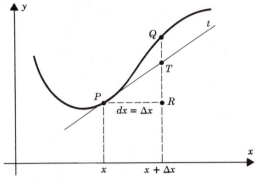

Figure 4.13

and $dx \neq 0$. Also, by multiplying both sides of this equation by dx, we obtain

$$\Delta y \simeq f'(x)\, dx.$$

Let us agree to define $f'(x)\, dx$ as a new dependent variable dy, that is, $dy = f'(x)\, dx$, and it follows that $\Delta y \simeq dy$. It is this value $dy = f'(x)\, dx$ that we will use to approximate Δy.

We now summarize these equations and introduce the usual terminology of differentials.

DEFINITION

Differential of x, dx; *Differential of y, dy.*   Let the function defined by $y = f(x)$ be differentiable in an interval containing the point x. Let dx be an independent variable called the *differential of x, dx = Δx. The differential of y at the point x, dy,* is defined by the equation $dy = f'(x)\, dx$. Notice that dy is a function of both dx and x.

Geometrical Interpretation of dx and dy

Returning to Figure 4.13, we notice that the slope of tangent line t is

$$m_t = f'(x) = \frac{\text{rise}}{\text{run}} = \frac{RT}{PR} = \frac{RT}{dx}$$

$$\therefore \frac{RT}{dx} = f'(x)$$

or

$$RT = f'(x)\, dx.$$

But

$$f'(x)\, dx = dy.$$
$$\therefore RT = dy$$

Geometrically:

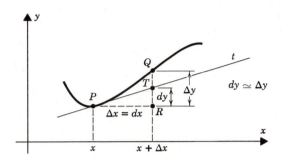

The differential has three aspects that we wish to consider:

1.   the use of differentials to approximate Δy, if Δx is "small,"
2.   the use of differentials to give increased meaning to the derivative, $\dfrac{dy}{dx}$,

 and

3.   the methods to obtain differentials and some formulas concerning differentials.

Example 1

Purpose   To show that the error introduced when dy is used to approximate Δy is small when compared with Δx:

Problem   The error E when dy is used to approximate Δy is $E = \Delta y - dy$.

 Show $\dfrac{E}{\Delta x} \to 0$ as $\Delta x \to 0$.

Solution

$$\lim_{\Delta x \to 0} \frac{E}{\Delta x} = \lim_{\Delta x \to 0} \frac{\Delta y - dy}{\Delta x} = \lim_{dx \to 0} \frac{\Delta y - f'(x)\,dx}{dx}$$

$$= \lim_{dx \to 0} \left[\frac{\Delta y}{dx} - f'(x) \right]$$

$$= \lim_{dx \to 0} \frac{\Delta y}{dx} - \lim_{dx \to 0} f'(x)$$

$$= f'(x) - f'(x)$$

$$= 0$$

 Since $\dfrac{E}{\Delta x} \to 0$ as $\Delta x \to 0$, E, which is equal to $\Delta y - dy$, $\to 0$ and thus dy is a good approximation to Δy.

Example 2

Purpose   To use the differential of y as an approximation of Δy:

Problem   Use differentials to approximate $\sqrt{26}$.

Solution   Since we wish to approximate a square root, we choose $y = f(x) = \sqrt{x}$.
Let $x = 25$, then $y = f(25) = \sqrt{25} = 5$
Let $\Delta x = dx = 1$, then $y + \Delta y = f(26) = \sqrt{26}$
or $\qquad\qquad\qquad\qquad 5 + \Delta y = \sqrt{26}$
Since $dy \simeq \Delta y$

$$5 + \Delta y \simeq 5 + dy$$
$$\sqrt{26} \simeq 5 + dy.$$

Find dy

$$dy = f'(x)\ dx = \frac{1}{2}x^{-1/2}\ dx = \frac{1}{2\sqrt{x}}\ dx.$$

When $x = 25$ and $dx = 1$, then $dy = \dfrac{1}{2\sqrt{25}}\ (1) = .1$

$$\therefore\ \sqrt{26} \simeq 5 + .1 = 5.1.$$

The second aspect that we wish to consider is the use of differentials to give added meaning to $\dfrac{dy}{dx}$. Beginning with the definition, $dy = f'(x)\ dx$, we may divide both sides of the equation by $dx \ne 0$ to obtain $\dfrac{dy}{dx} = f'(x)$. You may regard this as nothing new, since we have used this equality many times before this section. However, what is new is that this gives us another way to think of the derivative. The new interpretation of the derivative is as the quotient of two differentials, dy divided by dx. Thus, the first derivative becomes a fraction with numerator, dy and denominator, dx. This gives us a right, for example, to transform $\dfrac{dy}{dx} = 3x^2$ into $dy = 3x^2\ dx$ or $dy = 3x^2\ dx$ into $\dfrac{dy}{dx} = 3x^2$ if we wish. It also gives us an alternate method for finding differentials, that is, to find $\dfrac{dy}{dx}$ and then multiply both sides of this equation by dx, thus obtaining dy.

Example 3

Purpose   To illustrate the alternate method for finding the differential of y, dy:

Problem $y = \sqrt{x^3 - 2x + 4}$. Find dy.

Solution $\dfrac{dy}{dx} = \frac{1}{2}(x^3 - 2x + 4)^{-(1/2)}(3x^2 - 2)$
Therefore,

$$dy = \tfrac{1}{2}(x^3 - 2x + 4)^{-(1/2)}(3x^2 - 2)\ dx.$$

Formulas for finding the differential of a sum, product, and quotient of two functions can be derived. Fortunately, this is unnecessary, since the relationship $dy = f'(x)\, dx$ enables us to use our differentiation formulas to find dy. However, let us derive the formula for $d(u \cdot v)$ where u and v are functions of x, since this formula is of special interest in Chapter 6, Section 6.5.

Example 4

Purpose To derive a formula for $d(u \cdot v)$:

Problem Assume that there is an interval containing the point x where the derivatives of $u = g(x)$ and $v = h(x)$ exist.

$$\text{Prove } d(u \cdot v) = u \cdot dv + v \cdot du.$$

Solution Let $y = f(x) = u \cdot v$.

By definition, $\qquad\qquad dy = f'(x)\, dx$

or

$$d(u \cdot v) = \left[\frac{d}{dx}(u \cdot v) \right] dx$$

$$d(u \cdot v) = \left[u \cdot \frac{dv}{dx} + v \cdot \frac{du}{dx} \right] dx$$

$$d(u \cdot v) = u \cdot dv + v \cdot du$$

Exercise 4.9

In Problems 1 to 8, find dy for the given function.

1. $y = f(x) = x^2 - 4x + 2$

2. $y = f(x) = 2x^3 - 5x^2 - 4x$

3. $y = f(x) = (x^2 - 3)(2 - 3x)$

4. $y = f(x) = \dfrac{1}{x^2 - 1}$

5. $y = f(x) = \sqrt[3]{2x^3 - 3x}$

6. $y = f(x) = \dfrac{1}{(7 - x + 2x^2)^3}$

7. $y = f(x) = \dfrac{3x - 5}{x^2 + 2}$

8. $y = f(x) = \dfrac{4x}{\sqrt{2 + x^2}}$

In Problems 9 to 14, approximate by differentials.

9. $\sqrt{24}$

10. $\sqrt[3]{65}$

11. $\sqrt[3]{63}$

12. $\sqrt{9.5}$

13. $\frac{1}{101}$

14. $\frac{1}{99}$

15. If $f(x) = x^2 + 1$, find both Δy and dy when $x = 1$ and $dx = \Delta x = .5$. Make a careful sketch and identify both Δy and dy. Notice that $dy \simeq \Delta y$. What is the error, E, when dy is used to approximate Δy? What is the error when compared with Δx?

16. If $f(x) = \frac{1}{2}x^2$, find both Δy and dy when $x = 2$ and $dx = \Delta x = -.5$. Make a careful sketch and identify both Δy and dy. Notice that $dy \simeq \Delta y$. What is the error, E, when dy is used to approximate Δy?

17. The quantity demanded of the commodity x is $2x = 15 - 3y$, where y is the price in dollars of the commodity. If x changes from 6 to 7, find the corresponding change in y. Find both dy and Δy. Is there an error between Δy and dy? Discuss.

18. A factory finds its total cost can be determined by $C(x) = \$(75 + \frac{1}{2}x^2)$. Approximate the change in cost as x, the number of units, changes from 20 to 21.

19. The profit P for a corporation is given by $P(x) = \$[(400x - x^2) - (\frac{1}{2}x^2 - 76x + 3013)]$. Approximate the change in profit as x changes from 100 units to 105 units.

20. Approximate the cross-sectional area of a wall of an artery if the radius of the artery is 1 mm and the thickness of the wall is .05 mm.

21. The length of an edge of a cube may vary from 2 cm by $\pm.01$ cm. Use differentials to approximate the maximum possible error that can occur when calculating:
   (a) the volume of the cube
   (b) the surface area of the cube.
   (*Hint:* If the length of an edge of the cube is x, then $x = 2$ and $dx = .01$. The volume is $V = x^3$ and the surface area is $S = 6x^2$. Find dV and dS.)

22. The radius of a circle may vary from 3 feet by $\pm.1$ feet, or $r = 3 \pm .1$ feet. Use differentials to approximate the maximum possible error that can occur when calculating:
   (a) the area of the circle
   (b) the circumference of the circle.

23. A cubic wooden box has an interior volume of 64 cubic inches. The thickness of the wood is $\frac{1}{2}$ inch. Determine how much wood is used to make the box.

24. The yield of fruit from a restricted plot depends on the number of trees grown on the plot (because of crowding and the availability of sunlight). The annual yield of bushels of fruit can be found by the equation $Q(x) = 4.5x - .01x^2$ where x is the number of trees in the plot. By the use of differentials, approximate the change in yield of fruit if there are originally 200 trees and 10 more trees are planted.

■ **4.10 CHAPTER REVIEW**

Important Ideas

increasing function
decreasing function

relative maximum
relative minimum

absolute maximum   hypercritical number
absolute minimum   graphing
critical number   applied extrema problems
concave upward   differential of x, dx
concave downward   differential of y, dy
point of inflection

■ REVIEW EXERCISE (optional)

1. (a) What is meant by an increasing function?
   (b) What is meant by a decreasing function?

2. (a) What is meant by a relative maximum? An absolute maximum?
   (b) What is meant by a relative minimum? An absolute minimum?

3. From the graphs of the functions given below determine when the first derivative is positive, negative and zero.
   (a)

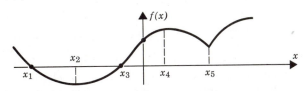

   (b)

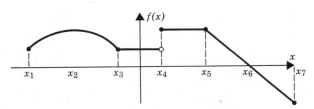

4. From the graph of f' determine when the function f is:
   (a) increasing
   (b) decreasing
   (c) attains a relative maximum or minimum.

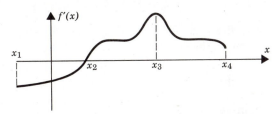

5. A psychologist is interested in studying the memory process. She gives 100 people identical lists of 10 words and asks them to memorize them.

She allows the first person 1 second, the second person 2 seconds, and so on. The results are shown on the following graph.

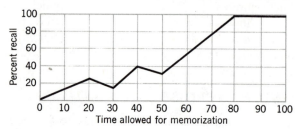

Fill in the following table:

Interval	Relative Extrema	Increasing or Decreasing
0–20		
21–30		
31–40		
41–50		
51–80		
81–100		

In Problems 6 to 21, find $f'(x)$ and $f''(x)$. Locate extrema (state whether a relative maximum or minimum) and point(s) of inflection. Indicate the intervals in which the function is increasing, decreasing, concave upward and concave downward. Graph each function.

6. $f(x) = 6x^2 - 11x - 7$

7. $f(x) = x(8 - x)$

8. $f(x) = x^3 - 9x$

9. $f(x) = x^3 - 3x^2 + 3$

10. $f(x) = (x + 1)^2(x - 3)$

11. $f(x) = x^3 - x^2 - 8x + 2$

12. $f(x) = x^3 - 6x^2 + 9x + 1$

13. $f(x) = 1 + 9x + 3x^2 - x^3$

14. $f(x) = (x - 2)^2(x + 2)^2$

15. $f(x) = x^4 - 2x^3$

16. $f(x) = x^4 + 8x^3 + 18x^2$

17. $f(x) = \sqrt{x + 2}$

18. $f(x) = \sqrt{1 - x^2}$

19. $f(x) = \dfrac{3}{x^2}$

20. $f(x) = x + \dfrac{4}{x}$

21. $f(x) = x^{1/3}$

22. Does the function $y = 2x$ have an absolute maximum and an absolute minimum (a) on the open interval $0 < x < 1$? (b) On the closed interval $0 \leqslant x \leqslant 1$?

23. Does the function $y = \dfrac{1}{x}$ have an absolute maximum and an absolute minimum on the open interval $0 < x < 1$?

In Problems 24 to 33 indicate and justify whether the statement is true or false.

24. $f(x) = 16 - 3x^3$ is never increasing.

25. $f(x) = x^4 - 4x^3$ has a relative maximum at $x = 0$.

26. $f(x) = \dfrac{1}{(x-3)^2}$ is never concave downward.

27. $f(x) = 5 - 2x$ is increasing for all x.

28. $f(x) = x^4$ has a relative extremum.

29. If $y = 3x$, then $dy = dx$.

30. If $f'(x_1) = 0$ and $f''(x_1) < 0$, then f has a relative maximum at $x = x_1$.

31. All critical numbers yield either relative maximum or relative minimum.

32. Given a function and its domain; the absolute extrema may occur at the end points of the domain.

33. A function is increasing on $x_1 < x < x_2$ if $x_1 < x_2$ and $f(x_1) < f(x_2)$.

34. The laboratory of a large pharmaceutical company runs a series of tests to determine the rate at which a new drug that is administered intramuscularly is absorbed into the bloodstream. The mathematical model constructed by the lab has the drug concentration Y as a function of time t (in hours) given by the equation $Y = \dfrac{2t}{t^2 + 4}$, $t \geq 0$.

    (a) Determine the percentage of the drug in the bloodstream:
        (1) at time of injection
        (2) 1 hour after injection
        (3) 2 hours after injection
        (4) 3 hours after injection.
    (b) Determine the time interval when the concentration is
        (1) increasing
        (2) decreasing.
    (c) Determine the peak concentration of the drug in the bloodstream and when it occurs.

35. Ms. X has determined that her pulse rate is affected by the amount of coffee she consumes each morning. Assume the model of this relationship is given by the equation $P = 70 + \frac{1}{8}x^3$ where the pulse rate is measured in pulses per minute at 11 a.m. each morning and x is the number of 6-ounce cups of coffee consumed between 7 a.m. and 11 a.m. in the morning.
    (a) Determine the possible maximum and minimum[1] pulse rates for Ms. X on any given morning at 11 a.m., assuming that Ms. X's only source of coffee is her 8-cup pot.
    (b) If Ms. X usually limits herself to four cups of coffee in the morning, approximate her pulse rate on those mornings when she drinks a fifth cup.

36. A football is thrown with a trajectory path equation: $y = \dfrac{90x - x^2}{45}$.

---

[1] Assume Ms. X is alive at the time the reading is taken!

(a) Graph the path of the football.
(b) Find the maximum height of the football.
(c) For what values of x is the football path increasing?
(d) For what values of x is the football path decreasing?

37. A man has 1000 shares of stock whose expected price per share over the next two weeks is given as a function of time, in days, by $P(t) = 2.5 + 20t - t^2$. If the man wishes to sell, when would be the best time for him to do so?

38. A psychologist gives the following function for the number of persons with a certain IQ as a function of the IQ: $N(x) = 2x^2 - 440x - 72$. Find the most probable IQ.

39. Two hundred apple trees are grown on a plot. The annual yield of apples is 500 bushels. For every 10 more trees planted and grown to maturity, the yield of the original 200 trees reduces by 20 bushels as a result of the availability of sunlight. The equation for the total number of bushels depending on the number of trees planted is

$$Q(x) = \frac{x}{200}\left[500 - \frac{20}{10}(x - 200)\right]$$

How many more trees should be planted for maximum yield?

40. The height of a BB shot t seconds after it is fired from a gun is given by $h(t) = -16t^2 + 96t + 7$. What time will the BB reach its maximum height? What is the maximum height?

41. Two brands of pain relievers are being tested for effectiveness. The relief of the two brands may be written as a function of time. The relief function of the first pain reliever is $R_1(t) = -4t^2 + 8t$, where t is the tth hour after the medication has been taken. The relief of the second brand may be found through the function $R_2(t) = -1.8t^3 + 5.4t$.
(a) At what time does the first pain reliever give maximum relief?
(b) At what time does the second brand give maximum relief?
(c) The first brand claims to have more strength and work faster than the second brand. At what time is this difference the smallest?

42. An object, thrown vertically upward from the ground, has an initial velocity of 16 feet per second. The equation of motion is $s = -4t^2 + 16t$, t the time in seconds elapsed after throwing the object and s the distance in feet that the object travels, the positive direction being upward. Find:
(a) The instantaneous velocity after 1 second; after 3 seconds.
(b) How long does it take the object to reach its highest point?
(c) How high does the object go?
(d) How many seconds elapse before the object reaches the ground?
(e) What is the instantaneous velocity when it reaches the ground?

43. A cost function is given to be $C(x) = \frac{1}{2}x^2 + 14x + 50$. For what value of x is the marginal average cost a minimum?

44. A manufacturer can sell x units per day of a certain commodity at a price of $\$(60 - .1x^2)$ per unit. The total cost for x units is $\$(1.2x + 40)$.
    (a) Show that the profit function $P(x) = -.1x^3 + 58.8x - 40$.
    (b) How many units should be produced daily for a maximum profit?

45. A manufacturer of ski equipment sells a set of boots, skis, and poles for $\$200$. It costs him $\$(120 + .02x^2)$ to manufacture x sets.
    (a) What is the profit if the number of sets sold is 2000?
    (b) How many sets of equipment need to be sold for maximum profit?
    (c) What is the maximum profit?

46. The National Forest Service finds that 16 trees planted on an acre of land will grow approximately 3 feet per year. For each additional tree planted, the growth will be reduced by $\frac{1}{2}$ inch. Find the number of trees per acre that will yield the largest amount of timber.

47. The Greyhound Bus Line has a special fall foliage tour in the New England states. The cost is $\$95$ per person for 75 passengers with a refund of $\$2$ for each 5 passengers in excess of 75. Find the number of passengers that will maximize the revenue received by the bus line.

48. The following graph shows both a cost and a revenue function in terms of x, the quantity supplied. How would the maximum profit be established? [Recall, the profit maximization will occur where $C'(x) = R'(x)$.]

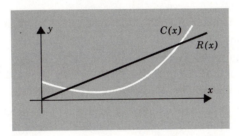

49. Most companies find that their sales volume is closely related to the price of their manufactured article—a lower price yields a higher volume. However, too low a price will diminish the profit regardless of the large sales volume. Therefore, a company must establish a wise and workable volume-profit relationship. If the following graph shows such a function, how would the maximum profit be determined?

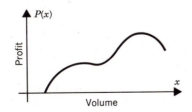

50. To determine a cost y for x units of an item produced, the function

$y = f(x)$ is called the cost function. If a cost function is given by $y = f(x)$, then

(1) the average cost, $\bar{y} = \dfrac{y}{x}$ or $\dfrac{f(x)}{x}$,

(2) the marginal cost is $\dfrac{dy}{dx}$ or $f'(x)$; and

(3) the marginal average cost is $\dfrac{d\bar{y}}{dx}$.

(a) Show that the average cost is a minimum when the average cost is equal to the marginal cost.
(b) Given a linear cost function $y = f(x) = ax + b$, find the average cost, marginal cost, and marginal average cost. Graph each of these for $a > 0$, $b > 0$.
(c) Given a quadratic cost function $y = f(x) = ax^2 + bx + c$, find the average cost, marginal cost, and marginal average cost. Minimize the average cost and show that the average cost is equal to the marginal cost at that value.

51. Find the dimensions of a rectangle of maximum area with a given perimeter P.

52. The perimeter of an isosceles triangle is 12 inches. Find the dimensions of the triangle that will maximize its area.

53. A rectangular swimming pool with a volume of 1350 cubic feet is to be constructed so that its length is 4 times its width. Find the dimensions that will minimize the surface area of the bottom and sides of the pool.

54. A man wishes to enclose a 600 square foot area of his backyard. Three sides of the enclosure will be made of fence which costs $5 per foot. The fourth side will be made of brick which costs $15 per foot. Find the dimensions of the enclosure that will minimize the cost of the building materials.

55. An oil drum is to hold 100 cubic feet of fuel oil. Find the radius of the drum that requires the least amount of metal.

56. What is the radius of a cylinder of minimum surface area and constant volume?

57. A company has found the relationship between total inventory of raw materials, y (in hundreds of dollars), and the production of x units of a manufactured product is $y = \frac{1}{2}\sqrt{x} + \frac{1}{4}x$. At what rate should the inventory be increased if there is a current demand for 100 units per month and this demand is increasing at a rate of 6 units per month? When the production reaches 400 units per month, at what rate should the inventory be increased?

58. In a biological experiment, a person's lung capacity is examined by the rate at which he can inflate a balloon. If the volume of the balloon is increasing at a rate of 9 cubic inches per minute, at what rate is the radius increasing when the radius is 3 inches?

59. Two bicyclists leave from a town at 8 a.m., one traveling east at 15 miles per hour and the other traveling south at 20 miles per hour. How fast is the distance between them changing at 12 noon?

60. Approximate by differentials
(a) $\sqrt[3]{28}$ (b) $\sqrt[3]{26}$

61. If $y = f(x) = 1 - x^2$ find the approximate change in y when x changes from 2 to 2.2.

62. Using differentials, approximate the change in surface area of a balloon as the radius increases from 4 inches to 4.25 inches.

63. The growth of a bacteria culture is given by $N = 100 + 100t + 5t^2$, where t is in hours. Approximate the change in the bacteria count from 5 hours to $5\frac{1}{2}$ hours.

64. If the weekly profit, in hundreds of dollars, from producing x units of a product is $P(x) = -x^2 + 42x - 221$, use differentials to approximate the change in profit if the production level changes from 15 to 16.

FIVE

Integration

5.1 INTRODUCTION

Chapters 2 to 4 were concerned with the derivative and its applications. The derivative is one of the two most important ideas in the calculus. The other important idea is that of the integral. Moreover, there exists a relationship between these two ideas. Thus, this chapter will introduce the integral, show the relationship between the derivative and the integral, and exhibit some of the many applications of the integral. We begin our study by examining the ideas of an antiderivative and an indefinite integral. It is the indefinite integral that will make the definite integral of later sections much easier to find. Our goal is to be able to evaluate integrals and to solve problems like the following.

The cost to produce the xth digital watch in a daily production run is given by $C(x) = \$\left(\dfrac{50}{\sqrt{x+1}}\right)$, $x \leqslant 200$. Approximate the total cost to produce the first 100 watches.

One way biologists study the growth of a bacteria culture is by observing the percentage of bacteria that regenerates in a given amount of time. If the equation of regeneration is found to be $B = 2x + \dfrac{40}{x^2}$, the number of bacteria with an a to b percent chance of regeneration is found by taking the definite integral of the equation of regeneration from a to b. Find the number of bacteria with 10 to 20 percent chance of regeneration.

A supply function is given to be $y = S(x) = 3 + x^2$ and the corresponding demand function for the same firm is $y = D(x) = -2x + 18$. Find both the consumers' surplus and the producers' surplus.

5.2 INDEFINITE INTEGRAL

In our previous work we began with a function and derived a new function called the derivative of the original function. Now we begin with the derivative and ask ourselves to find the original function. That is, we are putting our previous differentiation process into reverse. For example, given $f(x) = x^7$ find a function F such that its derivative $F'(x) = f(x) = x^7$. By inspection we see that F could be $F(x) = \frac{1}{8}x^8$, since $F'(x) = \dfrac{d}{dx}(\frac{1}{8}x^8) = x^7$. We will call $F(x) = \frac{1}{8}x^8$ an antiderivative of $f(x) = x^7$.

> **DEFINITION**
>
> *Antiderivative.* A function F is said to be an *antiderivative* of a function f on an interval I if $F'(x) = f(x)$ for all x in I.

Example 1

Purpose To illustrate the definition of the antiderivative:

Problem Show that the functions which are defined by $F(x) = \frac{1}{3}x^3$, $G(x) = \frac{1}{3}x^3 + 17$ and $H(x) = \frac{1}{3}x^3 - 64$ are antiderivatives of the function defined by $f(x) = x^2$.

Solution $F(x) = \frac{1}{3}x^3$ is an antiderivative of $f(x) = x^2$, since $\dfrac{d}{dx}[F(x)]$

$$= \frac{d}{dx}(\tfrac{1}{3}x^3) = x^2 = f(x).$$

$G(x) = \frac{1}{3}x^3 + 17$ is an antiderivative of $f(x) = x^2$,

$$\text{since } \frac{d}{dx}(\tfrac{1}{3}x^3 + 17) = x^2.$$

$H(x) = \frac{1}{3}x^3 - 64$ is an antiderivative of $f(x) = x^2$,

$$\text{since } \frac{d}{dx}(\tfrac{1}{3}x^3 - 64) = x^2.$$

$f(x) = x^2$ appears to have an unlimited set of antiderivatives. To form another antiderivative of $f(x) = x^2$, we simply add a different constant to $F(x) = \frac{1}{3}x^3$. We pause to question: "Do two antiderivatives for the same function differ by only a constant?" The answer is "yes."

When we wish to specify the entire set of functions that have x^2 for a derivative we use the following notation,

$$\int x^2\, dx = \tfrac{1}{3}x^3 + c.$$

The symbol, $\int$, is called the indefinite integral.

> **DEFINITION**
>
> *Indefinite Integral.* The set of all antiderivatives of f is the *indefinite integral of f* and is denoted by
>
> $$\int f(x)\, dx$$

Example 2

Purpose To find an indefinite integral:

Problem Find $\int \frac{1}{2}x^{-(1/2)} dx$.

Solution This indefinite integral stands for the set of functions such that the derivative of any element of this set is $\frac{1}{2}x^{-(1/2)}$. By inspection, $F(x) = x^{1/2}$ is a function whose derivative $F'(x) = \frac{1}{2}x^{-(1/2)}$. Therefore,

$$\int \frac{1}{2}x^{-(1/2)} dx = x^{1/2} + c.$$

The differential, dx, appears in the notation for the indefinite integral. It not only emphasizes the variable x, but $\int f(x)\, dx$ may be thought of as the set of all functions whose differentials are $f(x)\, dx$. In $\int f(x)\, dx$, $f(x)$ is called the *integrand* of the indefinite integral.

A Method for Obtaining $\int f(x)\, dx$

We wish to introduce certain techniques that will be an aid in finding indefinite integrals. After inspecting the following:

$$\int x\, dx = \frac{1}{2}x^2 + c$$

$$\int x^2\, dx = \frac{1}{3}x^3 + c$$

$$\int x^3\, dx = \frac{1}{4}x^4 + c$$

$$\int x^4\, dx = \frac{1}{5}x^5 + c,$$

it is hypothesized that the $\int x^n\, dx = \dfrac{1}{n+1}x^{n+1} + c, \qquad n \neq -1.$

THEOREM 5.2

For any real number $n \neq -1$, $\int x^n\, dx = \dfrac{1}{n+1}x^{n+1} + c$

Proof

Observe that

$$\int x^n\, dx \text{ will equal } \frac{1}{n+1}x^{n+1} + c \text{ if } \frac{d}{dx}\left(\frac{1}{n+1}x^{n+1} + c\right) = x^n.$$

Since

$$\frac{d}{dx}\left(\frac{1}{n+1}x^{n+1} + c\right) = \frac{1}{n+1}(n+1)x^n = x^n,$$

we have

$$\int x^n \, dx = \frac{1}{n+1}x^{n+1} + c, \qquad n \neq -1.$$

Example 3

Purpose To show the use of Theorem 5.2 to find the indefinite integral:

Problem A Find $\int x^{10} \, dx$.

Solution Using $\int x^n \, dx = \dfrac{1}{n+1}x^{n+1} + c$ where $n = 10$,

$$\int x^{10} \, dx = \tfrac{1}{11}x^{11} + c.$$

Problem B Find $\int (\sqrt{x})^3 \, dx$.

Solution $\displaystyle \int (\sqrt{x})^3 \, dx = \int x^{3/2} \, dx = \frac{1}{\frac{5}{2}}x^{5/2} + c = \tfrac{2}{5}x^{5/2} + c$

Problem C Find $\int \dfrac{1}{x^5} \, dx$.

Solution $\displaystyle \int \frac{1}{x^5} \, dx = \int x^{-5} \, dx = \frac{x^{-4}}{-4} + c = -\frac{1}{4x^4} + c$

To integrate more involved functions it is helpful to observe that the following properties yield answers consistent with the definition for the indefinite integral.

Properties of Indefinite Integrals

Property	*Example*
1. $\displaystyle \int k \, dx = kx + c$	1. $\displaystyle \int 7 \, dx = 7x + c$
2. $\displaystyle \int kf(x) \, dx = k \int f(x) \, dx$	2. $\displaystyle \int 4x^2 \, dx = 4 \int x^2 \, dx$
3. $\displaystyle \int [f(x) + g(x)] \, dx$	3. $\displaystyle \int (3x + x^3) \, dx$
$\displaystyle \quad = \int f(x) \, dx + \int g(x) \, dx$	$\displaystyle \quad = \int 3x \, dx + \int x^3 \, dx$

Cautions. 1. $\displaystyle\int f(x) \cdot g(x)\, dx \neq \int f(x)\, dx \cdot \int g(x)\, dx$

2. $\displaystyle\int \frac{f(x)}{g(x)}\, dx \neq \frac{\displaystyle\int f(x)\, dx}{\displaystyle\int g(x)\, dx}$

Example 4

Purpose To use the properties of indefinite integrals as an aid in integration:

Problem A Find $\displaystyle\int dx$.

Solution $\displaystyle\int dx = \int 1\, dx = x + c$ (Property 1 where $k = 1$).

Problem B Find $\displaystyle\int (2x^2 - 3 - 5\sqrt{x})\, dx$.

Solution $\displaystyle\int (2x^2 - 3 - 5\sqrt{x})\, dx = \int 2x^2\, dx - \int 3\, dx - \int 5x^{1/2}\, dx$

$$= 2\int x^2\, dx - 3\int dx - 5\int x^{1/2}\, dx$$

$$= 2 \cdot \frac{x^3}{3} - 3x - \frac{5x^{3/2}}{\frac{3}{2}} + c$$

$$= \frac{2}{3}x^3 - 3x - \frac{10}{3}x^{3/2} + c$$

(Theorem 5.2 and properties 1, 2, and 3)

Problem C Find $\displaystyle\int \left(\frac{x^4 + 2x^2 + 1}{x^2}\right) dx$.

Solution $\displaystyle\int \left(\frac{x^4 + 2x^2 + 1}{x^2}\right) dx = \int (x^2 + 2 + x^{-2})\, dx$

$$= \frac{x^3}{3} + 2x + \frac{x^{-1}}{-1} + c$$

$$= \frac{1}{3}x^3 + 2x - x^{-1} + c$$

Notice that you can *always* check your answer, $F(x) + c$, and know you are correct if $\dfrac{d}{dx}[F(x) + c] = f(x)$ where $f(x)$ is your given integrand. That is

$$\int f(x)\, dx = F(x) + c \quad \text{if} \quad \frac{d}{dx}[F(x) + c] = f(x)$$

Example 5

Purpose To use integration in an applied problem:

Problem A The fixed cost of production of a firm is $800. The marginal cost is given by $C'(x) = .03x^2 - .12x + 5$. Find the total cost function.

Solution The marginal cost, $C'(x)$ is the derivative of the cost function, $C(x)$. Therefore, to find $C(x)$, we integrate the marginal cost function.

$$C(x) = \int (.03x^2 - .12x + 5) \, dx$$

$$= .03 \int x^2 \, dx - .12 \int x \, dx + 5 \int dx$$

$$= \frac{.03}{3} x^3 - \frac{.12}{2} x^2 + 5x + c$$

$$= .01x^3 - .06x^2 + 5x + c$$

With no production ($x = 0$), the firm has a fixed cost of $800.

$$800 = .01(0)^3 - .06(0)^2 + 50(0) + c$$

$$c = 800$$

$$\therefore C(x) = .01x^3 - .06x^2 + 5x + 800.$$

Problem B The slope of the tangent line at a point on a curve is $\dfrac{dy}{dx} = 2x - 7$. If a point on the curve is (2,6), find the function defining the curve.

Solution The slope of the tangent line at a point on a curve is found by taking the derivative of the function defining the curve. Therefore, to find $y = f(x)$, integrate $\dfrac{dy}{dx} = 2x - 7$.

$$\frac{dy}{dx} = 2x - 7$$

$$dy = (2x - 7) \, dx$$

Integrate.

$$\int dy = \int (2x - 7) \, dx$$

$$y = 2 \cdot \tfrac{1}{2}x^2 - 7 \cdot x + c$$

$$y = x^2 - 7x + c$$

Since (2,6) is a point on the curve,

$$6 = (2)^2 - 7(2) + c$$

$$c = 16$$

and

$$y = f(x) = x^2 - 7x + 16.$$

Exercise 5.2

In Problems 1 to 22, find the indefinite integral.

1. $\displaystyle\int 10x \, dx$

12. $\displaystyle\int \frac{2}{7x^2} \, dx$

2. $\displaystyle\int dx$

13. $\displaystyle\int \left(\frac{4}{x^3} + \frac{5}{x^2} + 20\right) dx$

3. $\displaystyle\int (3x - 4) \, dx$

14. $\displaystyle\int \left(\frac{4}{5x^5} - \frac{5}{4}x^5\right) dx$

4. $\displaystyle\int (3x + 4) \, dx$

15. $\displaystyle\int \left(\frac{2}{3x^2} - \frac{3}{2x^3}\right) dx$

5. $\displaystyle\int (5 - 2x) \, dx$

16. $\displaystyle\int \frac{x^3 - 2x^2 + 7x}{x} \, dx$

6. $\displaystyle\int (3x^2 - 2x + 1) \, dx$

17. $\displaystyle\int \frac{x^3 + x^2 - 1}{x^2} \, dx$

7. $\displaystyle\int (3x^2 - x + 2) \, dx$

18. $\displaystyle\int (\tfrac{1}{3}x^{5/3} - 2x^{-(5/3)}) \, dx$

8. $\displaystyle\int (6x^5 - 2x^3 + 4x - 1) \, dx$

19. $\displaystyle\int 2\sqrt{x} \, dx$

9. $\displaystyle\int (3x + 1)^2 \, dx$

20. $\displaystyle\int \left(3\sqrt{x} + \frac{1}{2\sqrt{x}}\right) dx$

10. $\displaystyle\int (17 - 6x^2 + 3x^{-2}) \, dx$

21. $\displaystyle\int (\sqrt{x^3} + \sqrt[3]{x^2}) \, dx$

11. $\displaystyle\int \frac{x^3 - 6x^2 + 6x - 2}{3} \, dx$

22. $\displaystyle\int \left(\frac{2x^2 - 5x + \sqrt{x}}{\sqrt{x}}\right) dx$

In Problems 23 to 30, state whether the integration is true or false. You can *always* check: $\int f(x) \, dx = F(x) + c$ if $D_x[F(x) + c] = f(x)$.

23. $\displaystyle\int (x^2 + 3) \, dx = \frac{x^3 + x}{3} + c$

24. $\displaystyle\int \tfrac{1}{3}x^3 \cdot \sqrt[3]{x} \, dx = \tfrac{1}{13}x^4 \cdot \sqrt[3]{x} + c$

25. $\displaystyle\int \sqrt{x + 5} \, dx = \tfrac{2}{3}(x + 5)^{3/2} + 5 + c$

26. $\displaystyle\int (x^4 + 7)^2(4x^3) \, dx = \frac{(x^4 + 7)^3}{3} + c$

27. $\displaystyle\int \left(\frac{1}{2}\sqrt{x} + \frac{1}{2} + \frac{1}{2\sqrt{x}}\right) dx = \frac{1}{3}x^{3/2} + x^{1/2} + c$

28. $\int \left(\frac{1}{2}\sqrt{x} + \frac{1}{2} + \frac{1}{2\sqrt{x}} \right) dx = \frac{1}{3}x^{3/2} + \frac{1}{2}x + x^{1/2} + c$

29. $\int \sqrt{x^2 + 3} \, dx = \frac{x^2}{2} + \sqrt{3x} + c$

30. $\int \frac{2}{(x + 2)^3} \, dx = \frac{-1}{(x + 2)^2} + c$

31. The marginal cost for a factory is $C'(x) = .6x^2 + 2x + 3$ with an initial fixed production cost of \$420. Find the total cost function. (See Example 5, Problem A.)

32. The marginal cost for production is given by $C'(x) = 18\sqrt{x} + 4$. If the fixed cost is \$50, write a function for the total cost.

33. The marginal profit equation for an item is $P'(x) = 20 - .6x^2$. The profit when two items are sold is \$4.20. What is the profit equation?

34. A company has found that by observing production and sales, its marginal profit is given by $P'(x) = 28 - .09x^2$ when x units are sold. The company realizes a loss of \$500 when no units are sold. Find the profit function.

35. The change in average total cost per unit change in output, the marginal average cost, $\overline{C}'(x)$, is found to be $\$\left(\frac{1}{4} - \frac{17}{x^2} \right)$, find:

(a) the average cost function, $\overline{C}(x)$, given $\overline{C}(4) = 57$

(b) the cost function, $C(x)$. $\left[\text{Recall } \overline{C}(x) = \frac{C(x)}{x}. \right]$

36. The slope of the tangent line to a curve at any point on the curve is $3x^2 - 5$. Find an equation of the curve if it contains the point $(3,8)$. (See Example 5, Problem B.)

37. The slope of the tangent line to a curve at any point on the curve is $\frac{1}{2\sqrt{x}}$.

If a point on the curve is $(1,1)$, find the equation of the curve.

38. An oil well is producing oil at the rate of $R'(t) = 70 + 4.5t - .5t^2$, t in hours. Find a formula for the total output of oil after t hours.

39. A velocity function $v(t) = t^2 - 2t + 4$. Find the distance function, $s(t)$, given the initial condition $s(0) = 0$. (Recall, $v = \frac{ds}{dt}$; therefore, to find s, integrate the velocity function.)

40. An acceleration function is $a = 4 - 66t$. Find the velocity function under the condition that the initial velocity, $v(0)$, is zero. (Recall, $a = \frac{dv}{dt}$; therefore, to find v, integrate the acceleration function.)

41. The breaking distance for a vehicle may be found by solving the differential equation $\dfrac{W}{g}\dfrac{d^2x}{dt^2} = -fW$ where W is the weight of the vehicle, f is the coefficient of friction, g is a gravitational constant, x is distance, and t is time. $\left(\text{Recall that } \dfrac{d^2x}{dt^2} \text{ is acceleration and } \dfrac{dx}{dt} \text{ is velocity.}\right)$

(a) Integrate $\dfrac{W}{g}\dfrac{d^2x}{dt^2} = -fW$ to obtain the velocity function. Assume that when $t = 0$, $\dfrac{dx}{dt} = v_0$.

(b) Use the velocity function to determine how much time elapses before velocity reaches 0. In other words, find t when $\dfrac{dx}{dt} = 0$.

(c) Integrate the velocity function to obtain the distance function. Impose the condition that when $t = 0$, $x = 0$.

(d) Breaking distance is the distance traveled during the time it takes velocity to reach 0. Express breaking distance in terms of v_0, f, and g.

42. If a chemical process is taking place at a constant volume the enthalpy is given by $\Delta H = \int Cv\, dt$, where Cv is defined as the heat capacity at a constant volume. If $Cv = 2t^2 + 3t + 7$, find ΔH.

43. Theorem 5.2 states that $\displaystyle\int x^n\, dx = \dfrac{x^{n+1}}{n+1} + c$ where $n \neq -1$. Show that integration by this formula is impossible when $n = -1$.

5.3 AREAS BY SUMMATION, THE DEFINITE INTEGRAL

Let us illustrate the procedure for constructing a definite integral by first relating it to a concept that you are familiar with—that of an area. We will find areas under irregularly shaped curves by using the sum of the areas of certain rectangles. First, we investigate the use of summation notation.

Let us agree that n is a counter, increasing by one with every revolution of a machine. Set n initially equal to 1. Then with successive revolutions of the machine, n takes on the values: 1, 2, 3, 4, 5, 6, 7, 8, ... Let us take j to be a similar counter restricted to taking on values 1 to 5. Then j equals successively: 1, 2, 3, 4, 5.

Inspect the sum, $1^2 + 2^2 + 3^2 + 4^2 + 5^2$, and adopt a short method for writing this sum, called *summation notation,* by making use of the Greek letter Σ (Sigma). We will agree that

$$1^2 + 2^2 + 3^2 + 4^2 + 5^2 = \sum_{j=1}^{5} j^2.$$

In a similar manner,

$$1^3 + 2^3 + 3^3 + 4^3 = \sum_{i=1}^{4} i^3$$

and

$$1^3 + 3^3 + 5^3 + 7^3 + 9^3 = \sum_{n=1}^{5} (2n - 1)^3.$$

This notation is particularly useful when we have the sum of many terms. For example, instead of: $1^2 + 2^2 + 3^2 + 4^2 + \cdots + 100^2$, we will write this sum as $\sum\limits_{n=1}^{100} n^2$.

Also, it can be proved, by a technique known as mathematical induction, that

$$\sum_{k=1}^{n} k = 1 + 2 + \cdots + n = \frac{n(n + 1)}{2} \tag{1}$$

and

$$\sum_{k=1}^{n} k^2 = 1^2 + 2^2 + \cdots + n^2 = \frac{n(n + 1)(2n + 1)}{6}. \tag{2}$$

In addition, such results as $\sum\limits_{k=1}^{n} c = cn$ and $\sum\limits_{k=1}^{n} ca_k = c \sum\limits_{k=1}^{n} a_k$ can be proved, where c represents a constant.

Example 1

Purpose To illustrate the use of summation notation:

Problem A Use summation notation to express: $\sqrt{2} + 2 + \sqrt{6} + 2\sqrt{2}$.

Solution $\sqrt{2} + \sqrt{4} + \sqrt{6} + \sqrt{8} = \sum\limits_{n=1}^{4} \sqrt{2n}$

Problem B Write $\sum\limits_{i=1}^{n} 2f(x_i)$ without summation notation.

Solution $\sum\limits_{i=1}^{n} 2f(x_i) = 2f(x_1) + 2f(x_2) + \cdots + 2f(x_n).$

Problem C Show $\sum\limits_{k=1}^{10} k = 55$, using $\sum\limits_{k=1}^{n} k = \frac{n(n + 1)}{2}.$

Solution $\displaystyle\sum_{k=1}^{10} k = 1 + 2 + 3 + \cdots + 10 = \frac{(10)(10 + 1)}{2} = 55.$

Problem D   Does $\displaystyle\sum_{k=1}^{3} \frac{1}{n}k^2 = \frac{1}{n}\sum_{k=1}^{3} k^2$?

Solution $\displaystyle\sum_{k=1}^{3} \frac{1}{n}k^2 = \frac{1}{n}1^2 + \frac{1}{n}2^2 + \frac{1}{n}3^2 = \frac{1}{n}(1^2 + 2^2 + 3^2).$

$\displaystyle\frac{1}{n}\sum_{k=1}^{3} k^2 = \frac{1}{n}(1^2 + 2^2 + 3^2).$

Yes! $\displaystyle\sum_{k=1}^{3} \frac{1}{n}k^2 = \frac{1}{n}\sum_{k=1}^{3} k^2$

The problem of finding the area between $y = f(x) = 2x$ and the x axis, where this area is also bounded by $x = 0$ and $x = 1$ (see Figure 5.1) is as follows.

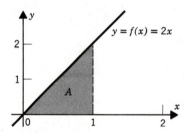

Figure 5.1

First we determine the approximate area, A, by dividing the interval $0 \leqslant x \leqslant 1$ into 4 equal subintervals. Then we construct rectangles on each sub-interval using, for the rectangle's height, the y value at the right end point of that subinterval (see Figure 5.2).

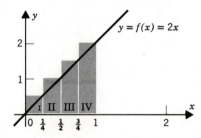

Figure 5.2

The area for each rectangle is developed in the following table. The width or base of each rectangle is $\frac{1}{4}$.

Rectangle	Height	Area
I	$f(\frac{1}{4}) = 2(\frac{1}{4}) = \frac{1}{2}$	$(\frac{1}{4})(\frac{1}{2}) = \frac{1}{8}$
II	$f(\frac{1}{2}) = 2(\frac{1}{2}) = 1$	$(\frac{1}{4})(1) = \frac{1}{4}$
III	$f(\frac{3}{4}) = 2(\frac{3}{4}) = \frac{3}{2}$	$(\frac{1}{4})(\frac{3}{2}) = \frac{3}{8}$
IV	$f(1) = 2(1) = 2$	$(\frac{1}{4})(2) = \frac{1}{2}$

The area, A, is approximately equal to the sum of the areas of the rectangles I, II, III, and IV. Therefore, $A \simeq \frac{1}{8} + \frac{1}{4} + \frac{3}{8} + \frac{1}{2} = \frac{10}{8} = 1\frac{1}{4}$.

How can we improve on this approximation? Investigate Figure 5.3. Does it not seem that for $n > 4$ the sum of the area of these n rectangles would yield a closer approximation to the true area A?

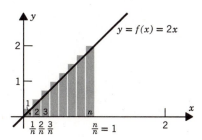

Figure 5.3

We have divided $0 \leqslant x \leqslant 1$ into n equal subintervals, each of width $\frac{1 - 0}{n} = \frac{1}{n}$. The rectangle is then built on each subinterval using, for the rectangle's height, the y value of the right end point of that subinterval.

The area for each rectangle is developed in the following table. The width or base of each rectangle is $\frac{1}{n}$.

Rectangle	Height	Area
1	$f\left(\frac{1}{n}\right) = 2 \cdot \frac{1}{n}$	$\left(\frac{1}{n}\right)\left(2 \cdot \frac{1}{n}\right)$
2	$f\left(\frac{2}{n}\right) = 2 \cdot \frac{2}{n}$	$\left(\frac{1}{n}\right)\left(2 \cdot \frac{2}{n}\right)$
3	$f\left(\frac{3}{n}\right) = 2 \cdot \frac{3}{n}$	$\left(\frac{1}{n}\right)\left(2 \cdot \frac{3}{n}\right)$
$\vdots$	$\vdots$	$\vdots$
n	$f\left(\frac{n}{n}\right) = 2 \cdot \frac{n}{n}$	$\left(\frac{1}{n}\right)\left(2 \cdot \frac{n}{n}\right)$

The area is approximately equal to the sum of the areas of the n rectangles.

Therefore,

$$A \simeq \left(\frac{1}{n}\right)\left(2 \cdot \frac{1}{n}\right) + \left(\frac{1}{n}\right)\left(2 \cdot \frac{2}{n}\right) + \left(\frac{1}{n}\right)\left(2 \cdot \frac{3}{n}\right) + \cdots + \left(\frac{1}{n}\right)\left(2 \cdot \frac{n}{n}\right)$$

$$= \frac{1}{n}\left(2 \cdot \frac{1}{n} + 2 \cdot \frac{2}{n} + 2 \cdot \frac{3}{n} + \cdots + 2 \cdot \frac{n}{n}\right)$$

$$= \frac{2}{n}\left(\frac{1}{n} + \frac{2}{n} + \frac{3}{n} + \cdots + \frac{n}{n}\right)$$

$$= \frac{2}{n^2}(1 + 2 + 3 + \cdots + n)$$

$$= \frac{2}{n^2} \sum_{k=1}^{n} k \qquad \text{See Formula (1), page 235.}$$

$$= \frac{2}{n^2} \cdot \frac{n(n + 1)}{2}$$

$$= \frac{n + 1}{n}$$

$$= 1 + \frac{1}{n}.$$

$$\therefore A \simeq 1 + \frac{1}{n}.$$

Most students intuitively suspect, and correctly so, the more rectangles that are used, the closer the approximated value, A, will be to the true area A. Examine the behavior of our approximation of A as n, the number of rectangles, increases without bound.

$$A \simeq 1 + \frac{1}{n}$$

$$\lim_{n \to \infty} \left(1 + \frac{1}{n}\right) = 1 + 0 = 1$$

To determine if this is a "good" estimate of A, let us calculate A by noting that it is a triangle. The area of a triangle is given by $A = \frac{1}{2}b \cdot h$ (b, base; h, height). For our problem $b = 1$, $h = 2$ and $A = \frac{1}{2}(1)(2) = 1$. Since 1 is the true area and we also obtained this answer by the limit method, we have

$$A = \lim_{n \to \infty} \left(1 + \frac{1}{n}\right) = 1.$$

It is possible to show that this limit method does give results that are in agreement with those obtained by area formulas and are consistent with certain desirable properties of areas. Rather than belabor this point, let us agree that if $f(x) \geq 0$ then the above limit method does yield the true area.

The Definite Integral

It is possible to use this same type of procedure for a function f defined on $a \leq x \leq b$ without ever mentioning the concept of area. We do this to define the *definite integral*. This particular definite integral, useful in many applied problems, is called the *Riemann integral*.

The *definite integral* is defined by the following procedure:

1. Let f be defined by $y = f(x)$ for $a \leq x \leq b$.

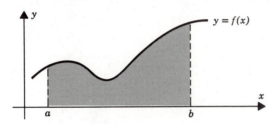

2. Divide the interval from a to b into n subintervals of equal length. The length of each subinterval, Δx, is then given by $\Delta x = \dfrac{b - a}{n}$.

3. Let x_1 be the x value of the right end point of the first subinterval; let x_2 be the x value at the right end point of the second interval; . . . and x_n the x value of the right end point of the nth subinterval.

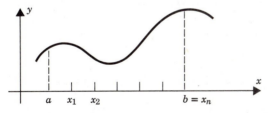

4. Corresponding to each point $x_1, x_2, \ldots, x_n$, find $f(x_1), f(x_2),$ $\ldots, f(x_n)$.

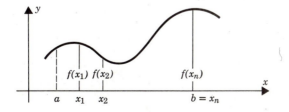

5. Form the sum, $f(x_1) \, \Delta x + f(x_2) \, \Delta x + f(x_3) \, \Delta x + \cdots + f(x_n) \, \Delta x$. This sum is called a Riemann sum and may be represented by

$$\sum_{i=1}^{n} f(x_i) \, \Delta x.$$

6. Take the limit of this Riemann sum as the number of subintervals increases without bound, $n \to \infty$. If this limit exists, it is called the definite integral of f from a to b and is denoted by $\int_a^b f(x) \, dx$.

Repeating the definition for definite integral:

DEFINITION

Definite Integral. Let f be defined for $a \leqslant x \leqslant b$. Then if the limit exists,

$\lim_{n \to \infty} \sum_{i=1}^{n} f(x_i) \, \Delta x = \int_a^b f(x) \, dx$. Here, x_i is the right end point of the ith subinterval and $\Delta x = \dfrac{b - a}{n}$.

Notes on the above Definition

1. If $\int_a^b f(x) \, dx$ exists it is a unique real number. This is true since

$\int_a^b f(x) \, dx = \lim_{n \to \infty} \sum_{i=1}^{n} f(x_i) \, \Delta x$ and, when this limit exists, the limit is a unique real number. See Appendix II, page 393, for a discussion of this new kind of limit.

2. The $\int_a^b f(x) \, dx$ exists if the defining limit exists. A large set of functions for which the limit exists is the set of continuous functions. It is proved in advanced courses of analysis that if f is continuous on $a \leqslant x \leqslant b$, then $\int_a^b f(x) \, dx$ exists.

3. x is called a dummy, since $\int_a^b f(x) \, dx = \int_a^b f(y) \, dy$. For example,

$$\int_0^1 x^2 \, dx = \int_0^1 t^2 \, dt = \int_0^1 z^2 \, dz, \quad \text{etc.}$$

4.   Since x_i is the right end point of the ith subinterval, $x_1 = a + \Delta x = a + \dfrac{b-a}{n}$, $x_2 = a + 2\,\Delta x = a + 2\dfrac{b-a}{n}$, $\ldots$, $x_i = a + i\,\Delta x = a + i\dfrac{b-a}{n}$. Then, $f(x_1) = f(a + \Delta x) = f\!\left(a + \dfrac{b-a}{n}\right)$, $\ldots$, $f(x_i) = f\!\left(a + i\dfrac{b-a}{n}\right)$. Thus, $\displaystyle\sum_{i=1}^{n} f(x_i)\,\Delta x = \sum_{i=1}^{n} f\!\left(a + i\dfrac{b-a}{n}\right) \cdot \dfrac{b-a}{n}$. See Exercise 5.3, Problems 31 and 32 for a specific example of this type of problem.

5.   A more general definition for a Riemann sum permits x_i' to be selected as any point in the ith subinterval, that is, $x_{i-1} \leqslant x_i' \leqslant x_i$. Then, a Riemann sum is defined to be $\displaystyle\sum_{i=1}^{n} f(x_i')\,\Delta x$. Moreover, the definite integral is then defined to be $\displaystyle\int_{a}^{b} f(x)\,dx = \lim_{n \to \infty} \sum_{i=1}^{n} f(x_i')\,\Delta x$, if this limit exists. See Exercise 5.3, Problem 34 for a specific example of such a Riemann integral. An even more general method permits the widths of the subintervals to be of unequal lengths. However, once the limit of the Riemann sum is taken, our definition gives the same answer as the more general methods.

Example 2

Purpose   To find a definite integral by using the definition:

Problem   Find $\displaystyle\int_{0}^{3} x^2\,dx$.

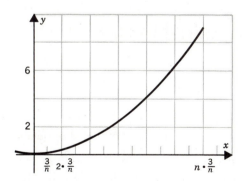

Solution   Divide $0 \leqslant x \leqslant 3$ into n equal subintervals, each of width $\dfrac{3}{n}$. Select x_i to be the right end point of each subinterval.

Form a Riemann sum.

$$\sum_{i=1}^{n} f(x_i)\, \Delta x = f(x_1)\, \Delta x + f(x_2)\, \Delta x + \cdots + f(x_n)\, \Delta x$$

$$= \left(\frac{3}{n}\right)^2\left(\frac{3}{n}\right) + \left(2 \cdot \frac{3}{n}\right)^2\left(\frac{3}{n}\right) + \cdots + \left(n \cdot \frac{3}{n}\right)^2\left(\frac{3}{n}\right)$$

$$= \frac{3^3}{n^3}(1^2 + 2^2 + \cdots + n^2)$$

$$= \frac{3^3}{n^3} \sum_{k=1}^{n} k^2 \qquad \text{See Formula (2), page 235.}$$

$$= \frac{3^3}{n^3}\left[\frac{(n)(n+1)(2n+1)}{6}\right]$$

$$= \frac{9}{2}\left[\left(\frac{n+1}{n}\right)\left(\frac{2n+1}{n}\right)\right].$$

By the definition of the definite integral,

$$\int_0^3 x^2\, dx = \lim_{n\to\infty} \sum_{i=1}^{n} f(x_i)\, \Delta x$$

or

$$\int_0^3 x^2\, dx = \lim_{n\to\infty} \left[\frac{9}{2}\left(\frac{n+1}{n}\right)\left(\frac{2n+1}{n}\right)\right]$$

$$= \lim_{n\to\infty} \left[\frac{9}{2}\left(1 + \frac{1}{n}\right)\left(2 + \frac{1}{n}\right)\right]$$

$$= \frac{9}{2}(1)(2) = 9$$

$$\int_0^3 x^2\, dx = 9.$$

In Example 2, area has never been mentioned in obtaining $\int_0^3 x^2\, dx = 9$. Of course, this particular problem can be rephrased as an area problem, and its solution will be exactly the same. The use of definite integrals to find an area is just one of the applications of the definite integral. As an example of a definite integral that is not an area, consider $\int_0^5 (x^2 - 5x)\, dx$. In the next section it is easily verified that $\int_0^5 (x^2 - 5x)\, dx = -\frac{125}{6}$. An area can never be negative.

However, if $f(x) \geq 0$, then $\int_a^b f(x) \, dx$ is equal to the area between $y = f(x)$, the x axis, and $a \leq x \leq b$.

The undesirable feature of our present position is that we have to find $\lim\limits_{n \to \infty} \sum\limits_{i=1}^n f(x_i) \, \Delta x$ to obtain $\int_a^b f(x) \, dx$. We must find a less involved approach for finding $\int_a^b f(x) \, dx$. This will be given in the next section.

Exercise 5.3

In Problems 1 to 8, expand the summation.

1. $\displaystyle\sum_{k=1}^4 \frac{1}{k}$

5. $\displaystyle\sum_{i=1}^5 f(x_i)$

2. $\displaystyle\sum_{k=2}^4 \frac{1}{k}$

6. $\displaystyle\sum_{i=1}^n f(x_i)$

3. $\displaystyle\sum_{n=1}^3 \frac{(n)(n+1)}{2}$

7. $\displaystyle\sum_{i=1}^n f(x_i) \, \Delta x$

4. $\displaystyle\sum_{n=1}^3 \frac{2}{(n)(n+1)}$

8. $\displaystyle\sum_{j=-1}^4 \sqrt{j+5}$

In Problems 9 to 19, write the following, using summation notation.

9. $2 + 4 + 6 + 8 + 10$

15. $x + x^3 + x^5 + \cdots + x^{2n-1}$

10. $2 + 4 + 6 + \cdots + 60$

16. $\left(\dfrac{1}{n}\right)^2 + \left(\dfrac{2}{n}\right)^2 + \cdots + \left(\dfrac{n}{n}\right)^2$

11. $x_1 + x_2 + x_3 + x_4$

17. $\dfrac{1}{1 \cdot 2} + \dfrac{1}{2 \cdot 3} + \dfrac{1}{3 \cdot 4}$

12. $x_1 + x_2 + \cdots + x_n$

18. $f(x_1) \, \Delta x + f(x_2) \, \Delta x + f(x_3) \, \Delta x$

13. $f(x_1) + f(x_2) + f(x_3) + f(x_4)$

19. $f(x_1) \, \Delta x + f(x_2) \, \Delta x + \cdots$
$\qquad\qquad\qquad\qquad\qquad + f(x_n) \, \Delta x$

14. $f(x_1) + f(x_2) + \cdots + f(x_n)$

20. A polynomial function may be expressed in Σ notation:

$$y = f(x) = a_0 + a_1 x + a_2 x^2 + a_3 x^3 + \cdots + a_{n-1} x^{n-1} + a_n x^n = \sum_{i=0}^n a_i \cdot x^i$$

Show that the derivative $f'(x) = \displaystyle\sum_{i=1}^n i \cdot a_i \cdot x^{i-1}$.

21. Verify:

(a) $\displaystyle\sum_{k=1}^{5} 13 = (13)(5)$

(c) $\displaystyle\sum_{i=1}^{4} 3x_i^2 = 3 \sum_{i=1}^{4} x_i^2$

(b) $\displaystyle\sum_{i=1}^{10} (ab) = 10ab$

(d) $\displaystyle\sum_{i=1}^{n} f(x_i)\,\Delta x = \Delta x \sum_{i=1}^{n} f(x_i)$

22. Cardiac output indicates the amount of blood pumped by the heart per unit time. The indicator-dilution method is one way of measuring cardiac output. This method involves injecting an indicator dye into a vein and then measuring the concentration of the dye at an observation point. The dye begins to show up at the observation point a few seconds after being injected, continues to increase to a maximum, and then begins to decrease. For simplicity, assume that recirculation of the dye does not occur. Cardiac output is equal to the amount of indicator injected divided by the area under the concentration versus time curve. Since the concentration curve does not have a known formula, the area is approximated by using Riemann sums.

A physician injects .5 mg of dye at time 0 and records the concentration of the dye at the observation point every second. The following data are obtained:

time (seconds)	0	1	2	3	4	5	6	7	8	9	10
concentration $\dfrac{mg}{1}$	0	.01	.05	.1	.5	1	2	1.1	.9	.3	.04

Using these data, determine cardiac output. It should be noted that cardiac output is usually expressed in liters per minute.

23. (a) Approximate the area bounded by the curve $y = 3x$, the x axis, and the interval $x = 0$ to $x = 1$. Follow the procedure developed in this section, which is: (1) Divide $0 \leq x \leq 1$ into four equal subintervals. (2) Construct rectangles on each subinterval using, for the rectangle's height, the y value at the right end point of that subinterval. (3) Find the sum of the areas of the rectangles.

(b) Find the true area bounded by the curve $y = 3x$, the x axis, and the interval $0 \leq x \leq 1$. Follow the procedure developed in this section, which is: (1) Divide $0 \leq x \leq 1$ into n equal subintervals. (2) Construct rectangles on each subinterval using, for the rectangle's height, the y value at the right end point of that subinterval. (3) Find the sum of the areas of the rectangles. (4) The true area is found by taking the limit as $n \to \infty$.

(c) Check your answer by finding the true area using elementary geometry (the area of a right triangle).

24. (a) Approximate the area bounded by the curve $y = x^2$, the x axis, and the interval $x = 0$ to $x = 3$, using six equal subintervals. (See Problem 23a for the procedure.)

(b) Find the true area bounded by the curve $y = x^2$, the x axis, and the interval $x = 0$ to $x = 3$. (See Problem 23b for the procedure.)

25. (a) Approximate the area bounded by the curve $y = x$, the x axis, and the interval $1 \leq x \leq 3$, using four equal subintervals. (See Problem 23a for the procedure.)

(b) Find the true area bounded by the curve $y = x$, the x axis, and the interval $1 \leq x \leq 3$. (See Problem 23b for the procedure.)

(c) Check your answer by finding the true area using elementary geometry [the area of a trapezoid which is $A = \frac{1}{2}(b + b')h$ where b and b' are the parallel bases and h is the height].

26. (a) Approximate the area bounded by the curve $y = \frac{1}{2}x^2 + 1$, the x axis and the interval $0 \leq x \leq 4$ using four equal subintervals. (See Problem 23a for the procedure.)

(b) Find the true area bounded by the curve $y = \frac{1}{2}x^2 + 1$, the x axis and the interval $0 \leq x \leq 4$. (See Problem 23b for the procedure.)

27. Find $\displaystyle\int_0^1 x \, dx$ by using the following procedure.

(a) Form a Riemann sum by dividing $0 \leq x \leq 1$ into n equal subintervals each of width $\dfrac{1}{n}$ and selecting x_i to be the right end point of each sub-interval. Thus $\displaystyle\sum_{i=1}^{n} f(x_i) \, \Delta x = f(x_1) \, \Delta x + f(x_2) \, \Delta x + \cdots + f(x_n) \, \Delta x$.

You should find the Riemann sum is equal to $\dfrac{1}{2}\left(\dfrac{n+1}{n}\right)$.

(b) By the definition of the definite integral,

$\displaystyle\int_0^1 x \, dx = \lim_{n \to \infty} \sum_{i=1}^{n} f(x_i) \, \Delta x$. Thus $\displaystyle\int_0^1 x \, dx = \lim_{n \to \infty} \frac{1}{2}\left(\frac{n+1}{n}\right)$. You should find $\displaystyle\int_0^1 x \, dx = \dfrac{1}{2}$.

28. Find $\displaystyle\int_0^1 (x + 2) \, dx$ by first forming a Riemann sum and then taking $\displaystyle\lim_{n \to \infty} \sum_{i=1}^{n} f(x_i) \, \Delta x$.

29. Find $\displaystyle\int_0^1 x^2 \, dx$. Riemann sum should equal $\left[\dfrac{1}{6}\left(\dfrac{n+1}{n}\right)\left(\dfrac{2n+1}{n}\right) \right]$.

30. Find $\displaystyle\int_0^1 2x \, dx$.

31. Find $\int_1^5 (x + 2)\, dx$ by using the following procedure.

(a) Form a Riemann sum by dividing $1 \leqslant x \leqslant 5$ into n equal subintervals each of width $\dfrac{4}{n}$. Select x_i to be the right end point of each subinterval; $\quad x_1 = 1 + \dfrac{4}{n}, \quad x_2 = 1 + 2 \cdot \dfrac{4}{n}, \ldots, x_i = 1 + i \cdot \dfrac{4}{n}, \ldots,$ $x_n = 1 + n\dfrac{4}{n}$. The $\displaystyle\sum_{i=1}^n f(x_i)\, \Delta x = \sum_{i=1}^n \left(1 + i \cdot \dfrac{4}{n} + 2\right) \cdot \left(\dfrac{4}{n}\right)$. You should find the Riemann sum equal to $12 + 8\left(\dfrac{n+1}{n}\right)$.

(b) By the definition of the definite integral,

$$\int_1^5 (x + 2)\, dx = \lim_{n \to \infty} \sum_{i=1}^n f(x_i)\, \Delta x = \lim_{n \to \infty} \left[12 + 8\left(\dfrac{n+1}{n}\right)\right].$$

32. Find $\int_1^4 (8 - 2x)\, dx$. You should find the Riemann sum equal to $18 - 9\left(\dfrac{n+1}{n}\right)$.

33. Find $\int_0^2 (\tfrac{1}{2}x + 2)\, dx$.

34. It can be shown that the definite integral can be found by selecting x_i to be any point of the ith subinterval. Find $\int_0^2 (\tfrac{1}{2}x + 2)\, dx$ by selecting x_i to be the left end point of each subinterval. The Riemann sum should be equal to $\left[4 + \left(\dfrac{n-1}{n}\right)\right]$.

5.4 WHY DWELL IN HELL? USE THE FUNDAMENTAL THEOREM OF INTEGRAL CALCULUS

The definite integral, $\int_a^b f(x)\, dx = \lim\limits_{n \to \infty} \sum_{i=1}^n f(x_i)\, \Delta x$, has a multitude of applications to problems of disciplines such as economics, management, science, biology, mathematics, psychology, physics, and chemistry. Since the definite integral is used frequently, can you imagine the hours that would be spent if the method of Section 5.3 were used every time a definite integral had to be evaluated?

As one speaker[1] at a mathematics convention recently said, "Why make the student dwell in hell by continuing to evaluate definite integrals as limits of Riemann sums?" Indeed, just as most derivatives were found by the use of shorter techniques, we have a quicker way to evaluate many definite integrals. The theorem that is used to evaluate definite integrals more quickly is a "five star" theorem that rates the colossal billing: The Fundamental Theorem of Integral Calculus. This theorem, when used with the following properties, makes the finding of definite integrals a much shorter and bearable procedure. These properties could be proved by using the definition of the definite integral.

Some Useful Properties of the Definite Integral

Let f be continuous and integrable for a given interval containing a, b, and c, then

1. $$\int_a^a f(x)\,dx = 0$$

   Example. $\displaystyle\int_5^5 x\,dx = 0$

2. $$\int_a^b f(x)\,dx = -\int_b^a f(x)\,dx$$

   Example. $\displaystyle\int_{-3}^1 (x^2 + 3x - 4)\,dx = -\int_1^{-3} (x^2 + 3x - 4)\,dx$

3. $$\int_a^b f(x)\,dx = \int_a^c f(x)\,dx + \int_c^b f(x)\,dx$$

   Example. $\displaystyle\int_{-2}^2 (x - 4)^2\,dx = \int_{-2}^0 (x - 4)^2\,dx + \int_0^2 (x - 4)^2\,dx$

4. $$\int_a^b kf(x)\,dx = k\int_a^b f(x)\,dx \quad (k \text{ any constant})$$

   Example. $\displaystyle\int_2^5 3x^2\,dx = 3\int_2^5 x^2\,dx$

5. $$\int_a^b k\,dx = k(b - a) \quad (k \text{ any constant})$$

   Example. $\displaystyle\int_1^6 4\,dx = 4(6 - 1) = 20$

---

[1] We are indebted to Jim Carney of Lorain County Community College for permission to use the title of his speech.

6. $\displaystyle\int_a^b [f(x) + g(x)]\, dx = \int_a^b f(x)\, dx + \int_a^b g(x)\, dx$

Example. $\displaystyle\int_0^3 (x^2 + \sqrt{3x + 1})\, dx = \int_0^3 x^2\, dx + \int_0^3 \sqrt{3x + 1}\, dx$

Cautions. 1. $\displaystyle\int_a^b f(x) \cdot g(x)\, dx \neq \int_a^b f(x)\, dx \cdot \int_a^b g(x)\, dx$

2. $\displaystyle\int_a^b \frac{f(x)}{g(x)}\, dx \neq \frac{\displaystyle\int_a^b f(x)\, dx}{\displaystyle\int_a^b g(x)\, dx}$

3. The variable x may never be placed in front of $\displaystyle\int_a^b$.

Example 1

Purpose To use the properties of the definite integral:

Problem Determine whether the following are true or false.

(A) $\displaystyle\int_3^3 4\, dx = 12$

(B) $\displaystyle\int_1^3 4x^2\, dx = 4 \int_1^3 x^2\, dx$

(C) $\displaystyle\int_{-1}^1 (2x - 7)\, dx = \int_{-1}^1 2x\, dx + \int_1^{-1} 7\, dx$

Solution (A) False. $\displaystyle\int_3^3 4\, dx = 0$. Property 1

(B) True. A *constant* factor may be placed in front of the integral symbol $\int$. Property 4. Thus, $\displaystyle\int_1^3 4x^2\, dx = 4 \int_1^3 x^2\, dx$.

(C) True. $\displaystyle\int_{-1}^1 (2x - 7)\, dx = \int_{-1}^1 2x\, dx - \int_{-1}^1 7\, dx$ by Property 6; and $\displaystyle\int_{-1}^1 2x\, dx - \int_{-1}^1 7\, dx = \int_{-1}^1 2x\, dx + \int_1^{-1} 7\, dx$ by Property 2.

So that you can see the power of the Fundamental Theorem, we immediately state and apply it. For those who are interested, in the next section, after some preliminary theorems are examined, a proof to The Fundamental Theorem of Integral Calculus is given. See Theorem 5.5d, page 257.

*****THE FUNDAMENTAL THEOREM OF INTEGRAL CALCULUS

Let the function f be continuous for $a \leqslant x \leqslant b$. Let F be any function whose derivative is f. Then

$$\int_a^b f(x)\ dx = F(b) - F(a).$$

To evaluate a definite integral, using the fundamental theorem of integral calculus, two steps are required.

1. Find a function F whose derivative is f, that is, find an antiderivative of f.
2. Evaluate $F(b) - F(a)$.

It is customary to use the symbol $\left[F(x) \right]_a^b$ or $F(x)\bigg]_a^b$ as another name for $F(b) - F(a)$.

This theorem is now used to find the definite integral.

Example 2

Purpose   To illustrate the use of the fundamental theorem of integral calculus to evaluate definite integrals:

Problem A   Find $\int_0^1 x^2\ dx$.

Solution $\int x^2\ dx = \frac{1}{3}x^3 + c$. Let $F(x)$ be any antiderivative, say, $\frac{1}{3}x^3 + 0 = \frac{1}{3}x^3$.

Then $\int_a^b f(x)\ dx = \left[F(x) \right]_a^b = F(b) - F(a)$ becomes

$$\int_0^1 x^2\ dx = \left[\tfrac{1}{3}x^3 \right]_0^1 = [(\tfrac{1}{3} \cdot 1^3) - (\tfrac{1}{3} \cdot 0^3)] = \tfrac{1}{3}.$$

Problem B   Find $\int_0^1 (3 - x)\ dx$.

Solution $\int (3 - x)\ dx = \int 3\ dx - \int x\ dx = 3x - \dfrac{x^2}{2} + c$.

Let $F(x) = 3x - \dfrac{x^2}{2} + 0$ or $F(x) = 3x - \dfrac{x^2}{2}$

Using: $\int_a^b f(x)\ dx = \left[F(x) \right]_a^b = F(b) - F(a)$,

$$\int_0^1 (3 - x)\ dx = \left[3x - \dfrac{x^2}{2} \right]_0^1 = \left[\left(3 - \dfrac{1}{2} \right) - (0 - 0) \right] = \dfrac{5}{2}.$$

Problem C Find $\int_1^2 (\sqrt{x})^3 \, dx$.

Solution $\int x^{3/2} \, dx = \frac{2}{5}x^{5/2} + c$. Select any antiderivative, say $\frac{2}{5}x^{5/2} + 0 = \frac{2}{5}x^{5/2}$.

Using: $\int_a^b f(x) \, dx = \left[F(x) \right]_a^b = F(b) - F(a)$,

$$\int_1^2 x^{3/2} \, dx = \left[\frac{2}{5}x^{5/2} \right]_1^2 = (\frac{2}{5}(2)^{5/2}) - (\frac{2}{5}(1)^{5/2}) = (\frac{2}{5} \cdot 2^2 \cdot 2^{1/2}) - (\frac{2}{5})$$

$$= \frac{8}{5}\sqrt{2} - \frac{2}{5} = \frac{8\sqrt{2} - 2}{5}.$$

Problem D $\int_1^8 (\sqrt[3]{x})^2 \, dx$

Solution $\int_1^8 (\sqrt[3]{x})^2 \, dx = \int_1^8 x^{2/3} \, dx$

$$= \frac{3}{5}x^{5/3} \Big]_1^8 = \frac{3}{5}(\sqrt[3]{x})^5 \Big]_1^8$$

$$= \frac{3}{5}\left[(\sqrt[3]{8})^5 - (\sqrt[3]{1})^5 \right] = \frac{3}{5}(2^5 - 1)$$

$$= \frac{3}{5}(32 - 1) = \frac{93}{5}$$

Problem E $\int_1^4 \left(\frac{2}{x^3} + \frac{1}{\sqrt{x}} \right) dx$.

Solution $\int_1^4 \left(\frac{2}{x^3} + \frac{1}{\sqrt{x}} \right) dx = \int_1^4 \frac{2}{x^3} \, dx + \int_1^4 \frac{1}{\sqrt{x}} \, dx$

$$= 2 \int_1^4 x^{-3} \, dx + \int_1^4 x^{-1/2} \, dx$$

$$= 2 \cdot \frac{1}{-2}x^{-2} \Big]_1^4 + \frac{1}{\frac{1}{2}}x^{1/2} \Big]_1^4$$

$$= \frac{-1}{x^2} \Big]_1^4 + 2\sqrt{x} \Big]_1^4$$

$$= \left[\left(-\frac{1}{16} \right) - (-1) \right] + [(2 \cdot 2) - (2 \cdot 1)]$$

$$= -\frac{1}{16} + 3 = \frac{47}{16}.$$

Example 3

Purpose To use the definite integral in an applied problem:

Problem From the results of a study in a certain small community, the number of robberies in the next year is expected to increase at a rate of $(4x + 5)$ robberies per month. Approximate the number of robberies expected to be committed in a year? During the last six months of the year?

Solution Let $R(x) = 4x + 5$. The number of robberies during the year is $R = R(1) + R(2) + \cdots + R(12)$. This sum can be approximated by an area.

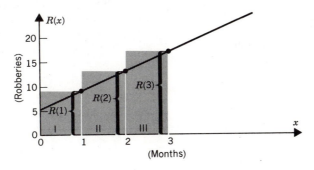

$$R(1) = A_1 = \text{area of rectangle I} = R(1) \cdot 1$$

$$R(2) = A_2 = \text{area of rectangle II} = R(2) \cdot 1$$

$$\vdots$$

$$R(12) = A_{12} = \text{area of 12th rectangle} = R(12) \cdot 1$$

$R = \sum_{i=1}^{12} R(i) \cdot (1) = \sum_{i=1}^{12} A_i$. The sum of the areas of the rectangles is approximately equal to the area under the curve from $x = 0$ to $x = 12$. This area is given by the definite integral:

$$R = \sum_{i=1}^{12} R(i) \cdot (1) \approx \int_0^{12} (4x + 5)\, dx = 2x^2 + 5x \Big]_0^{12}$$

$$= [2(12)^2 + 5(12)] - [0] = 348.$$

The number of robberies during the last six months of the year can be approximated by

$$\int_6^{12} (4x + 5)\, dx = 2x^2 + 5x \Big]_6^{12}$$

$$= [2(12)^2 + 5(12)] - [2(6)^2 + 5(6)]$$

$$= 348 - 102 = 246$$

The Fundamental Theorem of Integral Calculus not only makes integration simpler, but it embodies the relationship between a derivative, the limit of a quotient; and the integral, the limit of a sum. For the Fundamental Theorem states that the value of an integral, $\int_a^b f(x)\, dx$, may be found by finding a function F such that the derivative of F is equal to f. Then $\int_a^b f(x)\, dx = F(b) - F(a)$.

Another relationship between differentiation and integration is given in Theorem 5.5b. This theorem enables us to quickly find the derivative of an integral whose lower limit is any constant and whose upper limit is x. The theorem states that if f is continuous on an interval I, containing $x = k$ (k, a constant) and if x is another number in I, then $\dfrac{d}{dx} \displaystyle\int_k^x f(t)\, dt = f(x)$.

Example 4

Purpose To find $\dfrac{d}{dx} \displaystyle\int_k^x f(t)\, dt$ quickly:

Problem Find $\dfrac{d}{dx} \displaystyle\int_2^x t^{10}\, dt$ and $\dfrac{d}{dx} \displaystyle\int_{-20}^x t\sqrt[3]{t^7 - 8}\, dt$

Solution Use the theorem $\dfrac{d}{dx} \displaystyle\int_k^x f(t)\, dt = f(x)$.

$$\frac{d}{dx} \int_2^x t^{10}\, dt = x^{10}$$

$$\frac{d}{dx} \int_{-20}^x t\sqrt[3]{t^7 - 8}\, dt = x\sqrt[3]{x^7 - 8}$$

If time does not permit you to consider the proof of these extremely important theorems (Section 5.5), you may proceed directly to Exercise 5.4–5.5

5.5 PROOF OF THE FUNDAMENTAL THEOREM OF INTEGRAL CALCULUS (Optional)

We first consider the proofs of certain theorems necessary to prove the fundamental theorem of integral calculus. The section begins with a discussion of the mean value theorem for integrals and concludes with the proof of the fundamental theorem of integral calculus.

THEOREM 5.5a

The Mean Value Theorem for Integrals. If f is continuous in $a \leqslant x \leqslant b$ then there exists a value of $x = c$, $a < c < b$, such that

$$\int_a^b f(x)\, dx = f(c)(b - a).$$

The formal proof of this theorem will not be given. However, by interpreting the definite integral as an area, an intuitive approach to the theorem is presented. This does not take the place of a formal proof; it is merely stated to give substance to the theorem.

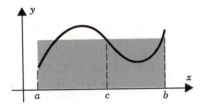

Assuming $f(x) \geqslant 0$, $\int_a^b f(x)\, dx$ has, as one of its interpretations, the area bounded by $y = f(x)$, $x = a$, $x = b$, and the x axis.

There appears to be a rectangle with base of length $(b - a)$ whose area is numerically equal to $\int_a^b f(x)\, dx$, that is, the same as the area under the curve.

Let $f(c)$ be the height of the desired rectangle. The area of the desired rectangle is then equal to its base times its height or $(b - a) \cdot f(c)$.

The area under the curve equals the area of the rectangle, or

$$\int_a^b f(x)\, dx = f(c)(b - a).$$

Thus, the mean value theorem for integrals guarantees that if f is continuous on $a \leqslant x \leqslant b$ then a value $x = c$, where $a < c < b$, may always be found such that

$$\int_a^b f(x)\, dx = f(c)(b - a).$$

Example 1

Purpose To use the mean value theorem:

Problem For the function given by $f(x) = x^2 + 1$ and the interval $1 \leqslant x \leqslant 3$, find the number c guaranteed by the mean value theorem for integrals.

Solution   Use the theorem $\int_a^b f(x)\ dx = f(c)(b - a)$.

$$\int_1^3 (x^2 + 1)\ dx = (c^2 + 1)(2)$$

$$\left[\frac{x^3}{3} + x\right]_1^3 = 2c^2 + 2$$

$$\frac{32}{3} = 2c^2 + 2$$

$$\frac{13}{3} = c^2$$

$$\pm\sqrt{\frac{13}{3}} = c$$

Only the value $\sqrt{\dfrac{13}{3}}$ is in the interval $1 < x < 3$, and this is the value guaranteed by the mean value theorem.

We now investigate the proof for $\dfrac{d}{dx}\displaystyle\int_k^x f(t)\ dt = f(x)$. We make use of properties 2 and 3 of the definite integral, page 247,

Property 2. $\displaystyle\int_a^b f(x)\ dx = -\int_b^a f(x)\ dx,$

Property 3. $\displaystyle\int_a^c f(x)\ dx + \int_c^b f(x)\ dx = \int_a^b f(x)\ dx.$

Watch for their use in this proof.

THEOREM 5.5b

If f is continuous on an interval, I, containing $x = k$ (k, a constant) and if x is another number in I, then

$$\frac{d}{dx}\int_k^x f(t)\ dt = f(x).$$

GIVEN. f is continuous on the interval, I. This condition guarantees that the definite integral exists.

TO PROVE. $\dfrac{d}{dx}\displaystyle\int_k^x f(t)\ dt = f(x).$

Proof

Find the derivative by forming a difference quotient and take its limit.

1. Define a function g:
$$y = g(x) = \int_k^x f(t)\, dt$$

2. Find $g(x + \Delta x)$
$$g(x + \Delta x) = \int_k^{x+\Delta x} f(t)\, dt$$

3. $\Delta y = g(x + \Delta x) - g(x)$
$$\Delta y = \int_k^{x+\Delta x} f(t)\, dt - \int_k^x f(t)\, dt$$

4. Apply properties 2 and 3 of the definite integral.
$$\begin{cases} \Delta y = \displaystyle\int_k^{x+\Delta x} f(t)\, dt + \int_x^k f(t)\, dt \\[2mm] \Delta y = \displaystyle\int_x^{x+\Delta x} f(t)\, dt \end{cases}$$

5. Form the difference quotient, $\dfrac{\Delta y}{\Delta x}$
$$\frac{\Delta y}{\Delta x} = \frac{\displaystyle\int_x^{x+\Delta x} f(t)\, dt}{\Delta x}$$

6. Apply the mean value theorem for integrals to $\displaystyle\int_x^{x+\Delta x} f(t)\, dt$, where $a = x$ and $b = x + \Delta x$.

There exists a value c where $x < c < x + \Delta x$ such that
$$\int_x^{x+\Delta x} f(t)\, dt = f(c)(\cancel{x} + \Delta x - \cancel{x})$$
$$= f(c)\Delta x$$

7. Rewrite step 5.
$$\frac{\Delta y}{\Delta x} = \frac{f(c)\,\cancel{\Delta x}}{\cancel{\Delta x}} = f(c)$$

8. Take the limit as $\Delta x \to 0$.

$$\xrightarrow[\ \ x \quad c \quad x+\Delta x\ \ ]{\ \ |\quad|\quad\quad|\ \ }$$

As $\Delta x \to 0$, $c \to x$, since $x < c < x + \Delta x$.

$$\lim_{\Delta x \to 0} \frac{\Delta y}{\Delta x} = \frac{dy}{dx} = \lim_{\Delta x \to 0} f(c)$$

Since f is continuous,
$$\lim_{\Delta x \to 0} f(c) = f(x), \text{ and we obtain:}$$
$$\frac{dy}{dx} = \lim_{\Delta x \to 0} f(c) = f(x).$$

Or, in general, at any value of x in the interval:

$$\frac{d}{dx} \int_k^x f(t)\, dt = f(x).$$

Another theorem is still needed before the fundamental theorem can be stated and proved.

> **THEOREM 5.5c**
>
> Let the functions f and g be continuous on $a \le x \le b$. If $f'(x)$ and $g'(x)$ exist for all x in $a < x < b$ and $f'(x) = g'(x)$ for all x in $a < x < b$, then there is a constant k such that $f(x) = g(x) + k$ or $f(x) - g(x) = k$.

GIVEN. The functions f and g are continuous for all x in $a \le x \le b$; $f'(x)$ and $g'(x)$ exist for all x in $a < x < b$; also, $f'(x) = g'(x)$ for all x in $a < x < b$.

TO PROVE. $f(x) - g(x) = k$. (See Figure 5.4.)

Proof

1. Let d be a value of x, $a < d \le b$.
2. Let $F(x) = f(x) - g(x)$. $F(x)$ satisfies all hypotheses of the mean value theorem for derivatives in $a \le x \le b$. Therefore, there exists a value of x, call it c, where $a < c < d$ and

    $$\frac{F(d) - F(a)}{d - a} = F'(c).$$

3. But, $F'(c) = f'(c) - g'(c) = 0$. This yields $\dfrac{F(d) - F(a)}{d - a} = 0$, or $F(d) - F(a) = 0$, or $F(d) = F(a)$. Since d was any value of x, $a < x \le b$, $F(x)$ is always the same height above the x axis.
4. Therefore, $F(x) = k$ or $f(x) - g(x) = k$.

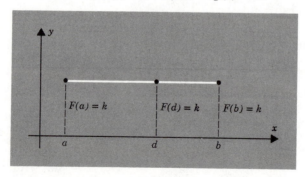

Figure 5.4

Example 2

Purpose To apply the previous theorem to two functions:

Problem Given $f'(x) = g'(x) = 5x^4$. What can be said about f and g?

Solution Let the function f be defined by $f(x) = x^5 + k$ and the function g be defined by $g(x) = x^5 + c$. Since $f'(x) = g'(x) = 5x^4$, the functions may differ by a constant.

We are now ready to state again and to prove the "five star" theorem—The Fundamental Theorem of Integral Calculus.

THEOREM 5.5d

The Fundamental Theorem of Integral Calculus. Let the function f be continuous for $a \leq x \leq b$. Let F be any one of the antiderivatives of f. Then

$$\int_a^b f(x) \, dx = F(b) - F(a).$$

GIVEN. f is continuous in $a \leq x \leq b$. F is an antiderivative of f.

TO PROVE. $\int_a^b f(x) \, dx = F(b) - F(a)$.

Proof

1. $\int_a^b f(x) \, dx$ exists, since f is given to be continuous over $a \leq x \leq b$. Also given, F is an antiderivative of f. We know another antiderivative of f, that is, $\int_a^x f(t) \, dt$. $\int_a^x f(t) \, dt$ is an antiderivative of f, since by Theorem 5.5b,

$$\frac{d}{dx} \int_a^x f(t) \, dt = f(x).$$

2. By Theorem 5.5c, since $\int_a^x f(t) \, dt$ and F are both antiderivatives of f, the two antiderivatives must satisfy $\int_a^x f(t) \, dt = F(x) + k$.

3. Let $x = a$,

$$\int_a^a f(t) \, dt = F(a) + k. \qquad \text{But,} \qquad \int_a^a f(t) \, dt = 0.$$

$$0 = F(a) + k \qquad \text{or} \qquad k = -F(a).$$

4. $\int_a^x f(t) \, dt = F(x) - F(a)$

5. Let $x = b$, $\int_a^b f(t) \, dt = F(b) - F(a)$.

6. Since t is a "dummy," we may replace it by any convenient letter. Let $t = x$ to obtain

$$\int_a^b f(x)\ dx = F(b) - F(a)$$

where F is an antiderivative of f.

Exercise 5.4–5.5

In Problems 1 to 12 use the properties of the definite integral to determine whether the following are true or false.

1. $\displaystyle\int_1^6 (x^2 + 7)\ dx = \int_1^3 (x^2 + 7)\ dx + \int_3^6 (x^2 + 7)\ dx$

2. $\displaystyle\int_2^2 (8 + 3x)\ dx = 0$

3. $\displaystyle\int_2^2 7(8 + 3x)\ dx = 7 \int_2^2 (8 + 3x)\ dx = 0$

4. $\displaystyle\int_1^{10} 5\ dx = 50$

5. $\displaystyle\int_0^7 3x^3\ dx = 3 \int_0^7 x^3\ dx$

6. $\displaystyle\int_{-1}^2 (9 + 7x - x^2)\ dx = \int_{-1}^2 9\ dx + \int_{-1}^2 7x\ dx - \int_{-1}^2 x^2\ dx$

7. $\displaystyle\int_3^4 \frac{1}{x}\ dx = -\int_4^3 x\ dx$

8. $\displaystyle\int_{-2}^4 3\ dx = 18$

9. $\displaystyle\int_0^4 (x + \sqrt{x})\ dx = \int_0^4 x\ dx + \int_0^4 \sqrt{x}\ dx$

10. $\displaystyle\int_0^4 x\sqrt{x}\ dx = \int_0^4 x\ dx \cdot \int_0^4 \sqrt{x}\ dx$

11. $\displaystyle\int_{-1}^{-2} (x + 1)^2\ dx = -\int_2^1 (x + 1)^2\ dx$

12. $\displaystyle\int_1^3 x\ dx = \int_1^4 x\ dx + \int_4^3 x\ dx$

In Problems 13 to 34, evaluate the definite integral by the Fundamental Theorem of Integral Calculus.

13. $\int_{1}^{4} 3 \, dx$

24. $\int_{1}^{3} (3x^2 + x - 2) \, dx$

14. $\int_{0}^{3} x \, dx$

25. $\int_{-1}^{1} (x^4 + x^2) \, dx$

15. $\int_{1}^{3} 3x \, dx$

26. $\int_{2}^{5} \frac{2}{7x^2} \, dx$

16. $\int_{2}^{5} (2x - 3) \, dx$

27. $\int_{1}^{2} \left(x^3 - \frac{1}{x^3} \right) dx$

17. $\int_{1}^{5} (7 - x) \, dx$

28. $\int_{1}^{4} \sqrt{x} \, dx$

18. $\int_{1}^{5} (7 + x) \, dx$

29. $\int_{0}^{9} \frac{1}{2}\sqrt{x} \, dx$

19. $\int_{-1}^{1} x^2 \, dx$

30. $\int_{1}^{9} (3\sqrt{x} - 2) \, dx$

20. $\int_{-2}^{0} 3x^2 \, dx$

31. $\int_{1}^{2} 6(x + \sqrt{x}) \, dx$

21. $\int_{1}^{3} (2x^2 - 5) \, dx$

32. $\int_{4}^{9} (6x^2 - 2x + 9\sqrt{x}) \, dx$

22. $\int_{-1}^{2} (6x^2 + 1) \, dx$

33. $\int_{1}^{4} \left(\frac{1}{2\sqrt{x}} + 3\sqrt{x} \right) dx$

23. $\int_{0}^{2} (2x - x^2) \, dx$

34. $\int_{1}^{3} \frac{x^3 + x^2 - 1}{x^2} \, dx$

35. A large city is hit by a flu epidemic and people are becoming ill at a rate of $270t - 9t^2$ people per day. Approximately how many people will get the flu between the first and twentieth days inclusive?

36. The number of tickets for a sporting event sold on a certain day is given by the equation $N = -3x + 150$, where x is the number of elapsed days since the ticket sale opened. Approximate the total number of tickets sold from the second through the fortieth day that the tickets were on sale?

37. It is estimated that a polluted lake contains 50,000 tons of pollutant. It is now planned to completely stop the dumping of all pollutants into the lake, and it is then estimated that the pollutant level will decrease because of natural diffusion of material. The rate of decrease is given by the following function: $P'(t) = 50,000 - 1500t - 90t^2$. Find the number of tons of pollution in the year 2000, taking the year 1985 as equal to 0.

38. A new factory is polluting a river according to $P(t) = t^{2/3} - 1$, where t is time in days and $P(t)$ is the amount of waste material dumped into the river on a given day. Approximate the total amount of waste dumped into the river during the first eight days of operation.

39. Assume that the unrestricted catch of fish from a certain lake would slowly decline over time (t in years) and that the quantity caught may be expressed as the rate (in thousands of pounds per year), $r = 37 - 2t$, $0 \leqslant t \leqslant 16$. Estimate the total catch in pounds over 16 years.

40. Economists are often interested in how a certain resource (money, land, food) is distributed among the population. One index that measures how equally a resource is distributed among the population is the Gini index. The Gini index, G, is calculated by $G = 1 - 2 \int_0^1 f(x)\, dx$ where $f(x)$ is a function relating a fraction of the population to the fraction of the resource accounted for. A value of G near 0 indicates that the resource is spread fairly equally among the population. In the following three instances, find the Gini index and determine the case in which the resource is the most equally distributed.
(a) $f(x) = \sqrt{x^3}$ (b) $f(x) = x^2$ (c) $f(x) = x^4$

41. A chemical reaction will occur only if the value for Gibbs free energy is negative.

$$\Delta G = \int_{P_1}^{P_2} V\, dP \qquad \text{where } V = \text{volume}$$
$$P = \text{pressure}$$
$$\Delta G = \text{change in Gibbs free energy}$$

If $V = 3P^2 + 2P + 4$, find ΔG when $P_1 = 3$ and $P_2 = 7$.

42. In chemistry the work obtained from a process can be found by the equation: Work $= \int P\, dV$, where P is equal to the pressure and V is the volume. If $P = 6V + 2V^2$, find the work when the volume changes from 1 to 6 ft³.

43. One way biologists study the growth of a bacteria culture is by observing the percentage of bacteria that regenerates in a given amount of time. If the equation of regeneration is found to be $B = 2x + \dfrac{40}{x^2}$, the number of bacteria with an a to b percent chance of regeneration is found by taking the definite integral of the equation of regeneration from a to b. Find the number of bacteria:
(a) with a 10 to 20 percent chance
(b) with a 20 to 40 percent chance.

44. Given that $\displaystyle\int_1^x t^2\, dt = \tfrac{1}{3}t^3 \Big]_1^x = \tfrac{1}{3}x^3 - \tfrac{1}{3}$:

(a) Justify that $g(x) = \displaystyle\int_1^x t^2\, dt$ represents a function.

(b) Find $\dfrac{d}{dx} \displaystyle\int_1^x t^2\, dt$ by finding $\dfrac{d}{dx}[\tfrac{1}{3}x^3 - \tfrac{1}{3}]$.

(c) Find $\dfrac{d}{dx} \displaystyle\int_1^x t^2 \, dt$ by using Theorem 5.5b. Compare your answers of (b) and (c).

In Problems 45 to 48, find the derivative of the function defined by a definite integral using Theorem 5.5b.

45. $\displaystyle\int_2^x \sqrt[3]{t} \, dt$

47. $\displaystyle\int_3^x \dfrac{1}{t^3} \, dt$

46. $\displaystyle\int_{22}^x \sqrt[3]{t} \, dt$

48. $\displaystyle\int_1^x (t^2 + t + 1) \, dt$

In Problems 49 and 50, find the number c guaranteed by the mean value theorem for integrals.

49. $\displaystyle\int_2^6 (x + 3) \, dx$

50. $\displaystyle\int_0^2 3x^2 \, dx$

In Problems 51 to 55, state whether the given statement is true or false.

51. If $f'(x) = g'(x) = 2x$, then $f(x)$ can equal x^2 and $g(x)$ can equal $x^2 + 16$.
52. If $f'(x) = g'(x) = 2x - 1$, then $f(x)$ can equal $x^2 - x + 1$ and $g(x)$ can equal $x^2 + x - 1$.
53. If $f'(x) = g'(x) = x^{-(1/2)}$, then $f(x)$ can equal $2\sqrt{x} + 7$ and $g(x)$ can equal $2\sqrt{x} - 7$.
54. If $f'(x) = g'(x) = 3(x^2 + 7)^2(2x)$, then $f(x)$ can equal $(x^2 + 7)^3 - 3$ and $g(x)$ can equal $(x^2 + 7)^3 + 30$.
55. If $f'(x) = g'(x) = \dfrac{-2}{x^3}$, then $f(x)$ can equal $\dfrac{1}{x^2} + 9$ and $g(x)$ can equal $\dfrac{1}{\sqrt{x}} + 9$.

5.6 APPLICATIONS OF THE DEFINITE INTEGRAL

Earlier in this chapter we mentioned one of the many applications of the definite integral. This specific application dealt with finding the area bounded by a curve, the x axis, and two vertical lines. This application is a special case of finding the area between two curves.

The concept of area between two curves will be extended to cases where one of the bounding curves may or may not be a coordinate axis. This extension of the concept of area must be used to find the area between two curves, $y = f(x)$ and $y = g(x)$. Not making use of the extension may lead to an incorrect area or to a negative number for an area. Remember, an area can never be negative!

The student is again cautioned that the definite integral is not always an area. It may in certain problems represent other quantities such as volume, future profit, quantity of bacteria present, and the like.

The Area Between Two Curves

Consider two functions f and g where the graph of f lies above the graph of g for values of x between a and b. Looking at the graphs separately we have

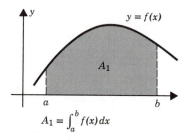

$$A_1 = \int_a^b f(x)\,dx$$

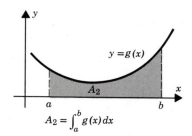

$$A_2 = \int_a^b g(x)\,dx$$

Put the graphs on the same coordinate system and concentrate on the area A between the two curves.

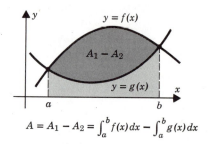

$$A = A_1 - A_2 = \int_a^b f(x)\,dx - \int_a^b g(x)\,dx$$

And from the integral properties,

$$A = \int_a^b [f(x) - g(x)]\,dx$$

Thus, the integrand is the function for the top curve minus the function for the bottom curve.

Example 1

Purpose To find the area between two curves:

Problem A Find the area between the curves, $y = 2x^2$ and $y = 3x$.

Solution First, find the points of intersection of $y = 2x^2$ and $y = 3x$ by solving the two equations simultaneously.

Substitute $y = 3x$ into $y = 2x^2$; then $3x = 2x^2$

$$2x^2 - 3x = 0$$
$$x(2x - 3) = 0$$
$$x = 0 \quad \text{or} \quad x = \tfrac{3}{2}$$

The points of intersection are $(0,0)$, $(\tfrac{3}{2}, \tfrac{9}{2})$.

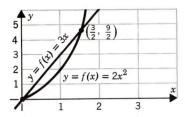

The area between the two curves $y = f(x) = 3x$ and $y = g(x) = 2x^2$ from $x = 0$ to $x = \tfrac{3}{2}$ is

$$A = \int_0^{3/2} [f(x) - g(x)]\, dx$$

$$= \int_0^{3/2} (3x - 2x^2)\, dx$$

$$= [\tfrac{3}{2}x^2 - \tfrac{2}{3}x^3]_0^{3/2} = [\tfrac{3}{2}(\tfrac{3}{2})^2 - \tfrac{2}{3}(\tfrac{3}{2})^3] - [\tfrac{3}{2}(0)^2 - \tfrac{2}{3}(0)^3]$$

$$= \tfrac{27}{8} - \tfrac{2}{3}(\tfrac{27}{8}) = \tfrac{1}{3}(\tfrac{27}{8})$$

$$= \tfrac{9}{8} = 1\tfrac{1}{8} \text{ square units.}$$

Problem B   Find the area bounded by the curves $y = -x^2 + 2x - 1$, $y = \tfrac{1}{2}x + 1$, $x = -1$, and $x = 2$.

Solution   From the sketch, the shaded region is the area to find.

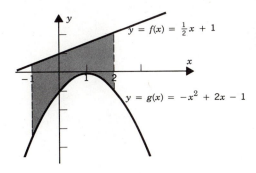

$$A = \int_{-1}^{2} [(\tfrac{1}{2}x + 1) - (-x^2 + 2x - 1)]\, dx$$

$$= \int_{-1}^{2} (x^2 - \tfrac{3}{2}x + 2)\, dx$$

$$= \frac{x^3}{3} - \frac{3}{4}x^2 + 2x \Big]_{-1}^{2}.$$

$$= (\tfrac{8}{3} - 3 + 4) - (-\tfrac{1}{3} - \tfrac{3}{4} - 2)$$

$$= \tfrac{27}{4} \text{ square units}$$

Problem C   Find the area of the region bounded by the curves $y = x^2$ and $y^2 = 8x$.

Solution

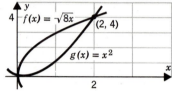

The points of intersection are $(0,0)$ and $(2,4)$.

$$y = f(x) = \sqrt{8x} = \sqrt{8}\sqrt{x} \quad \text{and} \quad y = g(x) = x^2$$

$$A = \int_{0}^{2} (\sqrt{8}\sqrt{x} - x^2)\, dx$$

$$= \left[\sqrt{8}\frac{2}{3}(x)^{3/2} - \frac{x^3}{3} \right]_{0}^{2} = \left[\left(\frac{16}{3} - \frac{8}{3} \right) - 0 \right]$$

$$= \tfrac{8}{3} \text{ square units}$$

This problem, like many others, can also be solved by an alternate procedure. If the curves enclosing the area can be represented by $x = f(y)$ and $x = g(y)$ and if $x = f(y)$ represents the right-hand curve, then the area between the two curves from $y = a$ to $y = b$ is $A = \int_{a}^{b} [f(y) - g(y)]\, dy$. This procedure is illustrated by reworking this example.

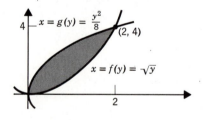

$$A = \int_{0}^{4} \left(\sqrt{y} - \frac{y^2}{8} \right) dy$$

$$= \left[\frac{2}{3}y^{3/2} - \frac{y^3}{24} \right]_{0}^{4}$$

$$= [(\tfrac{16}{3} - \tfrac{8}{3}) - (0)] = \tfrac{8}{3} \text{ square units.}$$

Problem D   Find the area between the curves, $y = \dfrac{1}{x^2}$ and $14x + 9y = 43$.

Solution

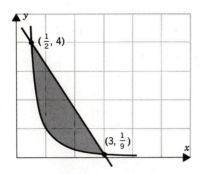

Solve $y = \dfrac{1}{x^2}$ and $14x + 9y = 43$ simultaneously to obtain their points of inter-section, which are $(\frac{1}{2}, 4)$ and $(3, \frac{1}{9})$.

Solution I (Integration with respect to x.)

Top curve is $y = \dfrac{43 - 14x}{9}$ and bottom curve is $y = \dfrac{1}{x^2}$

$$A = \int_{1/2}^{3} \left(\frac{43 - 14x}{9} - \frac{1}{x^2} \right) dx = \frac{43}{9}x - \frac{7x^2}{9} + \frac{1}{x} \Bigg]_{1/2}^{3}$$

$$= \left[\left(\frac{43}{3} - 7 + \frac{1}{3} \right) - \left(\frac{43}{18} - \frac{7}{36} + 2 \right) \right]$$

$$= \frac{125}{36} \text{ square units}$$

Solution II (Integration with respect to y.)

Curve on right is $x = \dfrac{43 - 9y}{14}$ and curve on left is $x = \dfrac{1}{\sqrt{y}}$

$$A = \int_{1/9}^{4} \left(\frac{43 - 9y}{14} - \frac{1}{\sqrt{y}} \right) dy = \frac{43}{14}y - \frac{9y^2}{28} - 2\sqrt{y} \Bigg]_{1/9}^{4}$$

$$= \left[\left(\frac{86}{7} - \frac{36}{7} - 4 \right) - \left(\frac{43}{126} - \frac{1}{252} - \frac{2}{3} \right) \right]$$

$$= \frac{125}{36} \text{ square units}$$

Example 2

Purpose To illustrate an area problem applied to economics:

Problem In Chapter 1, Section 6, page 29, equilibrium quantity and price are explained. The intersection of the demand function and supply function yields this equilibrium point. Of further interest to economists is both the consumers' surplus and the producers' surplus. When the market price is set at y_n, the consumer who is willing to pay more for a product will gain. The consumers' surplus, or total consumer gain, is represented by the area below the demand curve and above the line where y equals y_n, the equilibrium price. This area represents the difference between what the consumer should pay for the product and the actual expenditure for the product. Accordingly, the producer, who is willing to sell his product for less than y_n, will gain when the market price is set at y_n. The producers' surplus, or total gain by the producer, is represented by the area below the line where y equals the equilibrium price and above the supply curve. The graphical representation is given below.

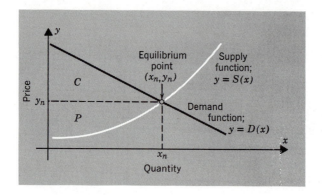

$$C = \text{consumers' surplus} = \int_0^{x_n} [D(x) - y_n]\, dx$$

$$P = \text{producers' surplus} = \int_0^{x_n} [y_n - S(x)]\, dx$$

If the supply function is found to be $y = S(x) = x^2 + 3x + 2$ and the demand function is $y = D(x) = -2x + 16$, find both the consumers' surplus and the producers' surplus.

Solution

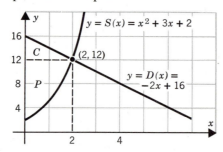

Solve $y = x^2 + 3x + 2$ and $y = -2x + 16$ algebraically to find the point of intersection $(2, 12)$.

$$C = \text{consumers' surplus} = \int_0^2 [(-2x + 16) - 12]\, dx$$

$$= \int_0^2 (-2x + 4) \, dx = -x^2 + 4x \Big]_0^2 = [(-4 + 8) - (0)] = 4$$

∴ consumers' surplus is $4.

$$P = \text{producers' surplus} = \int_0^2 [12 - (x^2 + 3x + 2)] \, dx$$

$$= \int_0^2 (-x^2 - 3x + 10) \, dx = \left[-\frac{x^3}{3} - \frac{3x^2}{2} + 10x \right]_0^2$$

$$= \left[\left(-\frac{8}{3} - 6 + 20 \right) - (0) \right] = \frac{34}{3} \approx 11.33$$

∴ producers' surplus is $11.33.

Probability for a Continuous Frequency Function (Optional)

Let x be a variable that can take on all values in an interval. Examples of such a variable are:

> x is the height of individuals in your math class;

or

> x is the exact annual debt of the United States Government from the year 1778 on;

or . . . .

A variable that can take on all the values in an interval is called a *continuous random variable*.

Associated with a continuous random variable is a function f, defined by $f(x)$, called its frequency function or probability density function. The frequency function of a continuous random variable has three properties:

1. $f(x) \geq 0$ for all x in the domain of f.
2. The total area between $y = f(x)$ and the x axis must be equal to one.
3. The probability that x takes on a value in the interval $c \leq x \leq d$ is given by
    $$\int_c^d f(x) \, dx.$$

These three properties require that the graph of $y = f(x)$ never dip below the x axis, the total area between $y = f(x)$ and the x axis to be equal to one, and the probability that x takes on values in $c \leq x \leq d$ to be equal to the area bounded by $x = c$, $x = d$, $y = f(x)$, and $y = 0$.

These requirements insure that the probability obtained will always be a number between zero and one.

Example 3

Purpose   The utilization of the definite integral to obtain probabilities:

Problem   In an electrical warehouse the proportion of orders filled per day has the frequency function f, given by $f(x) = 30(x^4 - x^5)$ for $0 \leqslant x \leqslant 1$.

(a)   Verify that this frequency function satisfies criteria 2, the total area under a frequency function must be equal to one.

(b)   What is the probability that less than 10 percent of the orders will be processed in one day?

(c)   What is the probability that between 90 percent and 100 percent of the orders will be processed in one day?

Solution   (a)   In this example we must verify that the total area between $y = f(x)$ and the x axis is equal to one.

Verify $\int_0^1 (30x^4 - 30x^5)\, dx = 1$.

$$\int_0^1 (30x^4 - 30x^5)\, dx = \left[6x^5 - 5x^6 \right]_0^1 = [(6 - 5)] - [0] = 1.$$

(b) $\int_0^{.1} (30x^4 - 30x^5)\, dx = \left[6x^5 - 5x^6 \right]_0^{.1} = [6(.1)^5 - 5(.1)^6] - 0$

$$= 6(\tfrac{1}{10})^5 - 5(\tfrac{1}{10})^6$$
$$= (\tfrac{1}{10})^5(6 - .5)$$
$$= (\tfrac{1}{10})^5(5.5)$$
$$= .000055$$

(This event is very unlikely to occur.)

(c) $\int_{.9}^1 (30x^4 - 30x^5)\, dx = \left[6x^5 - 5x^6 \right]_{.9}^1 = [1] - [6(.9)^5 - 5(.9)^6]$

$$= 1 + (.9)^5(-6 + 4.5)$$
$$\simeq 1 + (.59)(-1.5)$$
$$= .115$$

   Two additional properties useful in the study of distributions are the arithmetic mean and the variance. The arithmetic mean is a measure of the value at which data tend to center. The variance and its square root, which is called the standard deviation, are measures of the spread or scatter of a distribution.

DEFINITION

The kth Moment.   Given a frequency function, f, defined over the interval

$a \leqslant x \leqslant b$ where $\int_a^b f(x)\ dx = 1$. The *kth moment* with respect to the origin is defined to be

$$m_k = \int_a^b x^k f(x)\ dx.$$

Therefore, the first moment with respect to the origin, $m_1 = \int_a^b xf(x)\ dx$. The arithmetic mean, μ, is by definition the first moment about the origin. The second moment with respect to the origin is defined by $m_2 = \int_a^b x^2 f(x)\ dx$. In statistics, both variance and standard deviation are found with the aid of the second moment with respect to the origin.

Example 4

Purpose To find m_1, m_2 for a specific frequency function:

Problem Find m_1 and m_2 if $f(x) = 2x$ over the interval $0 \leqslant x \leqslant 1$.

Solution $m_1 = \int_0^1 xf(x)\ dx$

$$= \int_0^1 x(2x)\ dx = \int_0^1 2x^2\ dx$$

$$= \left[\frac{2}{3}x^3\right]_0^1 = \frac{2}{3}$$

$m_2 = \int_0^1 x^2 f(x)\ dx$

$$= \int_0^1 x^2(2x)\ dx = \int_0^1 2x^3\ dx$$

$$= \left[\frac{x^4}{2}\right]_0^1 = \frac{1}{2}$$

Exercise 5.6

In Problems 1 to 8, find the area of the region bounded by the x axis and the given curve. To show the problem geometrically, draw the graph of the given function.

1. $y = x^2 - 2x$

2. $y = 1 - x^2$

3. $y = 4x - x^2$

4. $y = x^2 - 6x$

5. $y = x^2 - 5x$

6. $y = x^2 - 5x + 4$

7. $y = -x^2 + 5x - 4$

8. $y = -x^2 + 5x - 6$

In Problems 9 to 16, find the area of the region bounded by the x axis, the given curve, and the interval on the x axis. Draw the graph of each function.

9. $y = x^2$, $0 \leqslant x \leqslant 3$

10. $y = x^2 - x - 2$, $-1 \leqslant x \leqslant 0$

11. $y = (x - 2)^2$, $0 \leqslant x \leqslant 4$

12. $y = x^3 + 1$, $-1 \leqslant x \leqslant 2$

13. $y = x^3 - 2x^2 + 3$, $-1 \leqslant x \leqslant 2$

14. $y = \sqrt{x}$, $1 \leqslant x \leqslant 9$

15. $y = \sqrt{x} - 2$, $4 \leqslant x \leqslant 16$

16. $y = \sqrt{x} + 3$, $1 \leqslant x \leqslant 4$

In Problems 17 to 22, find the area of the region bounded by the two given curves.

17. $y = x^2$; $y = x + 2$

18. $y = 9 - x^2$; $y = x^2 - 9$

19. $y = x^2 - x - 6$; $y = x + 2$

20. $y = x^2$; $y = -x^2 + 4$

21. $y = x^2 + 2x - 3$; $y = -x^2 + 9$

22. $y = x^3$; $y = 2x^2$

In Problems 23 to 25, find the area of the region bounded by the given curves and the interval on the x axis.

23. $y = x$; $y = -\frac{1}{4}x^2$; $1 \leqslant x \leqslant 4$

24. $y = \frac{1}{3}x^2$; $y = -x^2 + 4x - 4$; $0 \leqslant x \leqslant 3$

25. $y = x^3 + 3x^2 + 3x + 2$; $y = 3x - 1$; $-1 \leqslant x \leqslant 1$

In Problems 26 to 29, find the area of the region bounded by the pair of curves. Decide whether to integrate with respect x or with respect to y and, when possible, solve the problem both ways. Graph the curves.

26. $y = x^2$; $y^2 = 64x$

27. $y^2 = 4 - x$; $x = 0$

28. $y^2 = 4x$; $y = 2x - 4$

29. $x + y = 1$; $y^2 - x = 1$

30. Find the area of the region bounded by the curve $y = x^3 - 5x^2 + 3x + 9$ and the x axis.

31. Find the area in the first quadrant of the region bounded by the y axis and the curves $y = x^2 - 2x + 1$ and $x + 3y = 15$.

32. If the supply function is found to be $y = S(x) = \frac{3}{2}x + 3$ and the demand function is $y = D(x) = -\frac{5}{4}x + 14$, find both the consumers' surplus and the producers' surplus. (See Example 2, page 265.)

33. A supply function is found to be $y = S(x) = 3x + 5$ and the demand function is $y = D(x) = \frac{1}{4}x^2 - 4x + 18$, $(x \leqslant 8)$. Find both the consumers' surplus and the producers' surplus. (See Example 2, page 265.)

34. A manufacturer of electronic sphygmomanometers determines that the demand function for his product is given by $y = D(x) = (x - 8)^2$, and the supply function is given by $y = S(x) = 9x^2$. The price y is in dollars and the number of units produced is x.

(a) Determine the equilibrium price.

(b) Determine the consumers' surplus.

(c) Determine the producers' surplus.

35. A demand function is $D(x) = 4 - x^2$ where $D(x)$ is the price for x items bought. If the equilibrium price is \$3.00, sketch and find the consumers' surplus.

36. A greeting card supplier can supply x boxes of 100 greeting cards according to the supply equation: $S(x) = x^2 + 3x$. If the equilibrium price is \$10, sketch and find the producers' surplus.

37. One of the most important calculations in the study of thermodynamics is the calculation of the amount of work that can be obtained from a process. One of the easiest ways to calculate work is by finding the area under the curve of a pressure versus volume plot.

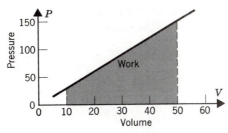

In this particular problem the pressure varies linearly with volume, $P = 3V + 2$. Find the work produced when the volume changes from 10 to 50.

38. The manager of a 60-unit apartment building renting each unit at \$75 per month will have no vacancies. Her marginal revenue function is determined to be $R'(x) = 225 - 5x$ (x, units).

(a) Find the revenue function.

(b) State the demand function.

(c) If she chooses to have only 56 units rented, what rent can she expect per unit?

39. (Optional) The proportion of orders processed per week by a mail-order catalogue firm has a frequency function $f(x) = 12(x^2 - x^3)$, $0 \leqslant x \leqslant 1$.

(a) What is the probability that, at most, 50 percent of the orders will be processed in one week?

(b) What is the probability that 75 percent to 100 percent of the orders will be processed in one week?

40. (Optional) Another mail-order catalogue firm finds that the proportion of orders processed per week has a frequency function $f(x) = c(x - x^2)$; $0 \leqslant x \leqslant 1$.

(a) Find c. $\left[\text{Use property 2, } \int_0^1 c(x - x^2) \, dx = 1 \right]$

(b) What is the probability that, at most, 50 percent of the orders will be processed in one week?

(c) What is the probability that 75 percent to 100 percent of the orders will be processed in one week?

41. (Optional) If the frequency function for a probability distribution over the interval $0 \leq x \leq 1$ is given by $f(x) = 3x^2$:
(a) Verify that the total area under the curve is 1.
(b) Find m_1, the first moment about the origin.
(c) Find m_2, the second moment about the origin.

42. (Optional) $f(x) = cx$ is a frequency function over $1 \leq x \leq 3$. Find c to guarantee $\int_1^3 cx \, dx = 1$. Find m_1 and m_2.

43. (Optional) Given the frequency distribution, $f(x)$ over the interval $a \leq x \leq b$ and $m_1 = \int_a^b xf(x) \, dx$ is the mean, μ. Show the variance, σ^2, is equal to the second moment about the origin minus the square of the first moment about the origin. The variance is defined to be the second moment about the mean; $\sigma^2 = \int_a^b (x - \mu)^2 f(x) \, dx$.

Therefore, show that

$$\sigma^2 = \int_a^b (x - \mu)^2 f(x) \, dx = \int_a^b x^2 f(x) \, dx - \mu^2$$

or

$$\sigma^2 = \int_a^b x^2 f(x) \, dx - \left[\int_a^b xf(x) \, dx \right]^2.$$

5.7 FURTHER TECHNIQUES OF INTEGRATION

Integration by Substitution

There still exist many other functions whose integrals cannot be found by the previous techniques. However, techniques of integration may be extended to a whole host of functions by making use of the formula:

$$\int u^n \, du = \frac{1}{n + 1} u^{n+1} + c, \qquad n \neq -1.$$

This powerful technique, called integration by substitution, is illustrated in the examples that follow.

Example 1

Purpose To illustrate the technique of integration by substitution:

Problem Find $\int \sqrt{x^2 + 1} \cdot 2x \, dx$.

Solution Let $u = x^2 + 1$

$du = 2x\, dx$

Note: $u = f(x)$ and by definition of the differential, $du = f'(x)\, dx$.

Upon substitution,

$$\int \sqrt{x^2 + 1} \cdot 2x\, dx = \int \sqrt{u}\, du$$

$$= \int u^{1/2}\, du$$

$$= \frac{u^{3/2}}{\frac{3}{2}} + c = \tfrac{2}{3} u^{3/2} + c.$$

Replace u by $x^2 + 1$ to obtain:

$$\int \sqrt{x^2 + 1} \cdot 2x\, dx = \tfrac{2}{3}(x^2 + 1)^{3/2} + c.$$

In addition, the technique of substitution may be extended by noting that $\int kf(x)\, dx = k \int f(x)\, dx$. Thus, if a good reason exists for doing so, $\int f(x)\, dx$ may be written as $\tfrac{1}{2} \int f(x) \cdot 2\, dx$, since this is equal to $\tfrac{1}{2} \cdot 2 \int f(x)\, dx = \int f(x)\, dx$. Therefore, if k is any nonzero constant, then

$$\int f(x)\, dx = \frac{1}{k} \int f(x) \cdot k\, dx.$$

Example 2

Purpose To show the procedure for inserting a constant into an integral when using the technique of substitution:

Problem Find $\displaystyle\int \sqrt{x^2 + 1}\, x\, dx$.

Solution Let $u = x^2 + 1$

$du = 2x\, dx$

$$\int \sqrt{x^2 + 1}\, x\, dx = \tfrac{1}{2} \int \sqrt{x^2 + 1}\, 2x\, dx$$

Upon substitution:

$$\tfrac{1}{2} \int \sqrt{x^2 + 1}\, 2x\, dx = \tfrac{1}{2} \int \sqrt{u}\, du = \tfrac{1}{2} \int u^{1/2}\, du$$

$$= \tfrac{1}{2} \frac{u^{3/2}}{\frac{3}{2}} + c = \tfrac{1}{3} u^{3/2} + c$$

$$= \tfrac{1}{3}(x^2 + 1)^{3/2} + c.$$

Caution: $\int \sqrt{x^2 + 1}\, dx$ cannot be solved by this procedure. It is illegal, immoral, and incorrect to insert a $2x$ and then write $\dfrac{1}{2x}$ in front of $\int$.

Example 3

Purpose To illustrate when integration by algebraic substitution can and cannot be used:

Problem Find (A) $\displaystyle\int (x^2 - 4)^2 \, dx$

(B) $\displaystyle\int (x^2 - 4)^2 x \, dx$

Solution (A) $\displaystyle\int (x^2 - 4)^2 \, dx$

Let $u = x^2 - 4$
$du = 2x \, dx$

No proper substitution can be made, since $2x$ is needed and only the factor 2 can be considered. Only a nonzero constant may be introduced into the numerator and denominator. In this problem the solution has to be obtained by another method.

$$\int (x^2 - 4)^2 \, dx = \int (x^4 - 8x^2 + 16) \, dx = \tfrac{1}{5}x^5 - \tfrac{8}{3}x^3 + 16x + c.$$

(B) $\displaystyle\int (x^2 - 4)^2 x \, dx$

Let $u = x^2 - 4$
$du = 2x \, dx$

$$\int (x^2 - 4)^2 x \, dx = \tfrac{1}{2} \int (x^2 - 4)^2 2x \, dx = \tfrac{1}{2} \int u^2 \, du$$

$$= \tfrac{1}{2} \cdot \frac{u^3}{3} + c$$

$$= \tfrac{1}{6}(x^2 - 4)^3 + c.$$

An alternate solution: Let $\int (x^2 - 4)^2 x \, dx = \int (x^5 - 8x^3 + 16x) \, dx$ [since $(x^2 - 4)^2 x = x^5 - 8x^3 + 16x$]. Finish this integration and check your results with the other solution. Notice that the first solution is easier and more expedient.

Example 4

Purpose To use the technique of substitution to evaluate a definite integral:

Problem Find $\displaystyle\int_0^2 \frac{x^2 \, dx}{\sqrt{x^3 + 1}}$.

Solution First, find $\displaystyle\int \frac{x^2 \, dx}{\sqrt{x^3 + 1}}$

Let $u = x^3 + 1$
$du = 3x^2 \, dx$

$$\int \frac{x^2 \, dx}{\sqrt{x^3 + 1}} = \frac{1}{3} \int \frac{3x^2 \, dx}{\sqrt{x^3 + 1}} = \frac{1}{3} \int \frac{du}{\sqrt{u}} = \frac{1}{3} \int u^{-1/2} \, du$$

$$= \frac{1}{3} \frac{u^{1/2}}{\frac{1}{2}} + c = \frac{2}{3} \sqrt{u} + c$$

or $\quad \int \frac{x^2 \, dx}{\sqrt{x^3 + 1}} = \frac{2}{3} \sqrt{x^3 + 1} + c$

Select $\dfrac{2}{3} \sqrt{x^3 + 1}$ as an antiderivative of $\dfrac{x^2}{\sqrt{x^3 + 1}}$. Then

$$\int_0^2 \frac{x^2 \, dx}{\sqrt{x^3 + 1}} = \left[\frac{2}{3} \sqrt{x^3 + 1} \right]_0^2 = \left[\left(\frac{2}{3} \sqrt{9} \right) - \left(\frac{2}{3} \sqrt{1} \right) \right] = \frac{4}{3}.$$

Note: In this example, the limits of integration are values of x. If the variable is changed to u, the limits of integration may be evaluated with respect to u. Therefore, by substitution in $u = x^3 + 1$, if $x = 0$, $u = 1$, and if $x = 2$, $u = 9$; and

$$\int_0^2 \frac{x^2 \, dx}{\sqrt{x^3 + 1}} = \frac{1}{3} \int_1^9 u^{-1/2} \, du = \left[\frac{2}{3} \sqrt{u} \right]_1^9 = \left[\left(\frac{2}{3} \sqrt{9} \right) - \left(\frac{2}{3} \sqrt{1} \right) \right] = \frac{4}{3}.$$

Example 5

Purpose To further illustrate a substitution technique:

Problem Find: (A) $\displaystyle\int x\sqrt{x + 3} \, dx$

(B) $\displaystyle\int_0^3 x^3 \sqrt{9 - x^2} \, dx$

Solution (A) $\displaystyle\int x\sqrt{x + 3} \, dx$

Let $u = x + 3$ $\qquad$ Solve $u = x + 3$ for x
$\quad du = dx$ $\qquad\qquad\quad x = u - 3$

Substituting for x, $x + 3$, and dx:

$$\int (u - 3)\sqrt{u} \, du = \int (u - 3)u^{1/2} \, du$$

$$= \int (u^{3/2} - 3u^{1/2}) \, du$$

$$= \tfrac{2}{5}u^{5/2} - 2u^{3/2} + c$$

$$= \tfrac{2}{5}(x + 3)^{5/2} - 2(x + 3)^{3/2} + c$$

(B) $\displaystyle\int_0^3 x^3\sqrt{9 - x^2}\, dx$

Let $u = 9 - x^2$ Solve $u = 9 - x^2$ for x^2
$du = -2x\, dx$ $x^2 = 9 - u$

If the variable is changed to u, the limits of integration may be evaluated with respect to u. Therefore, by substitution in $u = 9 - x^2$, when $x = 0$, $u = 9$; when $x = 3$, $u = 0$.

$$\int_0^3 x^3\sqrt{9 - x^2}\, dx = -\tfrac{1}{2}\int_0^3 x^2\sqrt{9 - x^2}\,(-2x\, dx)$$

$$= -\tfrac{1}{2}\int_9^0 (9 - u)\sqrt{u}\, du$$

$$= -\tfrac{1}{2}\int_9^0 (9u^{1/2} - u^{3/2})\, du$$

$$= -\tfrac{1}{2}\left[6u^{3/2} - \tfrac{2}{5}u^{5/2}\right]_9^0$$

$$= -\tfrac{1}{2}[0 - (162 - \tfrac{486}{5})] = \tfrac{162}{5}.$$

Example 6

Purpose   To use the substitution technique in applied problems:

Problem A   Find the area between the curves, $y = \sqrt{2x + 1}$ and $y = \tfrac{2}{3}x + \tfrac{1}{3}$.

Solution

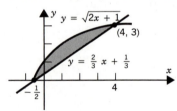

Solve $y = \sqrt{2x + 1}$ and $y = \tfrac{2}{3}x + \tfrac{1}{3}$ simultaneously to obtain their points of intersection, which are $(-\tfrac{1}{2},0)$ and $(4,3)$.

$$A = \int_{-1/2}^4 \left[\sqrt{2x + 1} - \left(\frac{2}{3}x + \frac{1}{3}\right)\right] dx$$

$$= \frac{1}{2}\int_{-1/2}^4 \sqrt{2x + 1}\cdot 2\, dx - \int_{-1/2}^4 \left(\frac{2}{3}x + \frac{1}{3}\right) dx$$

$$= \left[\frac{1}{3}(2x + 1)^{3/2} - \left(\frac{x^2}{3} + \frac{1}{3}x\right)\right]_{-1/2}^4$$

$$= [\tfrac{1}{3}(27) - (\tfrac{16}{3} + \tfrac{4}{3})] - [0 - (\tfrac{1}{12} - \tfrac{1}{6})] = [9 - \tfrac{20}{3}] - [\tfrac{1}{12}]$$

$$= \tfrac{9}{4}\ \text{square units}$$

Problem B   The cost to produce the xth digital watch in a daily production run is given by

$C(x) = \$\left(\dfrac{50}{\sqrt{x+1}}\right)$, $x \leq 200$. Approximate the total cost to produce the first

100 watches.

Solution   Let C = exact total production cost of the first 100 watches.
$C = C(1) + C(2) + \cdots + C(100)$.
This sum can be approximated by an area.

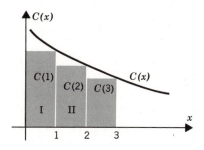

$$C(1) = A_1 = \text{area of rectangle I, which is } C(1) \cdot 1$$
$$C(2) = A_2 = \text{area of rectangle II, which is } C(2) \cdot 1$$
$$\vdots$$
$$C(100) = A_{100} = \text{area of 100th rectangle, which is } C(100) \cdot 1$$

$C = \displaystyle\sum_{j=1}^{100} C(j) = \sum_{j=1}^{100} A_j$. This sum of the areas of the rectangles is approximately

equal to the area under the curve from $x = 0$ to $x = 100$. This area is given by
a definite integral.

$$C = \sum_{j=1}^{100} C(j) \approx \int_0^{100} \frac{50}{\sqrt{x+1}}\, dx = 50 \int_0^{100} (x+1)^{-1/2}\, dx$$

$$= 50 \cdot 2(x+1)^{1/2} \Big]_0^{100} = 100\sqrt{x+1} \Big]_0^{100}$$

$$= 100\sqrt{101} - 100\sqrt{1} \approx 1005 - 100$$

$$C \approx \$905$$

Integration by Formula

If an integration problem cannot be easily done by the preceding techniques,
more sophisticated techniques must be developed, or one must refer to a table of
integrals. These tables are found in mathematical handbooks or in the appen-
dixes of many textbooks. Check the reference room of your library for them.

Before introducing seven formulas from the hundreds given in these tables,
we stress this fact: *your success with integration by formula depends on your
ability to match the formula exactly with your integration problem.*

Seven Integration Formulas

1. $\displaystyle \int \sqrt{ax + b}\ dx = \frac{2}{3a} \sqrt{(ax + b)^3} + c$

2. $\displaystyle \int \frac{x}{(ax + b)^3}\ dx = \frac{b}{2a^2(ax + b)^2} - \frac{1}{a^2(ax + b)} + c$

3. $\displaystyle \int x^2 \sqrt{ax + b}\ dx = \frac{2(15a^2x^2 - 12abx + 8b^2)\sqrt{(ax + b)^3}}{105a^3} + c$

4. $\displaystyle \int \frac{x}{\sqrt{ax + b}}\ dx = \frac{2(ax - 2b)}{3a^2} \sqrt{ax + b} + c$

5. $\displaystyle \int \frac{dx}{x\sqrt{ax^2 + bx}} = -\frac{2}{bx} \sqrt{ax^2 + bx} + c$

6. $\displaystyle \int \frac{dx}{x^2\sqrt{a^2 - x^2}} = \frac{-\sqrt{a^2 - x^2}}{a^2x} + c$

7. $\displaystyle \int \frac{x^3}{(a^2 - x^2)^3}\ dx = -\frac{1}{2(a^2 - x^2)} + \frac{a^2}{4(a^2 - x^2)^2} + c$

Some of the above formulas do involve derivations that you can work by previous methods.

Example 7

Purpose   To use formulas from a table of integrals in order to integrate:

Problem   Integrate $\displaystyle \int 5x^2\sqrt{3x + 2}\ dx$.

Analysis   This problem is similar to Formula 3 with $a = 3$ and $b = 2$.

Solution $\displaystyle \int 5x^2\sqrt{3x + 2}\ dx = 5 \int x^2\sqrt{3x + 2}\ dx$

$$= 5\left[\frac{2(15 \cdot 3^2 x^2 - 12 \cdot 3 \cdot 2x + 8 \cdot 2^2)\sqrt{(3x + 2)^3}}{105 \cdot 3^3} \right] + c$$

$$= \frac{2(135x^2 - 72x + 32)\sqrt{(3x + 2)^3}}{567} + c$$

Exercise 5.7

In Problems 1 to 26 integrate by substitution.

1. $\displaystyle \int (x^3 - 5)^4 \cdot 3x^2\ dx$

2. $\displaystyle \int \frac{3x^2}{\sqrt{x^3 - 5}}\ dx$

3. $\displaystyle \int_0^2 3x^2\sqrt{x^3 + 1}\ dx$

4. $\displaystyle \int_0^4 \frac{2x}{\sqrt{x^2 + 9}}\ dx$

5. $\displaystyle\int_{-2}^{0} \frac{dx}{(x-4)^2}$

16. $\displaystyle\int_{0}^{a} \sqrt{a^2 - x^2}\, x\, dx$

6. $\displaystyle\int \sqrt[3]{2x+3}\, dx$

17. $\displaystyle\int 5x^2(2x^3 + 5)^4\, dx$

7. $\displaystyle\int \frac{dx}{(2x+3)^3}$

18. $\displaystyle\int_{0}^{4} 9x\sqrt{9 + x^2}\, dx$

8. $\displaystyle\int (2x+3)^n\, dx,\ n \neq -1$

19. $\displaystyle\int (x^3 + 6x)^4(x^2 + 2)\, dx$

9. $\displaystyle\int (ax+b)^n\, dx,\ n \neq -1$

20. $\displaystyle\int \frac{(\sqrt{x} + 1)^2}{\sqrt{x}}\, dx$

10. $\displaystyle\int_{1}^{2} (x^2 - 1)^3 x\, dx$

21. $\displaystyle\int x\sqrt{2x+3}\, dx$

Let $u = 2x + 3$

11. $\displaystyle\int_{\sqrt{2}}^{3} \frac{x}{\sqrt[3]{x^2 - 1}}\, dx$

22. $\displaystyle\int 7x\sqrt{7x - 5}\, dx$

12. $\displaystyle\int \frac{4x^3 - 4}{\sqrt{x^4 - 4x + 4}}\, dx$

23. $\displaystyle\int 5x(7x + 5)^5\, dx$

13. $\displaystyle\int \frac{x^3 - 1}{\sqrt{x^4 - 4x + 4}}\, dx$

24. $\displaystyle\int 2x^3\sqrt{x^2 + 4}\, dx$

Let $u = x^2 + 4$

14. $\displaystyle\int_{0}^{2} \frac{x}{\sqrt{9 - x^2}}\, dx$

25. $\displaystyle\int_{0}^{2} x^3\sqrt{2x^2 + 1}\, dx$

15. $\displaystyle\int \frac{x}{\sqrt{a^2 - x^2}}\, dx$

26. $\displaystyle\int \frac{x}{\sqrt{5 - x}}\, dx$

27. Find the area of the region bounded by the curve $y = x\sqrt{x^2 - 9}$, the line $x = 5$, and the x axis.

28. Find the area of the region bounded by the curve $y = x\sqrt{16 - x^2}$ and the x axis.

29. Find the area bounded by the curves $x^2 = 4y + 5$ and $x - 2y = 1$.

30. Find the area bounded by the x axis, curve $y = \dfrac{1}{\sqrt[3]{2x + 1}}$, and the interval $x = 0$ to $x = 13$.

31. Find the area bounded by the curves $y = x$, $y = (2x - 1)^3$, $x = 1$, and the y axis.

32. A company finds that the production cost can be determined by the function $C(x) = \dfrac{60}{\sqrt{2x + 1}}$ where $C(x)$ is the cost to produce the xth item.

(a) Approximate the total cost to produce 84 items.

(b) Approximate the average cost of the company to produce 84 items. (Average cost = total cost of producing x items divided by the number of items.)

33. With environmental protection equipment operating, a steel mill is dumping pollutants into the Mahoning River during the xth year of operation according to this equation $y = \dfrac{10,000}{\sqrt[3]{(x + 1)^2}}$. Environmental standards require that the total pollutants dumped over the next 26 years be less than 80,000 tons. Approximate the total number of tons of pollutants dumped for this time period and observe whether the mill will meet environmental standards.

34. Find the consumers' surplus and the producers' surplus if the supply function is $y = S(x) = \frac{3}{2}x + 1$ and the demand function is $y = D(x) = \dfrac{100}{(2x + 1)^2}$. The equilibrium point is $(2,4)$. (See Section 5.6, Example 2.)

35. A factory manufacturing heavy machinery determines its marginal cost by the function $C'(x) = x\sqrt{x + 1}$ (x, pieces of heavy machinery) and the cost is \$7800 when 24 machines are produced. Find the cost function.

36. If the velocity function is given by $v = t\sqrt[3]{t^2 + 1}$, find the function, $s = f(t)$, where s is the distance an object moves along a straight line in time t with the initial condition, $t = 0$, $s = 0$.

In Problems 37 to 46, use the tables given on page 278 to integrate the following problems.

37. $\displaystyle\int \frac{1}{2}\sqrt{2x - 3}\, dx$

38. $\displaystyle\int x^2\sqrt{2 + x}\, dx$

39. $\displaystyle\int \frac{dx}{x^2\sqrt{25 - x^2}}$

40. $\displaystyle\int \frac{4x}{\sqrt{2x + 3}}\, dx$

41. $\displaystyle\int \frac{3x^3}{(9 - x^2)^3}\, dx$

42. $\displaystyle\int \frac{x}{(3x + 4)^3}\, dx$

43. $\displaystyle\int 7x^2\sqrt{x - 1}\, dx$

44. $\displaystyle\int \frac{-2x}{\sqrt{3x + 2}}\, dx$

45. $\displaystyle\int \frac{7}{x\sqrt{2x^2 + 7x}}\, dx$

46. $\displaystyle\int \frac{4x}{(4x - 5)^3}\, dx$

47. (Optional) The frequency function for a probability distribution over the interval $0 \leq x \leq 2$ is $f(x) = \dfrac{x}{\sqrt{2x^2 + 1}}$. Verify, the total area under the curve is 1.

5.8   SIMPSON'S RULE (Optional)

If no antiderivative of f is known then the fundamental theorem cannot be used to give the exact value of the definite integral. However, the definite integral may be *approximated* to any degree of accuracy by several techniques. One of these techniques, which gives very good results and is easily programmed for high-speed computers, is Simpson's rule.

We will utilize Simpson's rule without proof, but basically it is used to interpret the definite integral as an area under a curve, and then it approximates this area. The approximation is accomplished by dividing $a \leqslant x \leqslant b$ into an even number of equal subintervals by selecting $a = x_0, x_1, x_2, x_3, \ldots, x_n = b$.

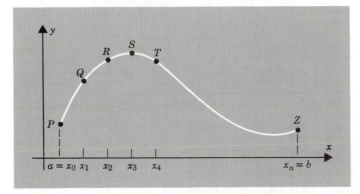

Let $P = (x_0, f(x_0))$; $Q = (x_1, f(x_1))$; $R = (x_2, f(x_2))$; .... Fit a parabola through P, Q, and R. The area under parabola PQR is taken to approximate the true area under $y = f(x)$ for $x_0 \leqslant x \leqslant x_2$. Repeat the process with R, S, and T. The area under parabola RST is approximately equal to the area under $y = f(x)$ for $x_2 \leqslant x \leqslant x_4$. Repeat the process until point Z is reached. Take the sum of these areas under successive parabolas to approximate the area under $y = f(x)$ for $a \leqslant x \leqslant b$. Then $\int_a^b f(x) \, dx$ will be approximated by this total area.

THEOREM 5.8

Simpson's Rule. Let f be a function continuous in the interval $a \leqslant x \leqslant b$. Divide $a \leqslant x \leqslant b$ into an even number of equal subintervals by selecting $a = x_0, x_1, x_2, \ldots, x_n = b$. Then:

$$\int_a^b f(x) \, dx \simeq \frac{b-a}{3n}[f(x_0) + 4f(x_1) + 2f(x_2) + 4f(x_3) + 2f(x_4) + \cdots$$
$$+ 2f(x_{n-2}) + 4f(x_{n-1}) + f(x_n)].$$

Example 1

Purpose To use Simpson's rule to approximate $\int_a^b f(x)\, dx$:

Problem Approximate $\int_0^2 \dfrac{dx}{1 + x^2}$ by using four equal subintervals.

Solution

x	$f(x) = \dfrac{1}{1 + x^2}$
0.0	1.0
0.5	0.8
1.0	0.5
1.5	0.3077
2.0	0.2

$$\int_0^2 \frac{dx}{1 + x^2} \simeq \frac{2 - 0}{3 \cdot 4}[1 + 4(.8) + 2(.5) + 4(.3077) + (.2)]$$

$$= \tfrac{1}{6}[1 + 3.2 + 1 + 1.2308 + .2]$$

$$= \tfrac{1}{6}[6.6308] \simeq 1.105$$

5.9 IMPROPER INTEGRALS (Optional)

In the fundamental theorem of integral calculus, the limits of integration, a and b, in $\int_a^b f(x)\, dx$, are real numbers. Also, f is a continuous function on $a \leqslant x \leqslant b$. If, in a certain application, it will be desirable for $a = -\infty$ or for $b = +\infty$ or f to be discontinuous for some x in $a \leqslant x \leqslant b$, then the definition of the definite integral and the techniques of integration must be extended to include these additional cases. Such integrals, where $a = -\infty$ or $b = +\infty$ or f is discontinuous on $a \leqslant x \leqslant b$, are called *improper integrals*. They do not always have a value assigned to them and, therefore, do not always exist. When they do exist, their value is attained by the following extensions of our definitions for a definite integral.

DEFINITIONS

Types of Improper Integrals. If the function f is continuous on the designated interval, then:

1. $\displaystyle \int_{-\infty}^b f(x)\, dx = \lim_{a \to -\infty} \int_a^b f(x)\, dx$ (if the limit is a real number).

2. $\displaystyle\int_a^\infty f(x)\ dx = \lim_{b\to\infty} \int_a^b f(x)\ dx$ (if the limit is a real number).

3. $\displaystyle\int_{-\infty}^\infty f(x)\ dx = \int_{-\infty}^c f(x)\ dx + \int_c^\infty f(x)\ dx$

$$= \lim_{a\to-\infty}\int_a^c f(x)\ dx + \lim_{b\to\infty}\int_c^b f(x)\ dx$$

(if the limits are real numbers). Here, c is any real number.

DEFINITION

Type of Improper Integral. If the function f is discontinuous at $x = d$ where $a \leqslant d \leqslant b$, then

$$\int_a^b f(x)\ dx = \int_a^d f(x)\ dx + \int_d^b f(x)\ dx$$

$$= \lim_{s\to d^-}\int_a^s f(x)\ dx + \lim_{t\to d^+}\int_t^b f(x)\ dx$$

(providing that both of these limits exist).

Example 1

Purpose   To illustrate the above definitions in the evaluation of improper integrals:

Problem   Find (if possible) (A) $\displaystyle\int_{-\infty}^{-2} \frac{1}{x^2}\ dx$

(B) $\displaystyle\int_3^\infty \frac{1}{x^2}\ dx$

(C) $\displaystyle\int_{-1}^1 \frac{1}{x^2}\ dx$

(D) $\displaystyle\int_5^\infty x\ dx$

Solution   (A) $\displaystyle\int_{-\infty}^{-2} \frac{1}{x^2}\ dx = \lim_{a\to-\infty}\int_a^{-2} x^{-2}\ dx = \lim_{a\to-\infty}\left[-x^{-1}\right]_a^{-2}$

$$= \lim_{a\to-\infty}\left[-(-2)^{-1} + a^{-1}\right]$$

$$= \lim_{a\to-\infty}\left[+\frac{1}{2} + \frac{1}{a}\right]$$

$$= \frac{1}{2} + 0 = \frac{1}{2}$$

(B) $\displaystyle\int_3^\infty x^{-2}\,dx = \lim_{b\to\infty}\int_3^b x^{-2}\,dx = \lim_{b\to\infty}\left[-x^{-1}\right]_3^b$

$$= \lim_{b\to\infty}\left[-\frac{1}{b}+\frac{1}{3}\right]$$

$$= 0 + \frac{1}{3} = \frac{1}{3}$$

(C) $f(x) = \dfrac{1}{x^2}$ is discontinuous at $x = 0$, which is within $-1 \le x \le 1$.

$\displaystyle\int_{-1}^1 \frac{1}{x^2}\,dx = \int_{-1}^0 \frac{1}{x^2}\,dx + \int_0^1 \frac{1}{x^2}\,dx$

$$= \lim_{s\to0^-}\int_{-1}^s x^{-2}\,dx + \lim_{t\to0^+}\int_t^1 x^{-2}\,dx$$

$$= \lim_{s\to0^-}\left[-x^{-1}\right]_{-1}^s + \lim_{t\to0^+}\left[-x^{-1}\right]_t^1$$

$$= \lim_{s\to0^-}\left[-s^{-1}+(-1)^{-1}\right] + \cdots$$

$$= \lim_{s\to0^-}\left[-\frac{1}{s}-1\right] + \cdots$$

This limit is not a real number so no value is assigned to this integral. In this case the definite integral is said to *diverge*. This is in contrast to parts (A) and (B) when the integrals are said to *converge*.

If you do not recognize that this is an improper integral and incorrectly apply the fundamental theorem, you obtain:

$$\int_{-1}^1 \frac{1}{x^2}\,dx = -\frac{1}{x}\bigg]_{-1}^1 = -2, \text{ a wrong answer!}$$

(D) $\displaystyle\int_5^\infty x\,dx = \lim_{b\to\infty}\int_5^b x\,dx = \lim_{b\to\infty}\left[\tfrac{1}{2}x^2\right]_5^b$

$$= \lim_{b\to\infty}\left[\tfrac{1}{2}b^2 - \tfrac{1}{2}(25)\right] = \infty.$$

This limit is not a real number, and no value will be assigned to this definite integral. The integral is called divergent.

Example 2

Purpose To use an improper integral to solve an applied problem:

Problem Find the area between the curve $y = f(x) = \dfrac{1}{\sqrt{x+4}}$ and the x axis over the interval $-4 < x \le 0$.

Solution   First sketch $y = \dfrac{1}{\sqrt{x+4}}$.

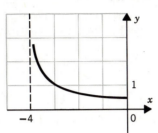

$$A = \int_{-4}^{0} \frac{1}{\sqrt{x+4}}\,dx$$

But, this function is not continuous at $x = -4$. Therefore, it must be evaluated, if possible, by one of the techniques discussed with improper integrals.

$$\int_{-4}^{0} \frac{1}{\sqrt{x+4}}\,dx = \lim_{a \to -4^{+}} \int_{a}^{0} (x+4)^{-(1/2)}\,dx$$

$$= \lim_{a \to -4^{+}} \left[\frac{(x+4)^{1/2}}{\frac{1}{2}} \right]_{a}^{0}$$

$$= \lim_{a \to -4^{+}} [4 - 2\sqrt{a+4}]$$

$$= 4 - 2 \cdot \lim_{a \to -4^{+}} \sqrt{a+4}$$

$$= 4 - 2 \cdot 0 = 4 \text{ square units}$$

Exercise 5.8–5.9

In Problems 1 to 12, if the improper integral exists, evaluate.

1. $\displaystyle\int_{1}^{\infty} \frac{1}{x^3}\,dx$

2. $\displaystyle\int_{0}^{1} \frac{1}{x^3}\,dx$

3. $\displaystyle\int_{-1}^{1} \frac{1}{x^3}\,dx$

4. $\displaystyle\int_{1}^{\infty} \frac{1}{\sqrt[3]{x}}\,dx$

5. $\displaystyle\int_{0}^{1} \frac{1}{\sqrt[3]{x}}\,dx$

6. $\displaystyle\int_{-1}^{1} \frac{1}{\sqrt[3]{x}}\,dx$

7. $\displaystyle\int_{0}^{2} \frac{4x}{(x^2 - 1)^2}\,dx$

8. $\displaystyle\int_{-\infty}^{0} \frac{-x}{(x^2 + 1)^2}\,dx$

9. $\displaystyle\int_{-2}^{2} \frac{1}{(x + 1)^2}\,dx$

10. $\displaystyle\int_{2}^{\infty} \frac{x}{(x^2 - 1)^{3/2}}\,dx$

11. $\displaystyle\int_{2}^{\infty} \sqrt{3x - 5}\,dx$

12. $\displaystyle\int_{2}^{\infty} \frac{1}{\sqrt{3x - 5}}\,dx$

13. Verify: $\int_1^\infty \frac{1}{x^n}\, dx$ (a) exists (converges) when $n > 1$.
                                             (b) does not exist (diverges) when $n < 1$.

14. Verify: $\int_0^1 \frac{1}{x^n}\, dx$ (a) exists (converges) when $n < 1$.
                                             (b) does not exist (diverges) when $n > 1$.

15. Find the area, if it exists, of the curve bounded by $y = \frac{1}{x^2}$, the x axis, and $x \geq 1$.

16. Find the area, if it exists, of the curve bounded by $y = \frac{1}{x^2}$, the x axis, and the interval from $x = 0$ to $x = 1$.

17. Use the properties of a continuous frequency function to find k for the frequency function $f(x) = \frac{k}{x^2}$ where $x \geq 1$.

18. The taxes per family paid in Camelot have a frequency function $f(x) = \frac{4x}{(2 + x^2)^2}$. (For $x \geq 0$ and x is in hundreds of dollars.)

    (a) Verify that the total area under the frequency function is equal to one.
    (b) What percentage of the families pay less than $100?
    (c) What proportion pay between $500 and $1000?
    (d) What percentage pay more than $2000?

19. The frequency function for a probability distribution function for $x \geq 1$ is given by $f(x) = \frac{c}{x^4}$.

    (a) Find c to guarantee that the total area under the curve is 1.
    (b) Find m_1, the first moment about the origin.
    (c) Find m_2, the second moment about the origin (see Section 5.6).

20. Use Simpson's rule to approximate the following integrals:

    (a) $\int_1^3 \frac{1}{x}\, dx$; $(n = 6)$ (b) $\int_0^2 \sqrt{4 - x^2}\, dx$; $(n = 4)$

21. A computer problem (optional). Use Simpson's rule to approximate the integrals:

    (1) $\int_1^3 \frac{1}{x}\, dx$ (2) $\int_0^2 \sqrt{4 - x^2}\, dx$

    for:

    (a) $n = 10$.
    (b) $n = 20$.
    (c) $n = 40$.

■ 5.10 CHAPTER REVIEW

Important Ideas

an antiderivative of a function
indefinite integral
properties of the indefinite integral

$$\int x^n \, dx = \frac{1}{n+1} x^{n+1} + c, \quad n \neq -1$$

summation notation
areas by summation
Riemann sum
definite integral
properties of the definite integral
the fundamental theorem of integral calculus
(mean value theorem for integrals)

$$\left(\frac{d}{dx} \int_k^x f(t) \, dt = f(x) \right)$$

(if $f'(x) = g'(x)$, then $f(x) - g(x) = k$)
application of the definite integral
integration by substitution
integration by formula
(Simpson's rule)
(improper integrals)

■ REVIEW EXERCISE (optional)

1. What is meant by a Riemann sum?

2. Define the definite integral.

3. What is meant by the indefinite integral?

4. State the fundamental theorem of integral calculus.
   (*Useless Hint*. Mrs. Whipkey once had a student answer this test question by the statement, "A friend in need is a friend indeed." Although the student at this point was truly in need of a friend, this is *not* the correct answer).

5. The function f is defined by $f(x) = \frac{1}{2}x$.
   (a) Divide the interval $0 \leqslant x \leqslant 2$ into four equal subintervals.
   Find $\sum_{i=1}^{4} f(x_i) \, \Delta x$ where x_i is the right end point of each subinterval.
   (b) Find $\int_0^2 \frac{1}{2}x \, dx$ by the definition of the definite integral.
   (c) Find $\int_0^2 \frac{1}{2}x \, dx$ using the fundamental theorem of integral calculus.

6. The function f is defined by $f(x) = x^2 - 1$.

(a) Divide the interval $1 \leqslant x \leqslant 3$ into six equal subintervals. Find $\displaystyle\sum_{i=1}^{6} f(x_i) \, \Delta x$ where x_i is the right end point of each subinterval.

(b) Find $\displaystyle\int_{1}^{3} (x^2 - 1) \, dx$.

In Problems 7 to 28, integrate.

7. $\displaystyle\int (2x^3 + 7x - 9) \, dx$

8. $\displaystyle\int (5 - x - x^2) \, dx$

9. $\displaystyle\int_{-2}^{5} (x + 2) \, dx$

10. $\displaystyle\int_{-1}^{2} (x^2 + 2x - 3) \, dx$

11. $\displaystyle\int_{0}^{3} (x + 2)^2 \, dx$

12. $\displaystyle\int \left(\frac{1}{2\sqrt{x}} - \frac{2}{x^3} \right) dx$

13. $\displaystyle\int \left(\frac{2}{\sqrt{x}} - \frac{1}{2x^2} \right) dx$

14. $\displaystyle\int \left(\sqrt[3]{x} + \frac{1}{\sqrt[3]{x}} - 3 \right) dx$

15. $\displaystyle\int_{-1}^{2} (2x^2 + 5)^3 4x \, dx$

16. $\displaystyle\int (x^3 - x + 1)^4 (3x^2 - 1) \, dx$

17. $\displaystyle\int_{-1}^{2} (2x^2 + 5)x \, dx$

18. $\displaystyle\int_{1}^{4} \frac{x^3 + x^2 - 1}{x^2} \, dx$

19. $\displaystyle\int_{1}^{3} \sqrt[3]{x^2 - 1} \, x \, dx$

20. $\displaystyle\int x\sqrt{2x + 1} \, dx$

21. $\displaystyle\int_{0}^{4} \frac{12x}{\sqrt{x^2 + 9}} \, dx$

22. $\displaystyle\int \frac{x}{\sqrt{3 - 2x}} \, dx$

23. $\displaystyle\int 4x^3\sqrt{16 - x^2} \, dx$

24. $\displaystyle\int x\sqrt{ax + b} \, dx$

25. $\displaystyle\int \frac{(\sqrt{x} - 1)^3}{\sqrt{x}} \, dx$

26. $\displaystyle\int x^2\sqrt{x - 1} \, dx$

27. $\displaystyle\int (x + 1)\sqrt{x - 1} \, dx$

28. $\displaystyle\int_{0}^{1} \frac{x}{\sqrt{x + 1}} \, dx$

29. Find the area bounded by the parabola $y = 3x - x^2$ and the x axis.

30. Find the area bounded by the parabola $y = x^2 - 3x$ and the x axis.

In Problems 31 to 34, find the area of the region bounded by the two given curves.

31. $y = x^2 - 2$; $y = x$

32. $y = x^2 + 2$; $y = 3x + 2$

33. $y = -x^2 - 2x + 1; 2x + y + 3 = 0$

34. $y = x^2 - 2; 8y = x^2$

35. Find the area of the region bounded by the curves $y = \sqrt{4x + 1}$ and $y = x + 4$ from $x = 0$ to $x = 2$.

36. In physics the potential energy of interaction between two bodies is defined as the integral of the force acting on the two bodies. If this force is given as $F(x) = 3x^3 + \dfrac{4}{x^2}$, find the potential energy of interaction.

37. A supply function is given to be $y = S(x) = 3 + x^2$ and the corresponding demand function for the same firm is $y = D(x) = -2x + 18$. Find both the consumers' surplus and the producers' surplus. (See Section 5.6, Example 2.)

38. The demand function for a type of art paper is $D(x) = (x - 4)^2$ where $D(x)$ is the price for every x units of 100 sheets. The paper producer is willing to supply paper according to $S(x) = x^2$. Sketch the graph.
(a) What is the equilibrium price?
(b) What is the value of the consumers' surplus?
(c) What is the equation for the change in price required to bring about a one unit change in quantity demanded?

39. What is the consumers' surplus and producers' surplus for the supply function $y = S(x) = 2x$ and the demand function $y = D(x) = (x - 4)^2$. $(0 \leqslant x \leqslant 4)$.

40. A supply function is given to be $y = S(x) = x^3 + 4x$. The equilibrium point is $(2,16)$. What is the producers' surplus for this function?

41. The supply and demand functions for a product are given to be $y = S(x) = x^4$ and $y = D(x) = 20 - x^2$.
(a) Graph these functions.
(b) What is the equilibrium point?
(c) What is the producers' surplus and the consumers' surplus?

42. A manufacturer determines that the demand function for a certain product is $D(x) = -2x + 8$, where x is the number of products in units of 100 and $D(x)$ is the price, in dollars, that the public would pay for the given quantity. The price is also determined by the supply function, $S(x) = x^2$ where x is the number of products in units of 100 and $S(x)$ is the price required for the seller to be willing to sell the quantity x.
(a) What price would the public pay if the number of goods supplied is 300 (three units of 100)?
(b) What would be the required price for the manufacturer to be willing to sell 100 units?
(c) What is the equilibrium quantity? What price would the public pay for that quantity?
(d) What is the value of the producers' surplus?

43. A furniture manufacturer determines that the cost to produce a living room set can be found by the function $C(x) = \dfrac{600}{\sqrt[3]{2x + 16}}$ where $C(x)$ is the cost to produce the xth set. Approximate the cost of producing the first 100 living room sets; of producing the next 100 sets.

44. A large water tank sprang a leak and is losing water at a rate of $4t + 30$ gallons per hour, where t is the time in hours. Approximate how much water will be lost from $t = 10$ to $t = 20$ (inclusive)?

45. The number of fish affected by a poisonous chemical in the water in the xth day can be found by the equation, $N = x^{-(1/2)}(x^{1/2} + 4)$. Find the total number of fish affected in the first nine days by integrating the function from $x = 0$ to $x = 9$.

46. An athlete exercises once every three days. The time span for his program is 2 years or 720 days. Since he exercises only every third day, he will exercise 240 times in the 2-year period. Assume that on the first day of exercising he does 16 exercises and he increases the number of exercises according to the formula $y = \dfrac{(x + 30)^2}{60}$. Therefore, the 2nd time he exercises he will do 17 exercises, the 10th time 27 exercises, the 100th time 282 exercises, and the 240th time 1215 exercises. Approximate the total number of exercises done between:
(a) the first day and 75th day of exercise
(b) the 50th and 100th day of exercise.

47. The slope of a tangent line to a curve is $\dfrac{dy}{dx} = 3x^2 + 2x - 1$. Find an equation of the curve if it contains the point $(-2, -1)$.

48. A contour chair company finds its marginal cost function to be $C'(x) = \$(14 + x)$ with a fixed production cost of $300 where x is the number of chairs produced daily. Find the total cost function.

49. A firm currently produces 50 units. Its marginal profit is given by $P'(x) = 10 + .4x + .09x^2$. Approximate the total profit earned from the sale of an additional 10 units. [Assume $P(0) = 0$.]

50. A business firm determines that its marginal cost is $\frac{1}{5}x^2 - 3x + 10$ dollars. The firm currently produces 10 units per hour. Approximate the total cost of producing 5 additional units. [Assume $C(0) = 0$.]

51. The rate of production in a factory at time t is $Q'(t) = .3t + 60$. Find the formula for the total production.

52. A physiologist conducting an experiment on exercise physiology finds a mouse's rate of change in its swim test performance with respect to time (in weeks) to be $P'(t) = 15 + 6t$. The first time the mouse is placed in the water ($t = 0$) it was able to swim for 6 minutes. Find $P(t)$.

53. The acceleration at time t is given to be $(12t + 2)$ ft/sec². Find the position function $s = f(t)$ if at time $t = 3$ sec, $v = 30$ ft/sec and $s = 5$ ft.

$$\left(Hint.\ a = \frac{dv}{dt} \text{ and } v = \frac{ds}{dt}.\right)$$

54. A car starts from rest and its velocity as a function of time is given by $v(t) = 4t + .3t^2$. Find a formula for the position of the car as a function of time.

55. The velocity of a particle moving in a straight line is given by $v = \dfrac{ds}{dt} =$

$(t^4 - 6t^2)^{1/2}(t^3 - 3t)$. Find the respective distance equation if $s = 10$ when $t = 0$. What is the total distance traveled when $t = 3$?

56. If a train starts from rest, what is the constant acceleration it must maintain to travel 500 feet in 200 seconds?

57. A motorcycle manufacturer claims that its bikes can accelerate from 0 to 60 miles per hour in 8 seconds. Assume constant acceleration and find the distance traveled in the 8 seconds.

In Problems 58 to 72, determine whether the statement is true or false and justify your answer.

58. $\displaystyle\int f(x)\ dx = F(x) + c$ if $F'(x) = f(x)$

59. $\displaystyle\int_a^b f(x)\ dx = F(x)\Big]_a^b = F(b) - F(a)$, where $F'(x) = f(x)$.

60. $\displaystyle\int_2^2 \frac{1}{x^2}\ dx = \frac{1}{4}$

61. $\displaystyle\int_2^5 3\ dx = 6$

62. $\displaystyle\int_6^x t^5\ dt = \frac{1}{6}x^6 - 6^5$

63. $\displaystyle\frac{d}{dx}\int_6^x t^5\ dt = x^5$

64. $\displaystyle\frac{d}{dx}\left(\frac{1}{6}x^6 - 6^5\right) = x^5$

65. $\displaystyle\sum_{i=1}^n x_i^2\ \Delta x = nx^2\ \Delta x$

66. $\displaystyle\sum_{k=1}^{16} k = 136$

67. $\displaystyle\int x\sqrt{x + 2}\ dx = \frac{2}{15}(x + 2)^{3/2}(3x - 4) + c$

68. If $\dfrac{dy}{dx} = 3x^2 + \dfrac{1}{x^2}$, then $y = x^3 - \dfrac{1}{x^3} + c$

69. If $v = \dfrac{t}{\sqrt{t^2 + 5}}$, then $s = f(t) = \sqrt{t^2 + 5} + c$

70. $\dfrac{1}{x^5} + 15$ is an antiderivative of $-4x^{-4}$

71. $\displaystyle\int_1^2 x^2 \, dx = \int_1^2 z^2 \, dz = \dfrac{7}{3}$

72. If $f'(x) = g'(x) = 3x^2 + 1$, then $f(x)$ can equal $x^3 + x + 1$ and $g(x)$ can equal $x^3 + 1$.

SIX

The Logarithmic and Exponential Functions; Review of the Great Ideas of the Calculus

6.1   INTRODUCTION

In this chapter two special functions are introduced—the logarithmic and the exponential. These functions deserve our careful attention because of their frequent use, not only in theoretical mathematics but also in applications taken from many fields.

In previous courses you have probably used logarithms as an aid in calculations. You may recall that the definition is: if $x > 0$, $b > 0$, $b \neq 1$, then $\log_b x = L$ if $b^L = x$. In your high school work the base $b = 10$ was usually selected. The system of logarithms with $b = 10$ is called the system of common logarithms. Common logarithms are seldom used in the calculus; that is, instead of common logarithms the calculus makes greater use of the natural logarithm. This new type of logarithmic function is developed in this chapter. Also, the inverse function, associated with our new logarithmic function, is discussed. Then, this inverse function, called the exponential function, is used in rate of growth problems involving bacteria, interest, population, and the like.

Additionally, this chapter affords a splendid opportunity to review the most important ideas of the calculus. Much is to be gained by encountering again the various ideas and techniques of previous chapters whose concepts may still be somewhat hazy. The review of previous work, as you progress through this chapter, will pay untold dividends in understanding the calculus and in giving cohesiveness to ideas that may appear to be totally unconnected.

Before we begin the development of the logarithmic and the exponential functions, let us examine a special limit that will be of great importance in our study of these functions.

The Number e

A hand-held calculator may be used to investigate the

$$\lim_{t \to 0} (1 + t)^{1/t}$$

Let us construct a table to indicate the behavior of $(1 + t)^{1/t}$ as t approaches zero.

t	$(1 + t)^{1/t}$	t	$(1 + t)^{1/t}$
1	2	-1	0
.1	2.59374	$-.1$	2.86797
.01	2.70481	$-.01$	2.73200
.001	2.71693	$-.001$	2.71964
.0001	2.71814	$-.0001$	2.71844
.00001	2.71828	$-.00001$	2.71828
.000001	2.71828	$-.000001$	2.71828
.0000001	2.71828	$-.0000001$	2.71828

Thus, it appears that $\lim\limits_{t \to 0} (1 + t)^{1/t}$ does exist, and we will call this limit the number e. This result could be proved, but rather than belabor this point we will accept the existence of a number e defined by

$$\lim_{t \to 0} (1 + t)^{1/t} = e \approx 2.71828.$$

Since e is a number that is approximately equal to 2.71828 (correct to five decimal places), it may be used to construct an exponential function, $y = e^x$, and a logarithmic function $y = \log_e x$. However, before we investigate these functions, it is advisable to review the properties of exponents and logarithms.

A Brief Review of the Properties of Exponents

Given $y = b^x$, $b \neq 0$.

1. If $x = 0$, define $y = b^0 = 1$.

   Examples. $10^0 = 1$; $\quad e^0 = 1$ (where $e \approx 2.71828$).

2. If $x = -n$ where $-n$ is a negative integer, define $y = b^{-n} = \dfrac{1}{b^n}$.

   Examples. $10^{-2} = \dfrac{1}{10^2} = \dfrac{1}{100}$; $\quad e^{-3} = \dfrac{1}{e^3}$.

3. Let x be a rational number represented by $\dfrac{m}{n}$. Then, define $b^{m/n} = \sqrt[n]{b^m} = (\sqrt[n]{b})^m$.

   Examples. $8^{2/3} = \sqrt[3]{8^2} = \sqrt[3]{64} = 4 \qquad$ or $\qquad 8^{2/3} = (\sqrt[3]{8})^2 = 2^2 = 4$;
   $e^{3/2} = \sqrt{e^3} = e\sqrt{e} \qquad$ or $\qquad e^{3/2} = (\sqrt{e})^3 = e\sqrt{e}$.

4. $b^m \cdot b^n = b^{m+n}$.

   Examples. $(3.1)^8(3.1)^{-2} = (3.1)^6$; $\quad e^4 \cdot e^3 = e^7$.

5. $(b^m)^n = b^{mn}$.

   Examples. $(10^2)^{-(1/2)} = 10^{-1} = \frac{1}{10}$; $\quad (e^2)^4 = e^8$.

6. $\dfrac{b^m}{b^n} = b^{m-n}$.

   Examples. $\dfrac{10^5}{10^2} = 10^3$; $\quad \dfrac{e^3}{e^6} = e^{-3} = \dfrac{1}{e^3}$.

7. Graph the exponential function, $y = b^x$ for selected values of b. A table of values is given for $y = 2^x$. The graphs of $y = 4^x$ and $y = (1.5)^x$ have the same basic shape. Notice also the graph of $y = .5^x$, which represents a decreasing function and the slopes of the tangent lines to the curve are

negative. This is in contrast to the other graphs which represent increasing functions and have slopes of tangent lines that are positive.

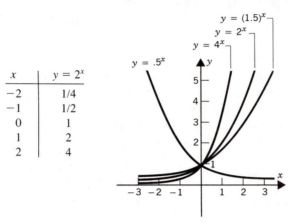

x	$y = 2^x$
-2	1/4
-1	1/2
0	1
1	2
2	4

When we filled in the points on the graph we used a continuous curve. This means that we are assuming that 2^x is defined when x is an irrational number. Such a definition could be made as follows: If $a > 0$ and x is a real number, then $a^x = \lim_{r \to x} a^r$, where r approaches x through rational values. Once such a definition is made, then: (1) a^x is defined for $a > 0$ when x is any real number, rational or irrational, and (2) properties 4, 5, and 6, which are stated above, could be proved to hold for m and n any real numbers, rational or irrational.

Example 1

Purpose   To further illustrate the properties of exponents:

Problem   State the properties that justify each of the following statements.

(A) $(e^2)^3 = e^6$

(B) $\sqrt{a^{10}} = a^5, \quad a > 0$

(C) $b^{-2} \cdot b^2 = 1$

(D) $\dfrac{10^0 \cdot 10^{-2}}{10^2} = \dfrac{1}{10^4}$

Solution   (A) $(e^2)^3 = e^6$ by property 5.

(B) $\sqrt{a^{10}} = a^{10/2} = a^5$ by property 3.

(C) $b^{-2} \cdot b^2 = b^0 = 1$ by properties 1 and 4.

(D) $\dfrac{10^0 \cdot 10^{-2}}{10^2} = 10^{-4} = \dfrac{1}{10^4}$ by properties 1, 2, and 6.

A Brief Review of the Properties of Logarithms

Given $y = \log_b x$.

1.   Definition: If $b > 0$, $b \neq 1$, $x > 0$, then $\log_b x = L$ if $b^L = x$.

Examples. $\log_{10} 100 = 2$ since $10^2 = 100;$

$$\log_2 \frac{1}{8} = -3 \quad \text{since} \quad 2^{-3} = \frac{1}{8};$$

$$\log_e \frac{1}{e} = -1 \quad \text{since} \quad e^{-1} = \frac{1}{e};$$

$$\log_b 1 = 0 \quad \text{since} \quad b^0 = 1;$$

$$\log_e 1 = 0 \quad \text{since} \quad e^0 = 1.$$

2. $\log_b b = 1.$

Examples. $\log_{10} 10 = 1$ since $10^1 = 10;$
$\log_e e = 1$ since $e^1 = e.$

3.   If m and n are positive real numbers, then

$$\log_b (m \cdot n) = \log_b m + \log_b n.$$

Example. $\log_e (2 \cdot 3) = \log_e 2 + \log_e 3.$

Also, $\log_b \left(\dfrac{m}{n} \right) = \log_b m - \log_b n.$

Example. $\log_{10} \left(\dfrac{3}{4} \right) = \log_{10} 3 - \log_{10} 4.$

4. $\log_b (m)^n = n \cdot \log_b m.$

Example. $\log_e 4^3 = 3 \cdot \log_e 4.$

Example 2

Purpose   To further illustrate the properties of logarithms:

Problem   State the properties that justify each of the following statements.

(A) $\log_3 81 = 4$ (C) $\log_e \sqrt{e} = \frac{1}{2}$

(B) $\log_{10} (35) = \log_{10} 5 + \log_{10} 7$ (D) $\dfrac{\log_b 5}{n} = \log_b \sqrt[n]{5}$

Solution   (A) $\log_3 81 = 4$ if $3^4 = 81$ by property 1.
(B) $\log_{10} (35) = \log_{10} (5 \cdot 7) = \log_{10} 5 + \log_{10} 7$ by property 3.
(C) $\log_e \sqrt{e} = \log_e e^{1/2} = \frac{1}{2} \log_e e = \frac{1}{2} \cdot 1 = \frac{1}{2}$ by properties 2 and 4.
(D) $\dfrac{\log_b 5}{n} = \dfrac{1}{n} \log_b 5 = \log_b 5^{1/n} = \log_b \sqrt[n]{5}$ by property 4.

Problem   Solve the given equations for x.

(E) $\log_3 9 = x$ (F) $\log_2 (x^2 + 3) - 2 \log_2 x = 2$

Solution   (E) $\log_3 9 = x$ if $3^x = 9$ by property 1. Therefore, $x = 2$.

(F) $\log_2 (x^2 + 3) - 2 \log_2 x = 2$

$$\log_2 \frac{x^2 + 3}{x^2} = 2 \text{ by properties 3 and 4.}$$

$$2^2 = \frac{x^2 + 3}{x^2} \text{ by property 1.}$$

$$4x^2 = x^2 + 3$$

$$x = 1; \ (x \neq -1, \text{ why?})$$

All of the above results can be proved by making use of the definitions of the logarithmic and the exponential functions. In addition, Appendix I contains a review of these concepts. Students not familiar with these properties are urged to examine the Appendix for more examples and problems.

Exercise 6.1

In Problems 1 to 10, simplify the expression.

1. $5^0 \cdot 5^3$

2. $\dfrac{5^0}{5^3}$

3. $9^{3/2}$

4. $(e^2)^{1/2}$

5. $\dfrac{10^2 \cdot 10^5}{10^{-2}}$

6. $a^9 \cdot a^{-4}$

7. $(64)^{2/3}$

8. $8^3 \cdot 8^{1/3}$

9. $\dfrac{e^3}{e^{-3}}$

10. $(7a^2b^3)^0$

In Problems 11 to 20, find the logarithms.
$\log_{10} 2 = .3010$, $\log_{10} 7 = .8451$, and $\log_{10} 10 = 1$.

11. $\log_{10} 14$

12. $\log_{10} 32$

13. $\log_{10} \frac{7}{2}$

14. $\log_{10} 70$

15. $\log_{10} 49$

16. $\log_{10} 200$

17. $\log_{10} \frac{1}{2}$

18. $\log_{10} 10^{10}$

19. $\log_{10} .016$

20. $\log_{10} \sqrt[3]{2}$

21. A decibel is a unit that measures the loudness of a sound. Decibel levels may be calculated by using the function, $D = 10 \log_{10} \left[\dfrac{\phi}{10^{-12} \text{ watts/m}^2} \right]$

where D is decibels and ϕ is the intensity of the sound. Determine the decibel levels for the following situations.
(a) A quiet office, $\phi = 10^{-8}$ watts/m^2.
(b) Conversation, $\phi = 10^{-6}$ watts/m^2.
(c) An automobile horn, $\phi = 1$ watt/m^2.

In Problems 22 to 31, solve the given equations for x.

22. $\log_e x = 1$ **27.** $\log_3 x = -2$

23. $\log_2 8 = x$ **28.** $\log_e x + \log_e (2 - x) = 0$

24. $\log_x 64 = 3$ **29.** $\log_{10} 5x - \log_{10} (x + 2) = 0$

25. $\log_7 x = 0$ **30.** $2 \log_5 x + \log_5 x^3 - \log_5 x^4 = 2$

26. $\log_{25} 5 = x$ **31.** $\log_2 \sqrt{x} + \log_2 (1 - 2x) = \frac{1}{2}\log_2 x^3$

In Problems 32 to 35, graph the given functions.

32. $y = 3^x$ **34.** $y = (\frac{1}{2})^x$

33. $y = 2^{-x}$ **35.** $y = \log_2 x$

6.2 THE NATURAL LOGARITHM FUNCTION AND ITS DERIVATIVE

In the previous section we reviewed logarithms and some of their properties. From the definition for logarithm, $y = \log_e x$, $x > 0$, if $x = e^y$. Thus we saw that

$$\log_e e = 1 \qquad \text{since} \qquad e^1 = e$$

$$\log_e \frac{1}{e} = -1 \qquad \text{since} \qquad e^{-1} = \frac{1}{e}.$$

A system of logarithms that uses e as the base is called a system of natural logarithms. This system is the most desirable system to use in the calculus because it simplifies the finding of derivatives and integrals. Since we use e as the base almost exclusively, we will abbreviate $\log_e x$ as $\ln x$ (read natural log of x).

DEFINITION

Natural Logarithm Function. The function defined by $y = f(x) = \ln x$, $x > 0$, is called the *natural logarithm function*. Moreover, $y = \ln x$ if $x = e^y$.

The problem of finding the derivative of $f(x) = \ln x$ involves the application of the delta process along with the definition $\lim_{t \to 0} (1 + t)^{1/t} = e$.

THEOREM 6.2α

If $y = f(x) = \ln x$, $x > 0$, then $\dfrac{d}{dx} (\ln x) = \dfrac{1}{x}$.

GIVEN. $y = f(x) = \ln x$, $x > 0$.

TO PROVE. $\dfrac{d}{dx} (\ln x) = \dfrac{1}{x}$.

Proof

$y = f(x) = \ln x$

$$\lim_{\Delta x \to 0} \frac{\Delta y}{\Delta x} = \lim_{\Delta x \to 0} \frac{f(x + \Delta x) - f(x)}{\Delta x}$$

$$= \lim_{\Delta x \to 0} \frac{\ln (x + \Delta x) - \ln x}{\Delta x}$$

$$= \lim_{\Delta x \to 0} \frac{1}{\Delta x} \ln \left(\frac{x + \Delta x}{x} \right)$$

$$= \lim_{\Delta x \to 0} \frac{1}{\Delta x} \ln \left(1 + \frac{\Delta x}{x} \right)$$

$$= \lim_{\Delta x \to 0} \frac{1}{x} \cdot \frac{x}{\Delta x} \ln \left(1 + \frac{\Delta x}{x} \right)$$

$$= \lim_{\Delta x \to 0} \frac{1}{x} \ln \left(1 + \frac{\Delta x}{x} \right)^{x/\Delta x}$$

Let $t = \dfrac{\Delta x}{x}$; as $\Delta x \to 0$, $t \to 0$

Therefore

$$\frac{d}{dx} (\ln x) = \lim_{t \to 0} \frac{1}{x} \ln (1 + t)^{1/t}$$

$$= \frac{1}{x} \ln e = \frac{1}{x} (1)$$

$$\frac{d}{dx} (\ln x) = \frac{1}{x}.$$

As an extension of this result, we use the chain rule to find the derivative of $y = \ln u$, where u represents a function of x which, in turn makes y a function of x. Recall that, under these conditions, the chain rule is

$$\frac{dy}{dx} = \frac{dy}{du} \cdot \frac{du}{dx}.$$

> **THEOREM 6.2b**
>
> If $y = \ln u$, $u > 0$ and u is a differentiable function of x, which makes y a function of x, then
>
> $$\frac{d}{dx}(\ln u) = \frac{1}{u} \cdot \frac{du}{dx}.$$

GIVEN. $y = \ln u$, $u > 0$ and u is a differentiable function of x, which makes y a function of x.

TO PROVE. $\dfrac{d}{dx}(\ln u) = \dfrac{1}{u} \cdot \dfrac{du}{dx}$.

Proof

By the chain rule, $\dfrac{d}{dx}(\ln u) = \dfrac{d}{du}(\ln u) \cdot \dfrac{du}{dx}$

or

$$\frac{d}{dx}(\ln u) = \frac{1}{u} \cdot \frac{du}{dx}$$

Example 1

Purpose To utilize the previous theorems to obtain derivatives of natural logarithmic functions:

Problem A $y = \ln(2x^2 + x)$, $(2x^2 + x) > 0$. Find $\dfrac{dy}{dx}$.

Solution If $\begin{cases} y = \ln u \\ \\ u = 2x^2 + x \end{cases}$, then $\begin{aligned} \frac{1}{u} &= \frac{1}{2x^2 + x} \\ \frac{du}{dx} &= 4x + 1 \end{aligned}$

and $\dfrac{dy}{dx} = \dfrac{1}{u} \cdot \dfrac{du}{dx} = \dfrac{1}{2x^2 + x}(4x + 1) = \dfrac{4x + 1}{2x^2 + x}$

Problem B $y = \ln(7x^3 - 6x + 9)$. Find $\dfrac{dy}{dx}$.

Solution $\begin{aligned} \frac{dy}{dx} &= \frac{1}{7x^3 - 6x + 9} \cdot \frac{d}{dx}(7x^3 - 6x + 9) \\ \\ &= \frac{21x^2 - 6}{7x^3 - 6x + 9} \end{aligned}$

Problem C $y = \ln \sqrt{15 + x^2}$. Find $\dfrac{dy}{dx}$.

Solution $\dfrac{dy}{dx} = \dfrac{1}{\sqrt{15 + x^2}} \cdot (\tfrac{1}{2})(15 + x^2)^{-1/2}(2x)$

$= \dfrac{x}{15 + x^2}$

See Example 2A for an alternate solution.

Problem D   A company finds that its total cost increases logarithmically. If the cost function is $C(x) = \dfrac{2x}{\ln x}$ (x, units of a commodity, $x > 1$; $C(x)$, dollars), find the marginal cost.

Solution   Recall the quotient rule!

$$C'(x) = \dfrac{\ln x \cdot \dfrac{d}{dx}(2x) - 2x \cdot \dfrac{d}{dx}(\ln x)}{(\ln x)^2}$$

$$= \dfrac{(\ln x)(2) - 2x \cdot \dfrac{1}{x}}{(\ln x)^2}$$

$$C'(x) = \dfrac{2 \ln x - 2}{(\ln x)^2} = \dfrac{2(\ln x - 1)}{(\ln x)^2}.$$

Problem E   From a study by the Planning Commission, the county's population P (in thousands) for each year since 1980 is approximated by the function, $P(t) = 40\left(t - t \ln \dfrac{t}{7}\right)$. The time t is in years where $t = 1$ corresponds to 1981, $t = 2$ corresponds to 1982, etc. In what year will the county's population be the largest?

Solution
$$P(t) = 40\left(t - t \ln \dfrac{t}{7}\right)$$

$$P'(t) = 40\left[1 - \left(t \cdot \dfrac{1}{t/7} \cdot \dfrac{1}{7} + \ln \dfrac{t}{7}\right)\right]$$

$$P'(t) = 40\left(1 - 1 - \ln \dfrac{t}{7}\right) = 40\left(-\ln \dfrac{t}{7}\right)$$

Set $P'(t) = 0$ and solve for t to find the critical number.

$$40\left(-\ln \dfrac{t}{7}\right) = 0$$

$$\ln \dfrac{t}{7} = 0$$

From the definition, $y = \ln x$ if $e^y = x$, it follows that $\ln \dfrac{t}{7} = 0$ is equivalent to

$$e^0 = \frac{t}{7}$$

$$1 = \frac{t}{7}$$

$$7 = t$$

The county's population will be the largest in 1987.

Problem F $y^3 \ln x - y^2 = 5$, $x > 0$. Find y'.

Solution Recall implicit differentiation!

$$y^3 \cdot \frac{d}{dx}(\ln x) + (\ln x) \cdot \frac{d}{dx}(y^3) - \frac{d}{dx}(y^2) = \frac{d}{dx}(5)$$

$$y^3 \cdot \frac{1}{x} + (\ln x)(3y^2 y') - 2yy' = 0$$

$$y'(3y^2 \ln x - 2y) = -\frac{y^3}{x}$$

$$y' = \frac{-y^3}{x(3y^2 \ln x - 2y)}$$

$$y' = \frac{-y^2}{x(3y \ln x - 2)}$$

Example 2

Purpose To use the properties of logarithms in differentiation:
The following properties may be used before you differentiate.

$$\ln (ab) = \ln a + \ln b$$

$$\ln \left(\frac{a}{b}\right) = \ln a - \ln b$$

$$\ln a^n = n \ln a$$

Problem A If $y = \ln \sqrt{15 + x^2}$, find $\dfrac{dy}{dx}$.

Solution $y = \ln \sqrt{15 + x^2} = \ln (15 + x^2)^{1/2} = \dfrac{1}{2} \ln (15 + x^2)$

$$\frac{dy}{dx} = \frac{1}{2} \cdot \frac{1}{15 + x^2} \cdot 2x = \frac{x}{15 + x^2}$$

Problem B If $y = \ln \left(\dfrac{x^2 + 6}{7 - x^3}\right)$, find $\dfrac{dy}{dx}$.

Solution $y = \ln \left(\dfrac{x^2 + 6}{7 - x^3} \right) = \ln (x^2 + 6) - \ln (7 - x^3)$

$$\frac{dy}{dx} = \frac{2x}{x^2 + 6} + \frac{3x^2}{7 - x^3}$$

Example 3

Purpose   To utilize natural logarithms in the solution of equations:

Problem   If P dollars is invested at a yearly rate of interest r, for t years, and interest is compounded k times per year, the amount A present at the end of t years is given by $A = P\left(1 + \dfrac{r}{k}\right)^{kt}$. Find how many years it will take a deposit of \$1000 to grow to \$1500 if it is invested at a yearly rate of 12%, compounded quarterly. ($\ln 1.03 = .02956$.)

Solution
$$A = P\left(1 + \frac{r}{k}\right)^{kt}$$

$$1500 = 1000\left(1 + \frac{.12}{4}\right)^{4t}$$

$$\frac{1500}{1000} = \frac{3}{2} = (1 + .03)^{4t}$$

Take ln of both sides of the equation:

$$\ln \frac{3}{2} = \ln (1.03)^{4t}.$$

From the properties of logarithms:

$$\ln 3 - \ln 2 = 4t \cdot \ln (1.03)$$
$$1.09861 - .69315 = 4t \cdot (.02956)$$
$$.40546 = .11824t$$
$$t = \frac{.40546}{.11824} \approx 3.43 \text{ years.}$$

The Graph of the Natural Logarithmic Function

Let us now examine the graph of $y = \ln x$. It has already been noted that x is always positive ($x > 0$) and when $x = 1$, $y = \ln 1 = 0$. Also, recall that when $x = e$, $y = \ln e = 1$.

Useful information is given by the derivatives of $y = \ln x$.

$y' = \dfrac{1}{x}$. Since $x > 0$, $y' > 0$ and the function is always increasing.

$y'' = -\dfrac{1}{x^2}$. Since $x > 0$, $y'' < 0$ and the curve is always concave downward.

From the table on page 427, the following *y* values are obtained (2-decimal-place accuracy).

x	.1	.5	1	2	2.72	4	8
y = ln *x*	−2.3	−.69	0	.69	1	1.39	2.08

By incorporating these results with the information given by the first and second derivative, the graph of the natural logarithmic function is obtained.

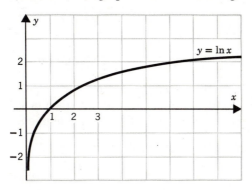

Since $\dfrac{dy}{dx} = \dfrac{1}{x}$ exists for all *x* > 0, that is, the function is differentiable for all *x* in its domain, it follows that the function is continuous for all *x* > 0. Since $\dfrac{dy}{dx} = \dfrac{1}{x} > 0$, the function is increasing. Also, the function has a special correspondence. This is called a one-to-one correspondence which is: to each *x* there is associated one and only one *y* and also to each *y* there is associated one and only one *x* value. Furthermore, if a function has the special property that it is a one-to-one correspondence, then the inverse relation, formed by interchanging the order of the members in each ordered pair, is also a function and is called the *inverse function*. Therefore, *y* = ln *x* represents a one-to-one function and has an inverse function which will be considered in the next section.

Exercise 6.2

In Problems 1 to 22, find the derivative of the function.

1. $y = \ln (3x^2 + 2x - 5)$

2. $y = \ln (7 - 3x + 2x^4)$

3. $y = \ln x^2$

4. $y = (\ln x)^2$

5. $y = \ln (x^2 - 7)^3$

6. $y = \ln (9 - 5x)^2$

7. $y = 5 \ln (x^2) + 3$

8. $y = 5 \ln (x^2 + 3)$

9. $y = \ln \dfrac{1}{x^2}$

10. $y = \ln \sqrt{x^2 + 14}$

11. $y = \ln \sqrt{4 - x^2}$

12. $y = x \ln \sqrt{x}$

13. $y = \sqrt{\ln (3x^2)}$

14. $y = \ln [(x^3 + 2)(x^2 + 3)]$

15. $y = \dfrac{\ln x^2}{x^2}$

16. $y = \dfrac{(\ln x)^2}{x^2}$

17. $y = \ln \left[\dfrac{x^4}{(x - 2)^2} \right]$

18. $y = \ln (\ln x)$

19. $y = \dfrac{\ln (x^2 - 1)}{\ln (x + 1)}$

20. $y \ln x - y^2 = 7$

21. $y^2 \ln (2x) + y \ln \sqrt{x} = 4$

22. $\ln (x + y) = \ln (xy) + 1$

23. Use logarithmic differentiation to prove if $y = x^n$, then $y' = nx^{n-1}$. (To use logarithmic differentiation: if $y = x^n$, then $\ln y = \ln x^n$; use the properties of logarithms and then differentiate implicitly with respect to x.)

In Problems 24 to 34, use the logarithmic properties and then find the derivative of the function.

24. $y = \ln (x^2 - 7)^3$

25. $y = \ln \sqrt{x^2 + 144}$

26. $y = \ln \sqrt{4 - x^2}$

27. $y = \ln \sqrt[3]{17 + x + x^3}$

28. $y = \ln [(x^3 + 2)(x^2 + 3)]$

29. $y = \ln [(2x + 3)^5(7 + 5x^3)]$

30. $y = \ln (x\sqrt{x^2 + 1})$

31. $y = \ln \sqrt{(2x + 1)(x^2 - 4)}$

32. $y = \ln \left[\dfrac{1}{(2x^2 + 13)^4} \right]$

33. $y = \ln \left[\dfrac{x^4}{(x - 2)^2} \right]$

34. $y = \ln \sqrt{\dfrac{x^2 - 7}{x^2 + 7}}$

35. A firm finds its sales increase logarithmically by $S = 1250 \ln x$ where x is the advertising expenditure. Find the rate of change of sales with respect to x.

36. A firm manufactures floppy discs for the personal computer. Their monthly revenue function is found to be $R(x) = 100(15x - 3x \ln x)$ where x is the selling price (in dollars) for a box of 10 floppy discs. At what price should the firm sell a box of floppy discs to maximize its monthly revenue? What is the maximum revenue?

37. A manufacturer determines that the revenue from the sales of x sets of dining room furniture increases logarithmically according to the function $R(x) = \ln (x^2 + 40x) + 200x$. What is the marginal revenue?

38. In chemistry and biology a very important concept is called entropy. Entropy is a measure of randomness and disorder in a system. Entropy, S, is related to probability, P, by the following equation where k is a constant $(k > 0)$: $S = k \ln P$.

(a) Find $\dfrac{dS}{dP}$

(b) Find $\dfrac{d^2S}{dP^2}$

39. (a) If $5000 is invested at a yearly rate of 6%, compounded semiannually, how many years will it take to double the initial investment? (See Section 6.2, Example 3, ln 2 = .69315; ln 1.03 = .02956.)
    (b) If N dollars is invested with the same conditions as part (a), how many years will it take to double the investment?

40. What will be the amount of an investment after 5 years, if $500 is invested at 8% compounded quarterly? What is the compound interest? (See Section 6.2, Example 3; ln 1.02 = .01980.)

41. How long will it take an investment of $3000 to grow to $4500, if it is invested at a yearly rate of 10% compounded annually?

42. How many years would it take $2000 invested at the yearly rate of 6%, compounded annually to mature to the following amounts? (ln 1.06 = .0583)
    (a) $2500
    (b) $3000
    (c) $4000

43. Find the equation of the tangent line to the curve $y = \ln 2x$ at $x = \frac{1}{2}$.

44. If $y = \ln (\frac{1}{2}x)$, find the equations of the tangent and normal lines to the curve at $x = 2$. (A normal line is perpendicular to a tangent line.)

45. Find the equation of the tangent line to the curve $y = x + 2 \ln x$ at the point $(1,1)$.

In Problems 46 to 50, find the first and second derivatives to determine when the function is increasing or decreasing, concave upward or concave downward, and graph the function.

46. $y = \ln (2x)$, $(x > 0)$ 49. $y = \ln |x|$, $(x \neq 0)$

47. $y = \ln (x^2 + 1)$ 50. $y = x \ln x$, $(x > 0)$

48. $y = \ln \left(\dfrac{1}{2x}\right)$, $(x > 0)$

51. (Optional) The natural logarithmic function is also defined as follows:
    $$\ln x = \int_1^x \frac{1}{t} \, dt, \ x > 0.$$
    (a) What is ln 1?
    (b) Using the theorem $\dfrac{d}{dx} \displaystyle\int_k^x f(t) \, dt = f(x)$, show that $\dfrac{d}{dx} (\ln x) = \dfrac{1}{x}$.

6.3   THE EXPONENTIAL FUNCTION

As shown in the last section, the natural logarithmic function has an inverse. This inverse, called the exponential function, is now defined.

DEFINITION

The Exponential Function. The *exponential function* is the inverse of the natural logarithmic function and is defined by $y = e^x$ if and only if $x = \ln y$.

The graph of $y = e^x$ is the same as the graph of $x = \ln y$ and is shown in Figure 6.1. The analysis of this graph is left to the student. See Exercise 6.3, Problem 21.

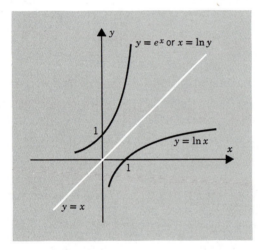

Figure 6.1

The graph for $y = e^x$ appears to have no sharp corners and to be the graph of a continuous function for all real numbers x. Assuming that this is the case, $y = e^x \leftrightarrow x = \ln y$ will have a derivative for all values of x. We now find that derivative.

THEOREM 6.3a

Assuming that the exponential function defined by $y = e^x$ has a derivative, then

$$\frac{d}{dx}(e^x) = e^x.$$

GIVEN. The exponential function, defined by $y = e^x$, has a derivative.

TO PROVE. $\dfrac{d}{dx}(e^x) = e^x$.

Proof

1. $y = e^x \leftrightarrow x = \ln y$
2. By implicit differentiation, differentiate $x = \ln y$ with respect to x.

   $1 = \dfrac{1}{y} \cdot y'$ or $y' = y$

3. But $y = e^x$ so $y' = e^x$
   Therefore,

$$\frac{d}{dx}(e^x) = e^x.$$

This is an interesting character—a function that is its own derivative. The more useful theorem for differentiating the exponential function is obtained by the chain rule.

THEOREM 6.3b

If $y = e^u$ and u represents a differentiable function of x, which causes y to be a function of x, then

$$\frac{d}{dx}(e^u) = e^u \cdot \frac{du}{dx}.$$

The proof is by the chain rule.

$$\frac{d}{dx}(e^u) = \frac{d}{du}(e^u) \cdot \frac{du}{dx} = e^u \cdot \frac{du}{dx}$$

Example 1

Purpose To utilize the previous theorem to find the derivatives of some exponential functions:

Problem A If $y = e^{x^2 - x}$, find $\dfrac{dy}{dx}$.

Solution If $\begin{cases} y = e^u \\ u = x^2 - x \end{cases}$, then $\begin{aligned} e^u &= e^{x^2 - x} \\ \frac{du}{dx} &= 2x - 1 \end{aligned}$

and $\dfrac{dy}{dx} = e^u \cdot \dfrac{du}{dx} = e^{x^2 - x}(2x - 1)$

Problem B If $y = x^2 e^{3x}$, find $\dfrac{dy}{dx}$.

Solution Use the product rule

$$\frac{dy}{dx} = x^2 \cdot \frac{d}{dx}(e^{3x}) + e^{3x} \cdot \frac{d}{dx}(x^2)$$

$$= x^2 \cdot 3e^{3x} + e^{3x}(2x)$$

$$\frac{dy}{dx} = 3x^2 e^{3x} + 2xe^{3x} = xe^{3x}(3x + 2)$$

Problem C If $y = \sqrt{e^x + 5}$, find $\dfrac{dy}{dx}$.

Solution $\dfrac{dy}{dx} = \dfrac{1}{2}(e^x + 5)^{-1/2}(e^x) = \dfrac{e^x}{2\sqrt{e^x + 5}}$

Problem D If $ye^x + y^3 + x^2 = 7$, find y'.

Solution Use implicit differentiation

$$y \cdot \frac{d}{dx}(e^x) + e^x \cdot \frac{d}{dx}(y) + \frac{d}{dx}(y^3) + \frac{d}{dx}(x^2) = \frac{d}{dx}(7) \quad (7)$$

$$ye^x + e^x y' + 3y^2 y' + 2x = 0$$

Solve for y'

$$y'(e^x + 3y^2) = -2x - ye^x$$

$$y' = \frac{-2x - ye^x}{e^x + 3y^2}$$

Example 2

Purpose To use the derivative of the exponential function in establishing a relationship:

Problem The number N of bacteria present at any time t is given by $N = 1000e^{.2t}$. Establish the relationship between N and the rate of change of N.

Solution The rate of change of N is given by $\dfrac{dN}{dt}$.

If $N = 1000e^{.2t}$, then $\dfrac{dN}{dt} = 1000(.2)e^{.2t}$.

By substitution: $\dfrac{dN}{dt} = .2(1000e^{.2t}) = .2N$.

Example 3

Purpose To show a use of the exponential function in compound interest where interest is compounded continuously:

Problem A P dollars is invested at $r\%$ (yearly) with interest compounded continuously. Establish a formula for the amount of money A present after t years.

Solution On page 294, it was shown that $\lim_{t \to 0} (1 + t)^{1/t} = e$; or using the variable s, $\lim_{s \to 0} (1 + s)^{1/s} = e$. Referring to Section 6.2, Example 3, $A = P\left(1 + \dfrac{r}{k}\right)^{kt}$, where k = number of times per year the interest is compounded. To compound interest continuously, let $k \to \infty$.

$$A = \lim_{k \to \infty} P\left(1 + \frac{r}{k}\right)^{kt}. \quad (1^\infty \text{ is a form whose answer we do not know.})$$

Let $s = \dfrac{r}{k}$ or $k = \dfrac{r}{s}$. As $k \to \infty$; $s \to 0$

$$A = \lim_{s \to 0} P(1 + s)^{(r/s)t}$$

$$= \lim_{s \to 0} P[(1 + s)^{1/s}]^{rt}$$

$$= P[\lim_{s \to 0} (1 + s)^{1/s}]^{rt}$$

$$A = Pe^{rt}$$

Problem B If \$1000 is invested at 5% compounded continuously, what is the amount present after 10 years?

Solution Use the formula $A = Pe^{rt}$

$$A = 1000e^{(.05)(10)}$$

$$= 1000e^{.5}$$

$$A \approx 1000(1.65) = 1650$$

Approximately \$1650 is present after 10 years.

Example 4

Purpose To use an exponential function in a problem related to biology:

Problem The number of bacteria in a controlled culture at t hours can be found by the equation $N = 150e^{t/3}$.

(A) What is the initial number of bacteria in the culture?
(B) What is the bacteria count at 1 hour?
(C) What is the instantaneous rate of growth at 1 hour?

Solution (A) $N = 150e^{t/3}$, when $t = 0$, $N = 150e^0 = 150(1)$
$$N = 150$$

(B) $N = 150e^{t/3}$, when $t = 1$, $N = 150e^{1/3} \approx 150(1.3956)$
$$N \approx 209$$

(C) $\dfrac{dN}{dt} = 150e^{t/3}(\tfrac{1}{3})$

$\dfrac{dN}{dt} = 50e^{t/3}$, when $t = 1$, $\dfrac{dN}{dt} = 50e^{1/3} \approx 50(1.3956)$

$$\frac{dN}{dt} \approx 70$$

A question that may have occurred to you is: Why not use base 10 for the logarithm and exponential functions? This is the procedure used in high school where logarithms to base 10 (common logarithms) were found. It is certainly admitted that 10 appears to be a more desirable number to work with than e. However, the formulas for finding the derivatives and integrals, when 10 is used as a base, are more complicated. This is one of the reasons that e is preferred as the base.

Exercise 6.3

In Problems 1 to 20, find the first derivative of the function.

1. $y = e^{2x}$

2. $y = e^{2x} + 3$

3. $y = e^{2x+3}$

4. $y = e^{2-x^2}$

5. $y = e^x \ln x^2$

6. $y = e^{x^2} \ln x$

7. $y = \dfrac{x^2}{e^x}$

8. $y = \dfrac{e^x - e^{-x}}{e^x + e^{-x}}$

9. $y = (e^{3x-1})^3$

10. $y = e^{3x} + 3x^e$

11. $y = e^{\sqrt{x+4}}$

12. $y = \sqrt{e^{x+4}}$

13. $y = \ln (e^x)$

14. $y = e^{\ln x}$

15. $y = \ln (xe^x)$

16. $y = e^{x \ln x}$

17. $y = e^{x\sqrt{x+1}}$

18. $y = \ln \left(\dfrac{e^{3x}}{x^3} \right)$

19. $\ln (xy) + xe^y = e^x$

20. $e^{xy} = e^x \, e^y$

21. Analyze the graph $y = e^x$ (see Section 6.3, Figure 6.1). Determine intercept(s), when the function is increasing or decreasing, and where the graph is concave upward or downward. Also investigate that $\lim_{x \to \infty} e^x = \infty$ and $\lim_{x \to -\infty} e^x = 0$.

In Problems 22 to 25, find the first and second derivatives, determine when the function is increasing or decreasing, when the curve is concave upward or downward, and graph the function.

22. $y = e^{-x}$

23. $y = e^{x^2}$

24. $y = x + e^x$

25. $y = xe^x$

26. The frequency function representing a certain normal distribution is

$$f(x) = \frac{1}{\sqrt{2\pi}} e^{-x^2}. \quad (\sqrt{2\pi} \approx 2.51)$$

(a) Where is $f \uparrow$ and $\downarrow$?

(b) Find the relative maximum for $f(x)$.

(c) Where is $f \cup$ and $\cap$?

(d) Graph f.

Observe that many students would like their grades based on this type of normal curve.

27. The concentration of a drug in the bloodstream is given by $K(t) = 5e^{-t}$ ($t \geq 0$), where t is the number of hours that have elapsed since an injection of 5 units of the drug.

    (a) Using the table on page 427, find the concentration $K(t)$ when $t = 0$, 1, 2, 3, 4, and 5 hours.

    (b) Plot the curve showing concentration $K(t)$ on the vertical axis and time t on the horizontal axis.

    If another injection of 5 units must be given whenever the concentration of the drug in the bloodstream decreases to .25 units, what length of time between injections is permissible?

    (c) Estimate the answer to the above question by inspecting the graph in (b).

    (d) Answer by solving the equation.

28. A discount appliance store is located in a community of 300,000 potential customers. For its gala grand opening, daily advertisements for a portable television set were placed in the local newspaper. Their primary aim, through this advertising campaign, was to maximize profit on the sales of the television sets and also to acquaint the community with their new store. They developed a response function to determine the proportion of potential buyers that would respond to their advertisement and purchase a set after 1 day, 2 days, and so forth; the function being $f(t) = 1 - e^{-.2t}$ (time, t, in days).

    (a) What percentage responded and bought after 1 day? 5 days? 10 days?

    (b) Graph this function and interpret the graph.

    The total advertising costs were found to be fixed costs of $5000 (artwork for layout, etc.) plus $1000 per day for newspaper space. Also, the markup per television set was $15.

    (c) Show that the profit exclusive of advertising costs is given by $4,500,000(1 - e^{-.2t})$.

    (d) Show that the profit function including advertising costs is

    $$P(t) = 4,500,000(1 - e^{-.2t}) - (5000 + 1000t).$$

    (e) Find the number of days the advertisement should run to maximize the profit.

29. A psychologist determines the following function that relates the extent of retardation of a child to the number of unpaired chromosomes he had inherited; $R(c) = c^2 e^{-c}$, where $R(c)$ is the amount of retardation and c is the number of unpaired chromosomes ($c \leq 2$). Draw the graph of this function and determine whether any relative extrema exist.

30. The number of cancer cells present at time t is given by $N = 100e^{.1t}$. Find the average rate of change in N over the interval $t = 0$ to $t = 10$.

31. The growth of a bacteria culture is observed to be $N = 100e^{.5t}$. What is the average rate of growth in the first hour observed (from $t = 0$ to $t = 1$)? In the second hour? What is the instantaneous rate of growth when $t = \frac{1}{2}$ hour? $t = 1\frac{1}{2}$ hour?

32. A stock market analyst gives the following function that relates the expected price of one share of common stock to the time, in months, at which it is sold. $E(t) = 75 + .4t^2 + 3e^{-t}$. Find $E'(t)$ and $E''(t)$.

33. Three banks, Pittsburgh National, Union National, and First National, each has a different savings program. Pittsburgh National's program consists of a yearly interest rate of 5%, compounded quarterly. Union National advertises a yearly rate of 4%, compounded continuously. First National boasts that its yearly rate is 6% compounded annually. If $1000 is to be placed in a savings account, which bank would yield the greatest return at the end of a 10-year period?

Hint. Use $A = P\left(1 + \dfrac{r}{k}\right)^{kt}$ for Pittsburgh National and First National and

$A = Pe^{rt}$ for the continual compounding of Union National. P is the amount originally invested, r is the yearly rate of interest, k is the number of times compounded yearly, and t is the number of years invested. A is the amount after t years.

34. An economist gives the following function for the Gross National Product, GNP, of the United States: $G(t) = 2.5 + .002t + .24e^{-t}$ where the GNP is in units of trillions of dollars and time t is in years. What is the percentage rate of growth or decline of the economy at $t = 3$?

(Optional) In Problems 35 to 38, find a value for x (call it $x = c$) guaranteed by the mean value theorem for derivatives given the following.

35. $y = \ln x$; $1 \leqslant x \leqslant e$ **37.** $y = e^x$; $0 \leqslant x \leqslant 1$

36. $y = \ln x^2$; $1 \leqslant x \leqslant e$ **38.** $y = e^{(1/2)x}$; $0 \leqslant x \leqslant 2$

39. (a) The derivative of the function defined by $y = a^x$, $a > 0$, can be determined by use of implicit differentiation and is found to be

$$y' = \frac{d}{dx}(a^x) = a^x \ln a.$$

Proof

$$y = a^x$$
$$\ln y = \ln a^x$$
$$\ln y = x \cdot \ln a, \qquad \text{differentiate}$$
$$\frac{1}{y} \cdot y' = 1 \cdot \ln a$$

$$y' = y \cdot \ln a$$
$$y' = a^x \ln a$$

Therefore, if $y = a^x$, then $y' = a^x \ln a$.

Example. $y = 3^x$
$$y' = 3^x \ln 3$$

(b) If $y = a^u$ and u is a differentiable function of x, which causes y to be a function of x, then

$$\frac{d}{dx}(a^u) = a^u \cdot \ln a \cdot \frac{du}{dx}.$$

Prove this by using the chain rule.

Example. $y = 3^{x^2}$
$$y' = 3^{x^2}(\ln 3)(2x) = 2x \cdot 3^{x^2} \ln 3$$

In Problems 40 to 47, find the first derivative. Refer to Problem 39.

40. $y = 5^x$

41. $y = a^{x^2}$

42. $y = 3^{\sqrt{x}}$

43. $y = (3e)^x$

44. $y = 10^{-x^2+x+7}$

45. $y = x \cdot a^x$

46. $y = 3^{\ln x}$

47. $y = \dfrac{3^{2x}}{x}$

In Problems 48 to 51, find the first and second derivatives, determine when the function is increasing or decreasing, when the curve is concave upward or concave downward, and graph the function ($\ln 2 = .69$, $\ln \frac{1}{2} = -.69$).

48. $y = 2^x$

49. $y = (\frac{1}{2})^x$

50. $y = 2^{x^2}$

51. $y = x + 2^x$

6.4 INTEGRALS THAT YIELD LOGARITHMIC AND EXPONENTIAL FUNCTIONS

Two basic formulas are obtainable by inspecting the formulas in the last sections, and by recalling that $\int f(x)\, dx = F(x) + c$ if $F'(x) = f(x)$.

THEOREM 6.4a

$$\int e^u\, du = e^u + c.$$

Proof

Since $\dfrac{d}{du}(e^u) = e^u$,

it follows that $\displaystyle\int e^u \, du = e^u + c$.

THEOREM 6.4b

$$\int \frac{du}{u} = \ln |u| + c$$

Proof

Case I.   If $u > 0$ then $|u| = u$.

$$\frac{d}{du} \ln |u| = \frac{d}{du} \ln u = \frac{1}{u}$$

Case II.   If $u < 0$ then $|u| = -u$ (where $-u$ is a positive number).

$$\frac{d}{du} \ln |u| = \frac{d}{du} \ln (-u) = \frac{1}{-u} \cdot \frac{d}{du}(-u) = \frac{1}{-u} \cdot (-1) = \frac{1}{u}$$

Therefore, in both cases $\dfrac{d}{du} \ln |u| = \dfrac{1}{u}$ and $\displaystyle\int \dfrac{du}{u} = \ln |u| + c$, provided

that u is in an interval which does not contain $u = 0$. This assumption will be
made in the remainder of the chapter.

Example 1

Purpose   To extend our techniques of integration to more complicated functions by utilizing these two new theorems:

Problem A   Find $\displaystyle\int \frac{4x + 7}{2x^2 + 7x + 3} \, dx$

Solution   Let $u = 2x^2 + 7x + 3$
$du = (4x + 7) \, dx$

$$\int \frac{4x + 7}{2x^2 + 7x + 3} \, dx = \int \frac{1}{u} \, du$$
$$= \ln |u| + c$$
$$= \ln |2x^2 + 7x + 3| + c$$

Problem B   Find $\displaystyle\int \frac{x^2 \, dx}{x^3 - 1}$

Solution   Let $u = x^3 - 1$

$du = 3x^2\, dx$

$$\int \frac{x^2\, dx}{x^3 - 1} = \frac{1}{3} \int \frac{3x^2\, dx}{x^3 - 1} = \frac{1}{3} \int \frac{du}{u}$$

$$= \frac{1}{3} \ln |u| + c$$

$$= \frac{1}{3} \ln |x^3 - 1| + c$$

Problem C   Find $\displaystyle\int_0^4 e^{(1/2)x}\, dx$

Solution   Let $u = \tfrac{1}{2}x$

$du = \tfrac{1}{2}\, dx$

Also, when $x = 0$, $u = 0$ and when $x = 4$, $u = 2$.

$$2 \int_0^4 e^{(1/2)x} \frac{1}{2}\, dx = 2 \int_0^2 e^u\, du$$

$$= 2e^u \Big]_0^2$$

$$= 2(e^2 - e^0) = 2(e^2 - 1).$$

Example 2

Purpose   To use integration in applied problems:

Problem A   Find the area of the region bounded by $y = e^{(1/2)x-1}$ and $y = \dfrac{-1}{x+1}$ from $x = 0$

to $x = 2$.

Solution

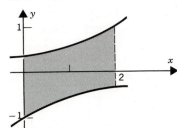

$$A = \int_0^2 \left(e^{(1/2)x-1} + \frac{-1}{x+1} \right) dx$$

$$= 2e^{(1/2)x-1} - \ln |x + 1| \Big]_0^2$$

$$= (2e^0 - \ln 3) - (2e^{-1} - \ln 1)$$

$$= 2 - \ln 3 - \frac{2}{e}$$

Problem B   The following function is important in psychology and is known as Weber's law: $\dfrac{\Delta l}{l} = k$.

l = amount of stimulation taken in by a subject.
Δl = the smallest change in stimulation necessary for the subject to notice a difference in stimulus.

If Δl is very small it can be approximated by dl. Substituting: $\dfrac{dl}{l} = k$. Now let us assume k is a function of the type of stimulus under consideration. This gives: $\dfrac{dl}{l} = ds$. Integrate this function as shown below.

Solution

$$\int_{l_1}^{l_2} \frac{dl}{l} = \int_{s_1}^{s_2} ds$$

$$\int_{l_1}^{l_2} \frac{dl}{l} = \int_{s_1}^{s_2} ds$$

$$\ln l \Big]_{l_1}^{l_2} = s \Big]_{s_1}^{s_2}$$

$$\ln l_2 - \ln l_1 = s_2 - s_1$$

$$\text{or } \ln \frac{l_2}{l_1} = s_2 - s_1$$

6.5   INTEGRATION BY PARTS

A type of integration that is often effective in reducing a given complicated integral to a simpler integral is integration by parts. The use of this technique depends on the wise use of the formula

$$\int u \, dv = uv - \int v \, du.$$

THEOREM 6.5

Integration by Parts.   If u and v are differentiable functions then

$$\int u \, dv = uv - \int v \, du.$$

GIVEN. u and v are differentiable functions.

TO PROVE. $\displaystyle\int u\,dv = uv - \int v\,du.$

Proof

Since u and v are differentiable functions, their product uv is also differentiable, and $d(uv) = u\,dv + v\,du$ (see Section 4.9, Example 4). Integrating both sides of this equation,

$$\int d(uv) = \int u\,dv + \int v\,du.$$

By our previous work, $\int d(uv)$ is a function whose differential is $d(uv)$. Therefore, $\int d(uv) = uv + c$. Indefinite integration and finding the differential are inverse processes of each other—the one process "undoes" the other. Returning to the above equation, we write

$$\int d(uv) = uv + c = \int u\,dv + \int v\,du.$$

Solve for $\int u\,dv$, and obtain

$$\int u\,dv = uv + c - \int v\,du.$$

Include the constant c with the arbitrary constant that will arise from integrating $\int v\,du$. Therefore, we have

$$\int u\,dv = uv - \int v\,du.$$

Example 1

Purpose To use the technique of integration by parts:

Problem A Find $\displaystyle\int x^2 \ln x\,dx$.

Solution Let $u = \ln x \qquad dv = x^2\,dx$

$$du = \frac{1}{x}\,dx \qquad v = \int x^2\,dx = \frac{x^3}{3}$$

$$\int u\,dv = uv - \int v\,du$$

$$\int x^2 \ln x\,dx = (\ln x)\left(\frac{1}{3}x^3\right) - \int \frac{x^3}{3}\cdot\frac{1}{x}\,dx$$

$$= \frac{1}{3}x^3 \ln x - \frac{1}{3}\int x^2\,dx$$

$$= \frac{1}{3}x^3 \ln x - \frac{1}{9}x^3 + c.$$

You should satisfy yourself that, if we had used $v = \dfrac{x^3}{3} + c$ instead of $v = \dfrac{x^3}{3}$, we would have obtained the same result.

Problem B Find $\displaystyle\int_0^1 xe^x \, dx$.

Solution First integrate by parts to find the indefinite integral, $\int xe^x \, dx$.

$$\text{Let } u = x \qquad dv = e^x \, dx$$

$$du = dx \qquad v = \int e^x \, dx = e^x$$

$$\int xe^x \, dx = xe^x - \int e^x \, dx$$

$$= xe^x - e^x + c$$

Return to the definite integral.

$$\int_0^1 xe^x \, dx = \left[xe^x - e^x \right]_0^1$$

$$= (1 \cdot e^1 - e^1) - (0 \cdot e^0 - e^0)$$

$$= e - e + 1 = 1$$

Exercise 6.4–6.5

In Problems 1 to 28, integrate.

1. $\displaystyle\int e^{2x} \, dx$

2. $\displaystyle\int 2xe^{x^2-1} \, dx$

3. $\displaystyle\int_0^2 e^{(1/2)x} \, dx$

4. $\displaystyle\int \left(\frac{e^x + e^{-x}}{2} \right) dx$

5. $\displaystyle\int \frac{-10}{x} \, dx$

6. $\displaystyle\int \frac{2}{x + 2} \, dx$

7. $\displaystyle\int_0^3 \frac{4}{2x + 1} \, dx$

8. $\displaystyle\int \frac{2x}{x^2 + 1} \, dx$

9. $\displaystyle\int_0^1 \frac{x}{x^2 + 1} \, dx$

10. $\displaystyle\int \frac{3x^2 - 3}{x^3 - 3x + 7} \, dx$

11. $\displaystyle\int \frac{x^2 - 1}{x^3 - 3x + 7} \, dx$

12. $\displaystyle\int \frac{dx}{1 - 3x}$

13. $\displaystyle\int \frac{e^x}{(1 + e^x)^3} \, dx$

14. $\displaystyle\int \frac{e^{\sqrt{x}}}{\sqrt{x}} \, dx$

15. $\displaystyle\int_1^2 \frac{\ln x}{x}\, dx$

16. $\displaystyle\int_0^3 \frac{e^x}{1 + e^x}\, dx$

17. $\displaystyle\int \frac{x^2 + 4x - 3}{x}\, dx$

18. $\displaystyle\int \frac{dx}{x \ln x}$

19. $\displaystyle\int \frac{(1 + \ln x)^2}{x}\, dx$

20. $\displaystyle\int_0^1 \frac{1 + e^x}{e^x}\, dx$

21. $\displaystyle\int x \ln x\, dx$

Let $u = \ln x;\ dv = x\, dx$

22. $\displaystyle\int xe^{-x}\, dx$

Let $u = x;\ dv = e^{-x}\, dx$

23. $\displaystyle\int \ln x\, dx$

24. $\displaystyle\int_0^2 x\, e^x\, dx$

25. $\displaystyle\int_1^2 x \ln x^2\, dx$

26. $\displaystyle\int (x + 1)^2 \ln x\, dx$

27. $\displaystyle\int_{-1}^1 x^3\, e^x\, dx$

28. $\displaystyle\int_1^e \sqrt{x}\, \ln x\, dx$

29. Find the area of the region bounded by $y = \dfrac{1}{x + 1}$, the x axis and the interval $1 \leqslant x \leqslant 3$.

30. Find the area of the region bounded by the curve $xy = 1$ and the line $2x + 3y - 7 = 0$.

31. Find the area of the region bounded by the curve $y = e^{2x}$, $x = 0$, $x = 2$, and the x axis.

32. Find the area of the region bounded by $y = e^x$, $y = e^{2x} - 2$, and the y axis. [*Hint.* First show that the point of intersection is $(\ln 2, 2)$.]

33. The marginal demand for a product is given by $-100\, e^{(1/3)x}$ where x is the selling price per unit. Find the decrease in demand if the selling price increases from \$3 to \$6 per unit.

34. The expected total expenditures, in dollars, for a business over the next 10 years can be estimated by the function $E(t) = 50{,}000\, e^{.5t}$ where t is in years. Approximate the total expenditures for years 3 through 6.

35. A market demand function for x units of a commodity is given as $D(x) = \dfrac{1000}{x}$. Find the area bounded by the demand curve and the x axis (quantity axis) from $x = 1$ to $x = 10$.

36. The marginal cost is given by $C'(x) = 150 + e^{.5x}$ where x is the number of units of a product in hundreds. If x increases from 200 to 400, find the total increase in cost.

37. The marginal revenue function is given by $R'(x) = -xe^{-x} + e^{-x}$. Find the total revenue function, $R(x)$, using integration by parts. (Revenue when $x = 0$ is zero.)

38. A supply function is found to be $y = 3e^{(1/2)x}$ and the demand function is $y = 24e^{-x}$.
(a) Verify that the equilibrium point is (ln 4, 6).
(b) Determine the consumers' surplus.
(c) Determine the producers' surplus.
(See Section 5.6, Example 2.)

39. A supply function is $y = S(x) = 10e^{.2x}$ with the equilibrium point $(5, 10e)$ or $(5, 27.18)$. What is the producers' surplus?

40. In studying the effects of a disease, researchers measure the number of times that a control population of rats showed the expected symptoms. The number of rats affected on the xth day after being exposed to the disease carrier can be estimated by the equation $\dfrac{x + 2}{x^2 + 4x + 4}$. Find the area under the curve from $x = 0$ to $x = 2$.

41. A biologist is studying the growth of bacteria. If $N = 2x + \dfrac{40}{x}$ represents the number of new bacteria on the xth day, approximate the total number of bacteria present on the twentieth day.

42. When a chemical process takes place at a constant pressure, the change in enthalpy, ΔH, is very important. $\Delta H = \int Cp \, dt$ where Cp is defined as the heat capacity at constant pressure. If Cp is given as $3t + 4e^t$, find ΔH.

43. Graph $f(x) = \dfrac{x^2 + 1}{x}$. Note extrema, point(s) of inflection, concavity, and when the function is increasing or decreasing. Find the area bounded by $f(x) = \dfrac{x^2 + 1}{x}$, the x axis, and $x = \frac{1}{2}$ to $x = 2$.

44. (a) Evaluate $\displaystyle\int x\sqrt{4 - x} \, dx$ by two different methods.
1. Let $w = 4 - x$. Then $x = 4 - w$ and $dx = -dw$. Finish this integration by substitution.
2. Let $u = x$, $dv = \sqrt{4 - x} \, dx$. Find du and v and substitute into the formula for integration by parts. Finish the integration and compare your answer to that obtained in part 1.

(b) Evaluate $\displaystyle\int x^3\sqrt{x^2 + 7} \, dx$ using both integration by substitution and integration by parts.

From a book of tables, some integration formulas are:

$$\int x^n \ln x \, dx = x^{n+1}\left[\frac{\ln x}{n + 1} - \frac{1}{(n + 1)^2}\right] + c; \qquad n \neq -1$$

$$\int x^n e^x \, dx = x^n e^x - n \int x^{n-1} e^x \, dx$$

$$\int \frac{dx}{x \ln x} = \ln |\ln x| + c$$

$$\int xe^{ax} \, dx = \frac{e^{ax}(ax - 1)}{a^2} + c; \qquad a \neq 0.$$

In Problems 45 to 52, integrate using these formulas.

45. $\displaystyle\int 5x^5 \ln x \, dx$

46. $\displaystyle\int \frac{dv}{v \ln v}$

47. $\displaystyle\int x^2 e^x \, dx$

48. $\displaystyle\int \sqrt{x} \ln x \, dx$

49. $\displaystyle\int 3ue^{2u} \, du$

50. $\displaystyle\int xe^{(-1/2)x} \, dx$

51. $\displaystyle\int \frac{dx}{x \ln (2x)}$

52. $\displaystyle\int_1^2 x^4 \ln x \, dx$

53. (Optional) Integrate if possible:

(a) $\displaystyle\int_1^\infty \frac{1}{x} \, dx$

(b) $\displaystyle\int_0^1 \frac{1}{x} \, dx$

54. (Optional) Find the area of the region bounded by the curve $y = xe^x$ and the x axis as x goes from $-\infty$ to 0. (Given $\lim\limits_{t \to -\infty} te^t = 0$.)

55. (Optional) A continuous frequency function is given to be $f(x) = e^{-x}$, $x > 0$.
(a) Verify that the total area between the given function and the x axis is equal to one. (Property 2 of a continuous frequency function)
(b) Find a number, call it x_0, such that the probability that x will exceed x_0 is $\frac{1}{2}$.

56. (Optional) Given the continuous frequency function $f(x) = kxe^{-x}$, $x \geq 0$.
(a) Find the value for k.
(b) Find the probability that $x < 1$.
(c) Find the probability that $1 < x < 2$.
Note. From an additional theorem on limits called L'Hospital's rule,

$$\lim_{x \to \infty} \left(\frac{x}{e^x} \right) = 0.$$

6.6 DIFFERENTIAL EQUATIONS

An equation that contains a derivative or differentials is called a *differential equation*. Differential equations differ greatly in their complexity and ease of solution. The most important technique used in solving differential equations is that of integration. Unfortunately, integration alone does not always yield a solution to a differential equation.

First Degree, Variables Separable Differential Equations

A differential equation is said to be of the first degree if any derivative that is present is raised to only the first power. Some (but not all) first degree differential equations have the additional special property that $\frac{dy}{dx} = f(x,y)$ may be written as $\frac{dy}{dx} = g(x) \cdot h(y)$. Here, $g(x)$ is a function of x alone and $h(y)$ is a function of y alone. From our previous work on differentials (see Section 4.9), $\frac{dy}{dx} = g(x) \cdot h(y) \leftrightarrow \frac{dy}{h(y)} = g(x)\, dx$, $h(y) \neq 0$. This form of the equation has the y and dy terms in one member and the x and dx terms in the other member. If a first degree differential equation can be written in the form $\frac{dy}{h(y)} = g(x)\, dx$, it is said to be a *separable equation*. By a *solution* to $\frac{dy}{dx} = f(x,y)$, we mean a function, $y = F(x)$, which does not involve $\frac{dy}{dx}$, but which causes $\frac{dy}{dx} = f(x,y)$ to be a true sentence.

Unfortunately, not all first-order equations are separable equations. For example:

	Rewritten	Variables Separable
$\frac{dy}{dx} = x^2 y$	$\frac{dy}{y} = x^2\, dx$	Yes
$x + y\frac{dy}{dx} = 4$	$y\, dy = (4 - x)\, dx$	Yes
$(3x - y)\, dx + (2x + y)\, dy = 4$	—	No

The technique for solving a variables separable equation is to separate the variables and then integrate both sides of the equation until no differentials are present. Recall that with every indefinite integration an arbitrary constant appears. Often the application that generated the differential equation imposes certain conditions that enable us to assign values to these arbitrary constants.

Example 1

Purpose   To solve first degree variables separable differential equations:

Problem   Solve:

(A) $\sqrt{x}\, dy = y\, dx$ $y > 0,\ x > 0$

(B) $y^2 \dfrac{dy}{dx} = x$ and when $x = 0,\ y = 1$.

Solution   (A) $\sqrt{x}\, dy = y\, dx$ or $\dfrac{dy}{y} = \dfrac{dx}{\sqrt{x}}$

$$\int \frac{dy}{y} = \int x^{-1/2}\, dx$$

$$\ln y + c_1 = 2x^{1/2} + c_2$$

Combine $c_2 - c_1 = c_3$ (Two constants may be combined to form a new constant.)

$$\ln y = 2\sqrt{x} + c_3$$
$$y = e^{2\sqrt{x}+c_3} = e^{2\sqrt{x}} \cdot e^{c_3} \qquad \text{Let } e^{c_3} = c$$
$$y = ce^{2\sqrt{x}}$$

(B) $y^2 \dfrac{dy}{dx} = x$ or $y^2\, dy = x\, dx$

$$\int y^2\, dy = \int x\, dx$$

$$\tfrac{1}{3}y^3 = \tfrac{1}{2}x^2 + c_1$$
$$2y^3 = 3x^2 + 6c_1$$

Call $6c_1 = c$

$$2y^3 = 3x^2 + c$$

This solution, containing c, is called the general solution. And, when $x = 0$, $y = 1$; then

$$2 \cdot 1^3 = 3 \cdot 0^2 + c$$
$$c = 2$$

The solution with c evaluated is called a particular solution and for this problem is

$$2y^3 = 3x^2 + 2 \qquad \text{or} \qquad y = \left(\frac{3x^2 + 2}{2}\right)^{1/3}$$

Example 2

Purpose   To solve an applied differential equation:

Problem   The rate of change of salt in a mixing tank is $\dfrac{dQ}{dt} = -\dfrac{3}{100}Q$. The initial amount of salt in the tank is Q_0. Find a formula for the amount of salt in the tank at any time t.

Solution
$$\frac{dQ}{dt} = -\frac{3}{100}Q \quad \text{or} \quad \frac{dQ}{Q} = -\frac{3}{100}\,dt$$

$$\int \frac{dQ}{Q} = -\frac{3}{100}\int dt$$

$$\ln Q = -\frac{3}{100}t + c$$

To solve for c, when $t = 0$, $Q = Q_0$

$$\ln Q_0 = -\frac{3}{100}(0) + c$$

$$\ln Q_0 = c$$

Therefore,
$$\ln Q = -\frac{3}{100}t + \ln Q_0$$
$$Q = e^{(-3/100)t + \ln Q_0}$$

or
$$Q = Q_0 e^{(-3/100)t}$$

6.7 GROWTH AND DECAY CURVES

A Growth Rate Principle

There exist certain quantities in biology, economics, business, and psychology that increase at a rate proportional to the amount of the quantity present at any time t. If Q represents the amount present at time t and if $\frac{dQ}{dt} > 0$, rate of growth is positive; then as time increases, Q increases and *growth* is taking place. However, if $\frac{dQ}{dt} < 0$, rate of growth is negative; as time increases Q decreases and *decay* is taking place.

We will say that the "growth rate principle" is operating if a quantity is increasing at a rate proportional to the amount of the quantity present. The growth rate principle will be illustrated by the following two examples.

Example 1

Purpose To use rate of growth and a differential equation to solve problems from biology:

Problem A The bacteria count in a culture rose from 100 units to 300 units in a 2-hour interval. Find an expression for the number of units of bacteria present at any time t. Assume that the rate of growth in the number of bacteria is proportional to the amount present.

Solution Let N = the number of units of bacteria present at any time t.

$$\frac{dN}{dt} = kN \qquad \text{or} \qquad \frac{dN}{N} = k \, dt$$

$$\int \frac{dN}{N} = \int k \, dt$$

$$\ln N = kt + c_1$$

In exponential form

$$N = e^{kt+c_1} = e^{kt} \cdot e^{c_1}$$

But e^{c_1} is a constant, call it c. Then

$$N = ce^{kt}$$

Substituting when $t = 0$, $N = 100$,

$$100 = ce^{k \cdot 0} = ce^0 = c$$
$$N = 100e^{kt}$$

When $t = 2$, $N = 300$,

$$300 = 100e^{k \cdot 2}$$
$$3 = e^{2k}$$
$$\ln 3 = \ln e^{2k} = 2k \cdot \ln e = 2k \qquad \text{Why?}$$
$$k = \frac{\ln 3}{2} = \frac{1}{2} \ln 3$$

Therefore, $\qquad N = 100e^{(1/2)(\ln 3)t}$

This represents the number of bacteria N present at any time t.

Problem B A biologist develops the following function for the rate of decay of a radioactive substance as a function of time: $\dfrac{ds}{dt} = -\dfrac{1}{4}s$. Find an expression for the number of grams of radioactive substance, s, at any time, t. If 100 grams of substance were present initially, how many grams are present after 2 hours?

Solution $\qquad \dfrac{ds}{dt} = -\dfrac{1}{4}s \qquad \text{or} \qquad \dfrac{ds}{s} = -\dfrac{1}{4} dt$

$$\int \frac{ds}{s} = -\frac{1}{4} \int dt$$

$$\ln s = -\frac{1}{4}t + c_1$$

$$s = e^{(-1/4)t+c_1}$$

or $\qquad$ $s = ce^{(-1/4)t}$ where $c = e^{c_1}$

when $t = 0$, $s = 100$

$$s = 100e^{(-1/4)t}$$

when $t = 2$, $s = ?$

$$s = 100e^{-1/2} = \frac{100}{\sqrt{e}}$$

$$s \approx 61 \text{ grams}$$

Example 2

Purpose   To use the rate of growth principle in problems from business:

Problem A $1000 is invested at a rate of 7% compounded continuously. What will be the amount of the investment at the end of 10 years? Assume that the rate of growth of the money is proportional to the amount present at any time t.

Solution   Let A = the amount of money present at any time t.

$$\frac{dA}{dt} = .07A$$

$$\frac{dA}{A} = .07 \, dt$$

$$\int \frac{dA}{A} = \int .07 \, dt$$

$$\ln A = .07t + c_1$$
$$A = e^{.07t + c_1} = e^{c_1} \cdot e^{.07t} = ce^{.07t}$$

At $t = 0$, $A = 1000$,

$$1000 = ce^0 = c$$
$$A = 1000e^{.07t}$$

At $t = 10$ years,

$$A = 1000e^{.07(10)} = 1000e^{.7}$$

From the tables, $e^{.7} = 2.0138$

$$A = 1000(2.0138) = \$2013.80$$

Graphically

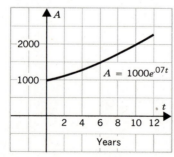

$$A = 1000e^{.07t}$$

Problem B   A company stops all advertising of a certain product. It is found that sales are declining at a rate proportional to the present number of sales. At the time advertising was stopped ($t = 0$), 500 products per day were sold. After 20 days, sales were down to 300 per day. Find a formula for the number of sales at any time.

Solution   The rate of growth (in this case, rate of decay) principle applies and $N = ce^{kt}$ where N is the number of products sold and t is time in days.

When $t = 0$, $N = 500$; when $t = 20$, $N = 300$

$$N = ce^{kt}$$
$$500 = ce^{k(0)}$$
$$500 = c$$
$$N = 500e^{kt}$$
$$300 = 500e^{k(20)}$$
$$\tfrac{3}{5} = e^{20k}$$
$$20k = \ln \tfrac{3}{5}$$
$$k = \tfrac{1}{20} \ln \tfrac{3}{5}$$

Therefore, $N = 500e^{(1/20)(\ln 3/5)t}$

or $N = 500e^{(.05 \ln .6)t}$

Graphically

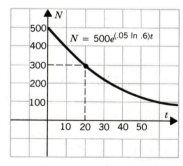

$N = 500e^{(.05 \ln .6)t}$ can also be expressed as $N = 500(.6)^{.05t}$. Do you see why? Recall, $\ln a^n = n \ln a$ and $.05t \ln(.6) = \ln(.6)^{.05t}$.   Also, $e^{\ln x} = x$ and $e^{\ln(.6)^{.05t}} = (.6)^{.05t}$.

6.8   A POPULATION MODEL (Optional)

Under the assumption that the rate of growth in the number, N, of a species present in a given population is proportional to the amount present, we have

$\dfrac{dN}{dt} = rN$. This is a separable equation and may be stated as

$$\frac{dN}{N} = r \, dt$$

$$\int \frac{dN}{N} = r \int dt$$

$$\ln N = rt + c_1$$

$$N = e^{rt+c_1} = e^{c_1} \cdot e^{rt}$$

$$N = ce^{rt}$$

At $t = 0$, assume the population had N_0 individuals present. This will enable us to evaluate c. For, at $t = 0$, $N_0 = c \cdot e^{r \cdot 0} = ce^0 = c \cdot 1 = c$. Therefore,

$$N = N_0 e^{rt}$$

is the population equation. The exact graph of this curve will depend on the values of N_0 and r. However, this is an exponential function, so the general shape of $N = N_0 e^{rt}$ will be related to the shape of $y = e^x$ as shown in the following drawing.

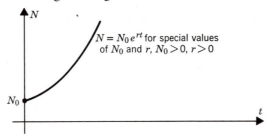

This particular result assumes that environment has little or no effect on the population. As such, it is more an indication of the potential of the population for a given species than an indication of what actually will happen. If bacteria divided every 20 minutes, according to this equation, within two days they would be a foot deep over the entire world. Also, within several thousand years any population of plants or animals that followed this law would weigh as much as the universe.

While populations obey the law $N = N_0 e^{rt}$ when environmental effects are negligible, as N increases the environment presents resistance that limits the growth of N and tends to keep the size of N under control. One example of a factor that would influence this environmental resistance is the size of the available food supply. This leads us to modify our original model to include the concept of carrying capacity, K, the maximum population that a system can support.

Example 1

Purpose   To modify our population model to include the concept of carrying capacity:

Problem   Assume that in a population of size N the rate of growth in N is given by

$\dfrac{dN}{dt} = rNF$, where F is a "correction factor." Establish that $F = \dfrac{K - N}{K}$, K represents carrying capacity, is a suitable choice for F by noting that as $N \to 0$, $F \to 1$, $\dfrac{dN}{dt} \to rN$ and as $N \to K$, $F \to 0$, $\dfrac{dN}{dt} \to 0$. Then, solve $\dfrac{dN}{dt} = rNF$.

Solution Investigate $F = \dfrac{K - N}{K}$.

When $N \to 0$, $F \to 1$. That is, when the population size is small, the old original rate of growth times F will be changed very little.

When $N \to K$, $F \to 0$. That is, when the population size is near to the carrying capacity K, the old rate of growth times F will be decreased. Therefore, F is the type of correction factor needed.

This leads us to the model:

$$\frac{dN}{dt} = rN\left(\frac{K - N}{K}\right) = \frac{r}{K} \cdot N(K - N)$$

$$\frac{dN}{N(K - N)} = \frac{r}{K}\, dt$$

It can be verified by addition that

$$\frac{1}{N(K - N)} = \frac{1}{KN} + \frac{1}{K(K - N)}$$

So, our model is

$$\left[\frac{1}{KN} + \frac{1}{K(K - N)}\right] dN = \frac{r}{K}\, dt$$

Integrate,

$$\int \frac{dN}{KN} + \int \frac{dN}{K(K - N)} = \int \frac{r}{K}\, dt$$

Since r and K are constants,

$$\frac{1}{K} \int \frac{dN}{N} + \frac{1}{K} \int \frac{dN}{K - N} = \frac{r}{K} \int dt$$

$$\int \frac{dN}{N} + \int \frac{dN}{K - N} = r \int dt$$

$$\ln N - \ln (K - N) = rt + c_1$$

$$\ln \left(\frac{N}{K - N}\right) = rt + c_1$$

$$\frac{N}{K - N} = e^{rt + c_1} = e^{c_1} \cdot e^{rt}$$

$$\frac{N}{K - N} = ce^{rt}$$

Solving for N, we obtain:

$$N(1 + ce^{rt}) = Kce^{rt}$$

Therefore

$$N = \frac{Kce^{rt}}{1 + ce^{rt}} = \frac{Kc}{e^{-rt} + c}$$

By assuming a value of K, r, and c, this graph can be shown to be a type called sigmoidal.

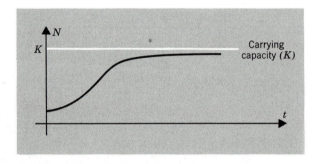

This is what we would expect, a rapid rise in population followed by a "leveling off" of the trend as $N \rightarrow K$.

The appropriateness of the sigmoidal growth pattern has been verified for several types of species in controlled laboratory experiments. The populations of yeast cells and water fleas grow at a rate that does not cause us to reject our sigmoidal model. However, when the sigmoidal model is applied to human communities, the model often fails to fit the data. This is because these communities are often too unstable to exhibit behavior in accordance to the sigmoidal law.

The population of the United States in the years 1790 to 1940 does fit more nearly the sigmoidal model than most populations. Population data for this period seem to suggest a carrying capacity for the United States of approximately 200 million. Since the 1970 United States population was more than 200 million, and until recently the United States population had been doubling every 35 years, these figures were frightening.

However, World War II has ruined the sigmoidal fit for the United States population. Therefore, a more refined model is needed both for population extrapolation and for estimating carrying capacity as a function of time. Students interested in this topic will find a bibliography and a more complex development in *Lectures on Mathematics in the Life Sciences,* The American Mathematical Society, 1968, Library of Congress Catalog Card No. 68-29411.

Exercise 6.6–6.8

In Problems 1 to 10, solve the differential equation.

1. $\dfrac{dy}{dx} = 2x + 2$ 6. $\dfrac{dy}{dx} = 3xy - y$

2. $\dfrac{dQ}{dt} = .5Q$ 7. $\dfrac{dy}{dx} = ye^x$

3. $y\,dx = x\,dy$ 8. $e^x\,dy + y\,dx = 0$

4. $dy = y^2\,dx$ 9. $\sqrt{x^3 + 3}\,dy = e^{-y}x^2\,dx$

5. $(x^2 - 1)\,y^2\,dx = x\,dy$ 10. $dy = xye^{x^2}\,dx$

11. In a legislative body, some members continue to serve as legislators whereas other members because of death, resignation, or political defeat are replaced by new individuals. Thus, the system can be viewed as starting at time 0 with a particular set of legislators and continuing until some other time when all of these particular legislators are out of office. The rate of change of the number of continuously serving members in the legislative body can be expressed as $\dfrac{dM}{dt} = -.09M$ where M is the number of continuously serving members and t is time in years. In 1965, there were 434 Democrats in the House of Representatives. Solve the preceding differential equation and calculate the number of these Democratic representatives that were predicted to be in office 10 years later. Compare this answer to the actual figure of 174.

12. In a certain tissue sample, cancer cells were found to increase at a rate of $20e^{(1/10)t}$. The time t is in days $(0 \leqslant t \leqslant 14)$, and 300 cancer cells were initially present. Find a function $N(t)$ that gives the number of cancer cells present at time t. Estimate the number of cells present after one week $(t = 7)$.

13. (a) Find the equation of the family of curves whose slope of the tangent line at a point on the curve is $3x^2 + 1$.
    (b) Find the equation of the particular curve with the above conditions that passes through the point $(1,4)$.
    (c) What is the slope of the tangent line to this particular curve at $(-1,0)$?

14. If the slope of the tangent line at a point on the curve is $\dfrac{2}{x^3}$, find:

    (a) The general equation for the curve.
    (b) The particular equation of the curve that passes through the point $(-1,0)$.
    (c) The particular equation of the curve that passes through $(1,-1)$.
    (d) Sketch the family of curves (as an aid in graphing, notice when the

function is increasing or decreasing and when it is concave upward or downward).

15. The slope of the tangent line at a point on a curve is $4e^{2x}$. Find:
(a) The general equation for the curve.
(b) The particular equation of the curve that passes through the point (0,5).

16. The following differential equation is used by sociologists to describe the rate at which news spreads through a population, $P'(t) = k[N - P(t)]$.
$P(t)$ = number of people who are aware of the news at any time t.
  k = constant.
  N = the total number of people.
Suppose that it is known that 10 percent of the population will become aware of the event within 1 hour. The formula for the number of informed people as a function of time is $P(t) = N(1 - e^{-kt})$. Find the time when 95% of the people will be informed.

17. Rumors, Rumors, Rumors! If you ask an older person who has served in the military, they will tell you that listening to rumors was a way of life. Surprisingly very few people own up to spreading them! We would like to construct a mathematical model for the diffusion of information. Our first attempt will be very simplistic: the more people are exposed to a rumor, the faster it will spread. Let P be the number of people exposed to a particular rumor, then $\dfrac{dP}{dt} = kP$ is our mathematical model for the diffusion.
(a) Assume 30 students in a mathematics class are told a rumor by their instructor. In one day the rumor spreads so that a total of 60 students will have heard the rumor. By using our simple model, predict the number of days necessary for the rumor to spread throughout the 1000 member campus community. (*Hint*. Let t be measured in days and let $P = 30$ when $t = 0$, $P = 60$ when $t = 1$.)
(b) You may wish to test this model using the following procedure: Generate a harmless rumor in your class and have all the members of the class living in a particular dormatory pass it on to exactly one other person in their dorm during the course of the day. Arrange for the polling of the residents of that dorm several days later to determine who has heard the rumor, and compare the results with those predicted by the model.

18. An object moving along a straight line at time t has an acceleration given by $6t - 18$. Find its position, s, at time t, if at $t = 2$ the velocity is -7 and its position is 4.
(Recall that if $s = f(t)$, then $\dfrac{ds}{dt}$ = velocity and $\dfrac{dv}{dt}$ = acceleration.)

19. A ball is thrown upward from a tower 75 feet high with an initial velocity of 64 ft/sec. Acceleration due to gravity is 32 ft/sec^2 downward. Find the velocity and position functions. How many feet has the ball traveled when $v = 0$?

20. If the rate of growth (or decay) is proportional to the amount of substance (call it x) present, then $\dfrac{dx}{dt} = kx$. Show $x = ce^{kt}$.

21. A bacteria culture grows from 100 to 750 in 1 hour. Assuming that the growth rate principle applies, when will the bacteria count be 400?

22. The rate of increase in a certain bacteria culture is 25%. If the number of cells is initially 200, how many cells will there be after 8 hours? Assume that the growth rate principle is operating.

23. A bacterial population increases at a rate proportional to its population where its population doubles in 1 hour. In how many hours will it be 100 times its original size?

24. If A dollars is invested at 5% compounded continuously, in how many years will the initial investment double? Triple?

25. At what yearly rate, compounded continuously, should 500 dollars be invested if it is expected to double in 10 years?

26. If a country's population is increasing at a continuous rate of 4% each year, in how many years will the country's population double? Assume that the growth rate principle is applicable.

27. The rate of people moving to a certain city is 10% of the present population each year. Assuming no one leaves the city, and the growth rate principle pertains, what is the population after 10 years if the initial population is 10,000 people?

28. Suppose it is known that the gross national product is presently 2.5 trillion dollars and is increasing at the rate of 3.5% per year. Find a formula for the gross national product at any year t.

29. Stock worth $10,000 when purchased in 1965 is worth $25,000 in 1985. If that same amount had been invested in a bank what interest rate compounded continuously would yield the same 1985 worth?

30. The National Bank advertises that it pays an interest rate of $5\frac{1}{2}$% compounded semiannually while The Community Bank advertises that it pays $5\frac{1}{4}$% compounded continuously. If you have $1000 to invest, what is the amount present after one year if you invest at the National Bank? If you invest at the Community Bank? Where will you place your money?

31. When dealing with radioactive substances it is usually true that the rate of decay is proportional to the amount of substance present. This allows us to write the following equation: $\dfrac{ds}{dt} = -ks$. An important radioisotope in medicine is I^{131}, which has a half-life of 8.08 days. This means that after 8.08 days exactly $\frac{1}{2}$ of the original amount of I^{131} will be present.

(a) Use the half-life to determine the proportionality constant k.

(b) How much I^{131} will be present after 25 days if 100 grams were initially present?

32. I^{131} is a radioactive element that is very important in chemotherapy. The rate of decay of I^{131} is proportional to the amount present at that time. A shipment of I^{131} leaves a laboratory at 8 a.m. on Tuesday. The radioactivity of this sample is 250 curies. The curie is a unit that measures the amount of radioactivity present in an element. If the shipment arrives at a hospital at noon the following day, what is the radioactivity when it arrives? Given the half-life of I^{131} is 8.08 days.

33. A patient is given .1 curies of I^{131}. The Atomic Energy Commission requires that any patient containing more than .03 curies of I^{131} be hospitalized. If the half-life of I^{131} is 8.08 days, when can the patient go home?

34. A doctor wishes to give a patient .005 curie of P^{32} on Tuesday at 11 a.m. A supply house will ship the exact amount you desire calibrated as of 8 a.m. on their regular shipping days of Tuesday and Friday. The shipments require 2 days for delivery. How much P^{32} should the doctor order so that he will have .005 curie at 11 a.m. Tuesday? (Half-life of $P^{32} = 14.3$ days.)

35. A certain drug manufactured by a pharmaceutical company is known to have a "half-life" of two years. (The strength of the drug is reduced 50% in two years.) It has been established that effective treatment cannot be accomplished if the strength of the drug is less than 80%. The medicine is labeled with its production date. How many months after the production date would the drug become ineffective?

36. A radioactive material has a half-life of 50 years. If there is originally 100 grams of this material, how much remains at the end of 75 years?

37. The controlled rate of growth of a species is proportional to the square root of the population. When $t = 1$, $N = 400$. When $t = 2$, $N = 625$. Find the equation for the population of the species with respect to time.

38. Phosphate compounds found in most detergents are highly soluble in water and act as a powerful fertilizer for algae. Assume that the growth principle is applicable and that there are 10 algae present in a certain area of a lake at time $t = 0$, and 2 hours later there are 20 algae present. Estimate the number of algae present in the lake area at the end of 30 hours.

39. A drought has struck a potato farm, and it is estimated that the number of healthy potato plants are decaying at a rate proportional to the total number of plants at any time. The drought has killed 40% of the original 200 plants in 25 days. Find a formula for the number of healthy plants at any time t.

40. From Section 6.8, Example 1, $N = \dfrac{Kce^{rt}}{1 + ce^{rt}}$. When $t = 0$ and $N = N_0$,

verify that a particular solution is

$$N = \frac{KN_0e^{rt}}{K + N_0(e^{rt} - 1)}.$$

41. From 1960 to 1970, the United States population rose from about 180 million to 205 million or at an average rate of 1.4% per year. Use the model $N = N_0e^{rt}$ and find the time t for the population to double itself.

■ 6.9 CHAPTER REVIEW

Important Ideas

$\log_b x = L$ if $b^L = x$

properties of exponents

properties of logarithms

natural logarithmic function

$\dfrac{d}{dx}(\ln x) = \dfrac{1}{x}$

$\dfrac{d}{dx}(\ln u) = \dfrac{1}{u} \cdot \dfrac{du}{dx}$

properties of the natural logarithm

$\qquad \ln (a^n) = n \cdot \ln a$

$\qquad \ln (ab) = \ln a + \ln b$

$\qquad \ln \left(\dfrac{a}{b}\right) = \ln a - \ln b$

exponential function

$\dfrac{d}{dx}(e^x) = e^x$

$\dfrac{d}{dx}(e^u) = e^u \cdot \dfrac{du}{dx}$

$\displaystyle\int e^u \, du = e^u + c$

$\displaystyle\int \dfrac{du}{u} = \ln |u| + c$

integration by parts

differential equations

growth rate principle

■ **REVIEW EXERCISE (optional)**

In Problems 1 to 6, solve the given equations for x.

1. $\log_x 8 = 3$

2. $\log_3 27 = x$

3. $\log_{16} x = \frac{1}{2}$

4. $(\ln x)^2 = \ln x^2$

5. $\log_3 (x^2 + 1) - \log_3 x = \log_3 2x$

6. $x = e^{3 \ln 2}$

7. A student invests \$2000 toward his college education. In how many years will the initial investment be worth \$5000 if it is invested at a rate of 6% compounded semiannually? ($\ln 1.03 = .02956$)

8. How many years will it take a deposit to double if it is invested at a yearly rate of 8% compounded quarterly? ($\ln 1.02 = .01980$)

9. A man invests \$500 in a bank that pays 8% interest, compounded quarterly. What is the compounded amount after four years?

10. Which investment would yield a higher total amount after five years?
(a) \$2000 invested at 12%, compounded semiannually or
(b) \$2500 invested at 8%, compounded annually.

In Problems 11 to 28 find the derivative of the function.

11. $y = \ln (7x^5 - 3x)$

12. $y = \ln (7x^5 - 3x)^4$

13. $y = e^{(7x^5 - 3x)}$

14. $y = e^{(7x^5 - 3x)^4}$

15. $y = \ln \sqrt{x}$

16. $y = \sqrt{\ln x}$

17. $y = e^{\sqrt{x}}$

18. $y = \sqrt{e^x}$

19. $y = \ln 2x + e^{2x}$

20. $y = \ln e^{3x^2}$

21. $y = e^x \ln x$

22. $y = \ln [(x^2 - x)(2 - x^3)]$

23. $y = e^x + x^e$

24. $y = \dfrac{x}{\ln x}$

25. $y = \dfrac{e^x - 1}{1 - e^{-x}}$

26. $y = \dfrac{\ln x^2}{e^{x^2}}$

27. $\ln xy = 1$

28. $x \ln y = e^x$

In Problems 29 to 40 integrate.

29. $\displaystyle \int \left(\frac{1}{x} + e^x \right) dx$

30. $\displaystyle \int \frac{2x + 7}{x^2 + 7x + 14} \, dx$

31. $\displaystyle \int_0^1 \frac{x + 3}{x^2 + 6x + 9} \, dx$

32. $\displaystyle \int \frac{3x^5 - x^3 + x^2}{x^3} \, dx$

33. $\displaystyle \int \frac{\sqrt{1 + \ln x}}{x} \, dx$

34. $\displaystyle \int xe^{x^2} \, dx$

35. $\displaystyle\int_1^2 e^{3x-1}\, dx$　　　　　　　　**38.** $\displaystyle\int_1^2 x \ln x\, dx$

36. $\displaystyle\int \frac{e^{2x}}{1 + e^{2x}}\, dx$　　　　　**39.** $\displaystyle\int xe^x\, dx$

37. $\displaystyle\int \frac{2e^x}{\sqrt{1 + e^x}}\, dx$　　　　**40.** $\displaystyle\int x^2 \ln x\, dx$

41. If $y = \ln 2x$, find the equation of the tangent line to this curve at $x = \frac{1}{2}$.

42. Find the area of the region bounded by the curve $y = e^x$, $x = -1$, $x = 1$ and the x axis.

43. Find the area bounded by $y = \dfrac{x}{x^2 + 2}$, the x axis, and the interval $0 \leqslant x \leqslant 2$.

44. During a Sell-O-Rama, an appliance store can expect additional revenue of \$2000 daily. After the sales promotion, the effect on sales, S, is $S = 2000e^{-x}$ (x, number of days after the Sell-O-Rama). What is the rate of change of sales with respect to x?

45. Show by graphing that $\ln x < x$ for $x > 0$. Use derivatives in your analysis.

46. The number of molecules per second colliding with a solid surface area 1 cm^3 is given by $\int v e^{-mv^2/2kT}\, dv$ where the only variable is the velocity v. Evaluate this integral.

47. A biologist observes that the number of bacteria in a controlled culture after t hours is given by $N = 100e^{.5t}$.
(a) What is the initial number of bacteria in the culture?
(b) At what time can the culture be expected to double?

48. Solve the differential equation.
(a) $(x^2 + 2)\, dx = xe^y\, dy$　　　　(b) $\dfrac{dA}{dt} = \dfrac{1}{2}A$

49. The slope of the tangent line at a point on the curve is $\dfrac{x}{x^2 + 1}$.
(a) Find the general equation for the curve.
(b) Find the particular equation of the curve that passes through the origin.

50. When dealing with the effect of gravity on gases the following equation applies: $-dP = \dfrac{PMg}{RT}\, dh$. P and h are the only variables. Separate these variables and find a formula for P as a function of h.

51. In a biology laboratory, bacteria cells increase from 100 to 400 in 2 hours. Assuming that the growth rate principle applies, how many bacteria cells are present after 1 hour?

52. Initially there are 500 bacteria in a certain culture. If the number of bacteria doubles in 10 minutes, when will 50,000 bacteria be present? Assume the growth rate principle is operating.

53. If $2000 is invested at 6% compounded continuously, what will be the amount of the investment after five years? How much interest has accrued?

54. If the population in a certain county increased from 50,000 to 70,000 in the years 1960 to 1980, what will be the population in the year 2000? Assume that the growth rate principle is operating.

55. In how many years would the population of a country double if the population increases at a continuous rate of 10%? Assume the growth rate principle is applicable.

56. The fruit fly population in a certain biology laboratory grows at a rate proportional to its size. At $t = 0$, 150 flies are present. In 12 days there are 1500 flies. Find a formula for the population at any time t.

57. Determine the domain for each of the following functions.
(a) $y = \ln (x - 4)$
(b) $y = \ln (x^2 - 9)$
(c) $y = e^{(1/2)x}$
(d) $y = e^{-2x}$

In Problems 58 to 68, determine whether the statement is true or false.

58. $\log_b N = L$ if $b^N = L$.

59. $\log_b 1 = 0$

60. $64^{2/3} = 4$

61. $\ln (4 + 3) = \ln 4 + \ln 3$

62. $\ln \sqrt{x} = \frac{1}{2} \ln x$

63. $\log_e x = \ln x$

64. $e^{\ln x} = x$

65. $\ln e^x = x$

66. $\dfrac{d}{dx} \left(\dfrac{1}{x} \right) = \ln x$

67. if $y' = e^x$, then $y = e^x + c$

68. $\displaystyle \int \ln x \, dx = \dfrac{1}{x} + c$

SEVEN

Partial Derivatives and Multiple Integrals

7.1   INTRODUCTION

Previous to this chapter we have considered the case where the dependent variable y is a function of only one independent variable x. With $y = f(x)$, we then worked in two-dimensional space, called E^2, with the derivative and integral of such a function.

Helpful as this case is, often the dependent variable is a function of more than one independent variable. For example, profit realistically has many independent variables affecting it. Also, the production function, which tells how much of the commodity will be produced, is often the function of two or more variables.

We expect that with more independent variables, the model and the solutions to problems will become more complicated. However, there is a definite "carry over" and extension of our previous work which will aid us in this chapter. For example, partial derivatives will play a role in this chapter comparable to the role played by the ordinary derivative in earlier chapters.

The partial derivative is used to maximize production and to minimize cost when the production and cost functions are given as functions of two or more independent variables. Also, in the biomedical field, partial derivatives are used in the study of diffusion, flow of electricity through tissues, bone stresses, and the interaction of sensory elements in the retina of the eye.

Before we can apply partial derivatives to problems, we must extend our previous definitions of limits and continuity and inspect the graph of a function where there is more than one independent variable. In so doing, we will concentrate on the model where there are two independent variables. This shifts our thinking to three-dimensional space called E^3.

7.2   THE THREE-DIMENSIONAL COORDINATE SYSTEM

Consider three mutually perpendicular axes intersecting in a common origin O. These axes divide three-dimensional space into eight octants. Of particular interest is Octant I where x, y, and z are nonnegative. For ease in understanding and in graphing, we shall often show only that portion of the graph that lies in Octant I. To draw the representation of a three-dimensional coordinate system in two-dimensional space introduces some distortion. To minimize this distortion, a unit on the x axis is drawn only about $\frac{2}{3}$ as long as a unit on the y and z axes.

To locate a point in this new three-dimensional coordinate system requires that an ordered triple, (x,y,z), be given. Thus, $(2,3,4)$ is a point P found by starting at the origin, O, moving 2 units along the positive x axis, 3 units to the right, and then 4 units up (see Figure 7.1).

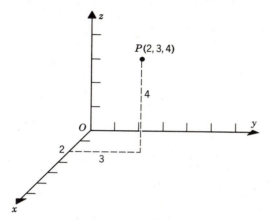

Figure 7.1

Points in other octants can be located in a similar manner. For example, the point $Q(-2,4,-5)$ is found by starting at the origin, O, moving 2 units along the negative x axis, 4 units to the right, and then 5 units down (see Figure 7.2).

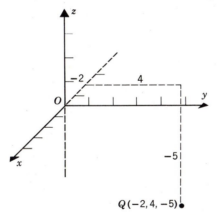

Figure 7.2

The surface where $z = 0$ is called the xy-plane. It is identical to two-dimensional space, E^2. The surface where $x = 0$ is called the yz-plane. A similar definition is given for the xz-plane.

7.3 DEFINITIONS OF FUNCTION AND LIMIT IN E^3 SPACE

Our concept of a function must now be extended to include the idea of two independent variables, x and y.

DEFINITION

Function. If a correspondence associates to each point in a set D of the xy-plane one and only one value z, then the correspondence is said to be a *function*. f is a function defined by $z = f(x,y)$ if to every point in D there is associated one and only one value of z. (Associations in Figure 7.3 shown only for P_1 and P_2.)

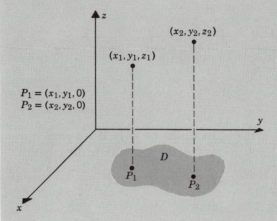

Figure 7.3

The concept of limit is also extended by the following working "definition" for the limit.

WORKING DEFINITION

Limit of a Function. We say that $\lim\limits_{(x,y)\to(a,b)} f(x,y) = L$ if the numbers $f(x,y)$ remain arbitrarily close to the real number L, whenever (x,y) is very near to (a,b).

The more elegant Cauchy definition of the limit is given in Appendix II, page 393.

In a similar manner, a corresponding definition to the E^2 definition is given for continuity.

DEFINITION

Function Continuous at (a,b). Let the function f be defined at $f(a,b)$. Then f is *continuous* at (a,b) if $\lim\limits_{(x,y)\to(a,b)} f(x,y) = f(a,b)$.

Example 1

Purpose To find the value of a function is E^3:

Problem A store's profit is dependent on the number of salesmen, x, and the amount of inventory, y (in thousands of dollars). If the profit is described by the function $P = f(x,y) = \$[1400 - (12 - x)^2 - (40 - y)^2]$, find the profit P: (A) when $x = 4$ and $y = 25$; (B) when $x = 6$ and $y = 30$.

Solution (A) When $x = 4$ and $y = 25$,

$$P = f(4,25) = \$[1400 - (12 - 4)^2 - (40 - 25)^2]$$
$$= \$1111$$

(B) When $x = 6$ and $y = 30$,

$$P = f(6,30) = \$[1400 - (12 - 6)^2 - (40 - 30)^2]$$
$$= \$1264.$$

7.4 GRAPHING IN E^3 SPACE

An equation of the type $y = f(x)$ in E^2 has for its graph a curve. However, it can be shown that an equation in E^3 space represents a *surface*. For example, it can be shown that the equation $ax + by + cz = d$ represents a plane in E^3.

Example 1

Purpose To show the technique for graphing a plane in E^3 space:

Problem Graph the portion of $2x + 3y + 4z = 12$ that occurs in Octant I.

Solution Find the traces in the coordinate planes, that is, find the curves in which the surface $2x + 3y + 4z = 12$ intersects the xy, xz, and yz planes.

Trace In	Trace Is
xy-plane	$2x + 3y + 4 \cdot 0 = 12$
Let $z = 0$	or a straight line in the xy-plane
	$2x + 3y = 12.$

yz-plane	Straight line in yz-plane
Let $x = 0$	$3y + 4z = 12$
xz-plane	Straight line in xz-plane
Let $y = 0$	$2x + 4z = 12$
	or
	$x + 2z = 6$

Graph these straight lines in their respective planes. These three straight lines are the boundaries of the portion of the plane that lies in Octant I.

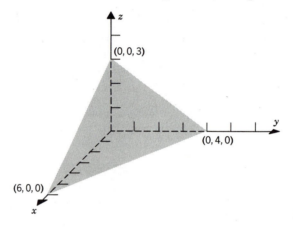

Figure 7.4

Example 2

Purpose To show the technique for graphing a surface in E^3 when one variable is missing from the equation:

Problem Graph $3x + 2y = 6$ in E^3.

Solution The trace on the xy-plane is the straight line $3x + 2y = 6$. Notice that the z variable is missing from the equation. All points on the surface $3x + 2y = 6$ will be above or below the line $3x + 2y = 6$ in the xy-plane. This is true since z can assume any value while the x and y coordinates of a point on the surface must satisfy $3x + 2y = 6$. The technique is to graph the trace $3x + 2y = 6$, in the xy-plane, and then to graph a line parallel to $3x + 2y = 6$. The lines are connected to show a portion of the plane satisfying $3x + 2y = 6$.

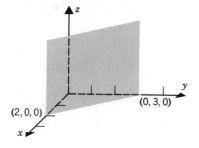

Figure 7.5

Notice that the surface is generated by a straight line marching along the line $3x + 2y = 6$ in such a way that it is always parallel to the z axis. Such a surface is an example of a cylinder. This cylinder also happens to be a plane.

Example 3

Purpose To picture a production function where there are two independent variables:

Problem The production function defined by $z = x^2 + y^2$ is given for a firm. Here, z measures the firm's yearly output, in thousands of units, x denotes the amount of labor used, in hours per week, and y is the capital input, in thousands of units per year. Sketch the first octant portion of the production function.

Solution

Trace In	Trace Is
xy-plane Let $z = 0$	$x^2 + y^2 = 0$ or the point $(0,0,0)$.
yz-plane Let $x = 0$	$z = y^2$ or a parabola in the yz-plane.
xz-plane Let $y = 0$	$z = x^2$ or a parabola in the xz-plane.

Also useful is the fact that the plane $z = k$, $k > 0$, which is parallel to the xy-plane, will intersect the surface $z = x^2 + y^2$ in the curve $k = x^2 + y^2$. This curve is a circle.

The first octant portion of the surface $z = x^2 + y^2$ is shown in Figure 7.6. Notice that this surface is not a plane.

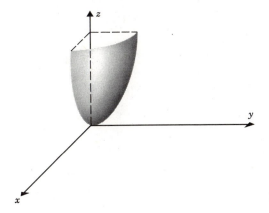

Figure 7.6

Example 4

Purpose To show the sketching of another surface in E^3 where the surface is not a plane:

Problem Sketch: $x^2 + y^2 + z^2 = 16$.

Solution

Trace In	Trace Is
xy-plane Let $z = 0$	$x^2 + y^2 + 0^2 = 16$ or the circle in the xy-plane with center at $x = 0$, $y = 0$ and radius, $r = 4$.
yz-plane Let $x = 0$	$y^2 + z^2 = 16$ or the circle in the yz-plane with center at $y = 0$, $z = 0$, and $r = 4$.
xz-plane Let $y = 0$	$x^2 + z^2 = 16$ or the circle in the xz-plane with center at $x = 0$, $z = 0$, and $r - 4$.

These traces suggest to us that the surface we are graphing is a sphere whose center is at $(0,0,0)$ and with radius equal to 4 units. The graph of this sphere is shown below. Notice that this is not the graph of a function. Why?

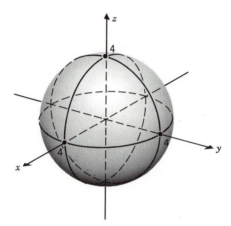

Figure 7.7
Graph of $x^2 + y^2 + z^2 = 16$.

As one might expect, with more involved equations the surfaces often are difficult to represent. One example of such a surface is the hyperbolic paraboloid or saddle. An equation of a hyperbolic paraboloid is

$$\frac{y^2}{b^2} - \frac{x^2}{a^2} = \frac{z}{c}.$$

This surface is sketched in Figure 7.8 for $c > 0$.

An important property of this hyperbolic paraboloid or saddle is observed from Figure 7.8. Imagine that you are standing at the origin $(0,0,0)$ and can look in only two directions—along the positive and negative x axes. Then you would conclude that you are at a point of relative maximum. However, standing at $(0,0,0)$ and being able to look only in the positive and negative y directions would lead you to the conclusion that you are at a point of relative minimum.[1]

---

[1]Definitions for relative maximum and relative minimum in E^3 are given in Section 7.7.

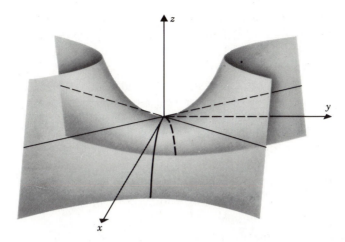

Figure 7.8
Graph of hyperbolic
paraboloid or saddle
with equation
$$\frac{y^2}{b^2} - \frac{x^2}{a^2} = \frac{z}{c}.$$

Such a point as $(0,0,0)$, which appears to be both a point of relative maximum and relative minimum, is an example of a "saddle point" (see Definition, page 359) and will not be called a relative maximum or a relative minimum.

To find true points of relative extrema on a surface, the technique of partial differentiation is needed. The next section deals with the procedure for finding partial derivatives.

Exercise 7.2–7.4

1. Plot, in E^3 space, the points whose coordinates are $(0,1,3)$, $(1,0,3)$, $(1,3,0)$, $(-2,-2,5)$, $(0,-3,0)$, $(4,1,0)$, $(2,5,-2)$.

2. Given, $f(x,y) = x^2 + xy + y^2$.
    Find: $f(1,2)$; $f(0,3)$; $f(2,2)$; $f(2,0)$.

3. Given, $f(x,y) = ye^x + x^2 + \ln y$.
    Find: $f(0,1)$; $f(1,2)$.

4. Given, $f(x,y) = \sqrt{xy} + \sqrt{x+y}$.
    Find: $f(0,1)$; $f(4,0)$; $f(9,16)$.

5. Given, $f(x,y) = x^{1/3}y^{1/2}$
    Find: $f(-1,4)$; $f(8,9)$; $f(64,64)$

6. The surface area of a rectangular solid with a square base can be determined from the equation $A = 2s^2 + 4sh$, where s is the length of one side of the square base and h is the height of the solid. Find the surface area for the following values of s and h.
    (a) $(s,h) = (3,4)$ (c) $(s,h) = (2,4)$
    (b) $(s,h) = (5,5)$ (d) $(s,h) = (10,20)$

7. It is possible to extend the distance formula to E^3. (See Review Exercise, Problem 4.) The distance between $P_1(x_1,y_1,z_1)$ and $P_2(x_2,y_2,z_2)$ is given by $d = \sqrt{(x_2 - x_1)^2 + (y_2 - y_1)^2 + (z_2 - z_1)^2}$. Find the distance between P_1

and P_2 for each of the following pairs of points:
(a) $P_1(1,5,7)$; $P_2(0,-3,3)$ (c) $P_1(-3,3,2)$; $P_2(1,1,-6)$
(b) $P_1(2,-4,5)$; $P_2(5,0,5)$ (d) $P_1(4,2,-7)$; $P_2(5,-3,-1)$

8. A total budget expenditure of \$3000 is to be made for three commodities whose prices are \$100, \$50, and \$60, respectively.
   (a) If x, y, and z represent the numbers of the commodities purchased, verify that the budget equation is $10x + 5y + 6z = 300$.
   (b) Sketch the surface of this equation. The graph is often referred to as a budget plane.

9. A manufacturing firm pays skilled workers \$16.50 per hour, semiskilled workers \$11.25 an hour, and unskilled workers \$4.75 an hour. Express the total hourly wages paid by the firm as a function of the skilled, semi-skilled, and unskilled employees.

10. The Bahia Boat Company has found that the price, z, in dollars for a Bahia yacht depends on the number of yachts, x, demanded and the price of inboard motors, y. The function is $z = f(x,y) = -\frac{1}{2}x + 3y + 20{,}000$.
    (a) Find the price of a yacht when $x = 50$ and $y = \$2700$.
    (b) Sketch the function.

11. The growth of a plant is dependent on the availability of sunlight, x, and the moistness of the soil, y. Assume that a plant's growth equation is $z = 2x - .2y^2 + 2y$. What would be the value of z, when $x = 10$, and $y = 5$? Sketch the function through the use of traces in the coordinate planes.

12. The learning of a laboratory animal is dependent on the amount of rein-forcement, x, used in a given trial and, the number of previous trials, y. Assume one such learning equation is $z = f(x,y) = 2\sqrt{x} + \frac{1}{2}\sqrt{y}$. What is the value of the function when $(x,y) = (1,4)$? Sketch the traces in the xz-plane and yz-plane.

In Problems 13 to 23, sketch the surface represented by the equation in E^3 space. Use traces in the xy, xz, and yz-planes to aid your graphing.

13. $x + 2y + 3z = 6$ 19. $3y + z = 12$

14. $6x + 4y + 3z = 12$ 20. $x^2 + y^2 + z^2 = 4$

15. $x = 3$ 21. $x^2 + y^2 = 4z$

16. $y = 2$ 22. $z^2 + y^2 = x$

17. $2x + 3y = 12$ 23. $z = x^2$

18. $2x + z = 12$

7.5 PARTIAL DERIVATIVES AND THEIR GEOMETRIC INTERPRETATION

With f a function of x and y, defined by $z = f(x,y)$, it is useful to speak of the exact rate of change in f with respect to x when y is held constant. Also of interest is the exact rate of change in f with respect to y when x is held constant. Such exact rates of change of the function f with respect to one of the independent variables, when all other independent variables are held constant, are called *partial derivatives*. Specifically, the exact rate of change of f with respect to x (y held constant) is called the first partial derivative of f with respect to x and is denoted by $\dfrac{\partial z}{\partial x}$ or $f_x(x,y)$. The first partial derivative of f with respect to y (x held constant) is denoted by $\dfrac{\partial z}{\partial y}$ or $f_y(x,y)$. The definitions of these partial derivatives are now given.

DEFINITIONS

Partial Derivative. Let f be a function of two variables, x and y, with f defined by $z = f(x,y)$. The *first partial derivative of f with respect to x* is

$$f_x(x,y) = \frac{\partial z}{\partial x} = \lim_{\Delta x \to 0} \frac{f(x + \Delta x, y) - f(x,y)}{\Delta x};$$

the *first partial derivative of f with respect to y* is

$$f_y(x,y) = \frac{\partial z}{\partial y} = \lim_{\Delta y \to 0} \frac{f(x, y + \Delta y) - f(x,y)}{\Delta y};$$

provided these limits exist.

The notations $\dfrac{\partial z}{\partial x}\bigg]_{(x_1,y_1)}$ and $f_x(x_1,y_1)$ are used to indicate the value of the first partial derivative of f with respect to x at the point (x_1,y_1). Correspondingly, $\dfrac{\partial z}{\partial y}\bigg]_{(x_1,y_1)}$ or $f_y(x_1,y_1)$ is the first partial derivative of f with respect to y evaluated at (x_1,y_1).

As is true with ordinary derivatives, the above definitions are not usually used to find the first-order partial derivatives. Rather than deal with the limits, one finds that the partial derivatives may be obtained by using his knowledge of ordinary derivatives along with the meaning of a partial derivative. This technique is illustrated in the following examples.

Example 1

Purpose   To find partial derivatives for certain functions:

Problem   Find $\dfrac{\partial z}{\partial x}$ and $\dfrac{\partial z}{\partial y}$ if:

(A) $z = f(x,y) = 2x^2y + y^2 + x^2$.
(B) $z = f(x,y) = ye^x + xe^{2y}$.

Solution   (A) $z = f(x,y) = 2x^2y + y^2 + x^2$

To find $\dfrac{\partial z}{\partial x}$, treat y as a constant.

$$f_x(x,y) = \frac{\partial z}{\partial x} = 4xy + 2x.$$

Notice that $\dfrac{\partial}{\partial x} y^2 = 0$.

To find $\dfrac{\partial z}{\partial y}$, treat x as a constant.

$$f_y(x,y) = \frac{\partial z}{\partial y} = 2x^2 + 2y.$$

(B) $z = f(x,y) = ye^x + xe^{2y}$

$$f_x(x,y) = \frac{\partial z}{\partial x} = ye^x + e^{2y}.$$

$$f_y(x,y) = \frac{\partial z}{\partial y} = e^x + 2xe^{2y}.$$

Example 2

Purpose   To find and interpret the meaning of the partial derivatives of a profit function:

Problem A   The Glover Company finds that their weekly profit is given by $P = f(x,y) = 10000 - 2x - .3y^2$. The profit, P, is in dollars while x represents the weekly cost of raw materials in dollars, and y represents the weekly labor costs in dollars. Find and interpret $\dfrac{\partial P}{\partial x}$ and $\dfrac{\partial P}{\partial y}$.

Solution $P = 10000 - 2x - .3y^2$.

$$\frac{\partial P}{\partial x} = -2$$

$$\frac{\partial P}{\partial y} = -.6y$$

The $\dfrac{\partial P}{\partial x} = -2$. This means that the exact rate of change in profit per unit change in raw material costs is -2 (when labor costs are held constant). The $\dfrac{\partial P}{\partial y} = -.6y$. The exact rate of change in profit per unit change in labor costs is $-.6y$ (when raw material costs are held constant).

Problem B Economists often regard the quantity produced Q as a function of capital K and labor L, which is denoted by $Q = f(K,L)$. If $Q = f(K,L) = K^{1/2}L^{1/2}$, find the marginal productivity of labor $\dfrac{\partial Q}{\partial L}$ or $f_L(K,L)$. Also, find $f_L(30,120)$.

Solution $Q = f(K,L) = K^{1/2}L^{1/2}$

$$\frac{\partial Q}{\partial L} = f_L(K,L) = K^{1/2}\left(\frac{1}{2}\right)L^{-1/2} = \frac{K^{1/2}}{2\,L^{1/2}}$$

$$f_L(30,120) = \frac{\sqrt{30}}{2\sqrt{120}} = \frac{1}{2}\sqrt{\frac{1}{4}} = \frac{1}{4} = .25$$

In order to give a geometric interpretation to $f_x(x,y)$ and $f_y(x,y)$, let us return to the portion of the sphere $x^2 + y^2 + z^2 = 16$ where $z \geq 0$, that is, where $z = \sqrt{16 - x^2 - y^2}$. The first octant portion of this hemisphere is shown in Figure 7.9.

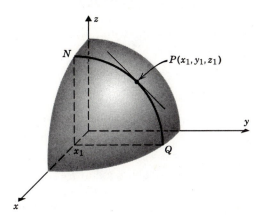

Figure 7.9

The plane $x = x_1$, $0 < x_1 < 4$, intersects the hemisphere in a curve whose first octant trace is NPQ, a curve in the plane $x = x_1$. P has coordinates (x_1,y_1,z_1) or $(x_1,y_1,\sqrt{16 - x_1^2 - y_1^2})$.

The $\dfrac{\partial z}{\partial y}\bigg]_{(x_1,y_1,z_1)}$ represents the slope of the tangent line drawn to curve NPQ at (x_1,y_1,z_1). A similar interpretation can be given for $\dfrac{\partial z}{\partial x}$ as the slope of the tangent line drawn to the curve that is the intersection of $z = f(x,y)$ and the plane $y = y_1$.

7.6 HIGHER ORDER PARTIAL DERIVATIVES

Assume the function f is defined by $z = f(x,y)$. If f is twice differentiable, there exist four partial derivatives that are known as *second-order partial derivatives*. These partial derivatives are defined by the following equations.

$$\frac{\partial}{\partial x} \left(\frac{\partial z}{\partial x} \right) = \frac{\partial^2 z}{\partial x^2} = f_{xx}(x,y)$$

$$\frac{\partial}{\partial y} \left(\frac{\partial z}{\partial y} \right) = \frac{\partial^2 z}{\partial y^2} = f_{yy}(x,y)$$

$$\frac{\partial}{\partial x} \left(\frac{\partial z}{\partial y} \right) = \frac{\partial^2 z}{\partial x \partial y} = f_{yx}(x,y)$$

$$\frac{\partial}{\partial y} \left(\frac{\partial z}{\partial x} \right) = \frac{\partial^2 z}{\partial y \partial x} = f_{xy}(x,y)$$

In these definitions your attention is called to the fact that:

1. $\dfrac{\partial^2 z}{\partial x \partial y} = f_{yx}(x,y)$ means, starting with $z = f(x,y)$, the partial derivatives are taken with respect to y, and then with respect to x *in that order*. $\dfrac{\partial^2 z}{\partial y \partial x} = f_{xy}(x,y)$ means the order for taking partial derivatives is first with respect to x, and then with respect to y.

2. At point $P_1(x_1,y_1)$, $\dfrac{\partial^2 z}{\partial x \partial y} = \dfrac{\partial^2 z}{\partial y \partial x}$ if the partial derivatives found are each continuous on a circular neighborhood of P_1.[1] This can be proved, but the proof will not be given here.

Example 1

Purpose To find second-order partial derivatives:

Problem Given: $z = f(x,y) = x^2 e^y + y^2 + x^3 + \ln x^2$.
Find: $f_{xx}(x,y)$, $f_{xy}(x,y)$, $f_{yy}(x,y)$, and $f_{yx}(x,y)$.

Solution $f_x(x,y) = \dfrac{\partial z}{\partial x} = 2xe^y + 3x^2 + \dfrac{2}{x}$

---

[1] A circular neighborhood of (x_1,y_1) is the set of points interior to a circle whose center is at (x_1,y_1).

$$f_{xx}(x,y) = \frac{\partial^2 z}{\partial x^2} = 2e^y + 6x - \frac{2}{x^2}$$

$$f_{xy}(x,y) = \frac{\partial^2 z}{\partial y \partial x} = \frac{\partial}{\partial y}\left(2xe^y + 3x^2 + \frac{2}{x}\right) = 2xe^y$$

$$f_y(x,y) = \frac{\partial z}{\partial y} = x^2 e^y + 2y$$

$$f_{yy}(x,y) = \frac{\partial^2 z}{\partial y^2} = x^2 e^y + 2$$

$$f_{yx}(x,y) = \frac{\partial^2 z}{\partial x \partial y} = \frac{\partial}{\partial x}(x^2 e^y + 2y) = 2xe^y$$

Notice that $f_{xy}(x,y) = f_{yx}(x,y)$.

Exercise 7.5–7.6

In Problems 1 to 8, find $f_x(x,y), f_y(x,y), f_{xx}(x,y), f_{xy}(x,y), f_{yy}(x,y)$, and $f_{yx}(x,y)$ for each function.

1. $f(x,y) = x^3 - 3x^2 y + 2y$

2. $f(x,y) = 3x^2 + x^2 y^2 - 2y^4$

3. $f(x,y) = e^{xy} + \ln(xy)$

4. $f(x,y) = \ln(2x^2 + y^2)$

5. $f(x,y) = xe^{x^2 - y^2}$

6. $f(x,y) = \dfrac{y - 2x}{xy}$

7. $f(x,y) = (x^3 + y^2)^4$

8. $f(x,y) = x^2 \ln \sqrt{y} + y^2 e^{\sqrt{x}} + y$

9. If $f(x,y) = \dfrac{x}{y} + y^2 e^{x^2}$, show that $f_{xy}(x,y) = f_{yx}(x,y)$.

10. If $f(x,y) = x^{1/5} y^{1/3}$, show that $f_{xy}(x,y) = f_{yx}(x,y)$.

11. If $f(x,y) = x \ln y^2 - \dfrac{y}{x^2 + 2}$, show that $f_{xy}(x,y) = f_{yx}(x,y)$.

12. If $z = x^2 y + x \ln y$, show that $x \cdot \dfrac{\partial^2 z}{\partial x^2} + y \cdot \dfrac{\partial^2 z}{\partial y \partial x} = 4xy + 1$.

13. If $z = 2 \ln(x^2 + y^2)$ show that $\dfrac{\partial^2 z}{\partial x^2} + \dfrac{\partial^2 z}{\partial y^2} = 0$.

14. If the joint cost of producing the quantities x and y of two commodities is $C = f(x,y)$, then $\dfrac{\partial C}{\partial x}$ and $\dfrac{\partial C}{\partial y}$ are called the *partial marginal costs* with respect to x and y, respectively. Find the partial marginal costs with respect to x and with respect to y if
(a) $C = x \ln(4 + e^y)$ (b) $C = 2x + 4y$.

15. The production of most commodities requires an input of two factors. For example, let K be the units of capital good inputs, and let L be the units of labor inputs. Then $Q = f(K,L)$ gives the total output produced and is called a *production function*. $\dfrac{\partial Q}{\partial K}$ is called the *marginal productivity of K* and represents the rate of increase of Q as input K is increased, assuming that input L remains constant. $\dfrac{\partial Q}{\partial L}$ is called the *marginal productivity of L* and represents the rate of increase of Q as input L is increased, assuming that input K remains constant. It should be noted that the inputs (K and L) must be nonnegative for a production problem to be meaningful.

   (a) Given the production function $Q = 10K^{2/3}L^{1/3}$, find the marginal productivity of K and the marginal productivity of L. Prove that if the quantities of both inputs are doubled, output will double.

   (b) Given the production function $Q = 3K^2L - 2L^2$, find the marginal productivity of K and the marginal productivity of L. Notice that if $L > \frac{3}{4}K^2$, then the marginal productivity of L is negative. Interpret this result.

16. The following production function gives the quantity of output produced when K units of capital and L units of labor are utilized. If $Q(K,L) = 3KL^2 + 6K + 4L$, find $\dfrac{\partial Q}{\partial K}$ and $\dfrac{\partial Q}{\partial L}$.

17. The total revenue of two products, x and y, can be expressed by the function $R(x,y) = 100y - 4y^2 + 90x - x^3$. Find the marginal revenue for product x and the marginal revenue for product y.

18. Assume that a store's profit is dependent on the number of salesmen, x, and the amount of inventory, y (in thousands of dollars). If the profit is described by the function $P = f(x,y) = 1400 - (12 - x)^2 - (40 - y)^2$, find $\dfrac{\partial P}{\partial x}$ and $\dfrac{\partial P}{\partial y}$. State, in words, the significance of these partial derivatives.

19. The pollution index, P, for a community is $P = f(x,y) = ax^2 + bxy + cy^2$ where a, b, and c are constants. Find $\dfrac{\partial P}{\partial x}$ and $\dfrac{\partial P}{\partial y}$.

20. A company has found its weekly sales, S, to be a function of the number of times, x, their television commercial is shown in a city and the number of cities, y, showing the commercial. If $S = f(x,y) = 47xy + 3x + 780$,
   (a) Find $f_x(x,y)$, $f_y(x,y)$.   (b) Interpret $f_x(x,y)$ and $f_y(x,y)$.

21. A firm that manufactures toaster ovens finds monthly production P depends on the number of employees x and the number of machines being used y. The production function is $P = f(x,y) = 2x^2 + 3xy + \frac{1}{2}y^2$. The

partial derivative P_x or $f_x(x,y)$ is called the marginal productivity of labor and P_y or $f_y(x,y)$, the marginal productivity of machinery (equipment). Find

(a) $f_x(x,y)$

(b) $f_y(x,y)$

(c) $f_x(x,12)$, which is the rate of change of production with respect to the number of employees when 12 pieces of machinery are being used.

(d) $f_y(20,y)$, which is the rate of change of production with respect to the number of pieces of machinery in use when 20 workers are employed.

(e) $f_x(20,12)$, which is the number of additional toaster ovens that would be produced when 12 units of machinery are in use and the number of employees is increased from 20 to 21.

(f) $f_y(20,12)$, which is the number of additional toaster ovens that would be produced with a work force of 20 and the number of units of machinery is increased from 12 to 13.

22. A psychologist has determined that the time it takes a monkey to respond to an electrical stimulus depends on two things:

(1) The strength of the shock.

(2) The number of shocks the monkey had received in the past.

She reports the following function: $T(V,P) = 3VP^2 + 4P + 2V^{-1}$ where V is the number of volts and P is the number of shocks received in the past.

(a) Find $\dfrac{\partial T}{\partial V}$ and $\dfrac{\partial T}{\partial P}$.

(b) State in words the significance of these partial derivatives.

23. The volume of a right circular cylinder is $V = \pi r^2 h$, where r represents the radius and h the height. Find $\dfrac{\partial V}{\partial r}$, $\dfrac{\partial V}{\partial h}$, $\dfrac{\partial^2 V}{\partial h \partial r}$, $\dfrac{\partial^2 V}{\partial r \partial h}$, $\dfrac{\partial^2 V}{\partial r^2}$, and $\dfrac{\partial^2 V}{\partial h^2}$.

24. The velocity of a moving particle is dependent on the force on the particle, F, and the time since the particle began moving. The velocity function for such a particle is $V = F(t^2 - 32t)$. What is $\dfrac{\partial V}{\partial t}$ and what does this represent in words? What is $\dfrac{\partial V}{\partial F}$ and what does this represent in words?

25. In thermodynamics the following formula is very important.

$$H = U + PV \qquad \text{where } H = \text{enthalphy}$$
$$U = \text{internal energy}$$
$$P = \text{pressure}$$
$$V = \text{volume}$$

If pressure is constant find $\dfrac{\partial H}{\partial U}$ and $\dfrac{\partial H}{\partial V}$.

26. A differential is said to be "exact" when it meets Euler's test for exactness. An example will illustrate this test:

$$\text{Given: } dz = \underline{4x^2y} \ dx + \underline{\tfrac{4}{3}x^3} \ dy$$

The procedure involves taking the partial derivative of each of the underlined portions with respect to its complimentary variable.

Let $4x^2y = M$ and $\tfrac{4}{3}x^3 = N$

Substituting: $dz = M \ dx + N \ dy$

Now find $\dfrac{\partial M}{\partial y}$ and $\dfrac{\partial N}{\partial x}$.

$$\frac{\partial M}{\partial y} = \frac{\partial(4x^2y)}{\partial y} = 4x^2$$

$$\frac{\partial N}{\partial x} = \frac{\partial(\tfrac{4}{3}x^3)}{\partial x} = 4x^2$$

Only if $\dfrac{\partial M}{\partial y} = \dfrac{\partial N}{\partial x}$ is the differential exact. In our example the differential dz is exact because both $\dfrac{\partial M}{\partial y}$ and $\dfrac{\partial N}{\partial x}$ equal $4x^2$. Use the above procedure to determine if the following differentials are exact.

(a) $dz = 2y^2x \ dy + \tfrac{2}{3}y^3 \ dx$ (*Hint.* Let $N = 2y^2x$ and $M = \tfrac{2}{3}y^3$.)

(b) $dr = 3s^2t \ dt + 2t^2s \ ds$ (*Hint.* Let $M = 3s^2t$ and $N = 2t^2s$.)

7.7 RELATIVE EXTREMA IN E^3 SPACE

There exist definitions for relative extrema in E^3 that correspond to the definitions given in E^2.

DEFINITIONS

Relative Extrema. The function f is said to have a *relative maximum at* (x_1, y_1) if there exists a circular neighborhood of (x_1, y_1) such that for all points (x, y) in this neighborhood

$$f(x, y) \leqslant f(x_1, y_1).$$

f is said to have a *relative minimum at* (x_1, y_1) if there exists a circular neighborhood of (x_1, y_1) such that for all points (x, y) in this neighborhood

$$f(x, y) \geqslant f(x_1, y_1).$$

It also seems logical that at a point of relative extrema, the two first partial derivatives (if they exist) must be equal to zero. This is suggested by noticing the following sketch.

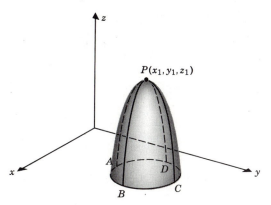

1. The surface $z = f(x,y)$ has a relative maximum at P. The plane $x = x_1$ intersects $z = f(x,y)$ in curve APC.

2. P is a relative maximum for curve APC. Therefore, the slope (if it exists) of curve APC at P must equal zero. The slope of curve APC at P is $\dfrac{\partial z}{\partial y}$ (see geometric interpretation of f_x and f_y). Hence $\dfrac{\partial z}{\partial y}\bigg]_P = 0$.

3. The plane $y = y_1$ intersects $z = f(x,y)$ in curve BPD.

4. P is a relative maximum for curve BPD. Therefore, the slope (if it exists) of curve BPD at P must equal zero. So, $\dfrac{\partial z}{\partial x}\bigg]_P = 0$.

However, it is possible for both $f_x(x_1,y_1) = 0$ and $f_y(x_1,y_1) = 0$ and still not have a relative extremum at (x_1,y_1). Such a point is called a saddle point.

DEFINITION

Saddle Point. The point (x_1,y_1,z_1) is called a *saddle point* of the surface $z = f(x,y)$ if $f_x(x_1,y_1) = 0$ and $f_y(x_1,y_1) = 0$ but (x_1,y_1,z_1) is not a point of relative extremum.

Therefore, even if $f_x(x_1,y_1) = 0$ and $f_y(x_1,y_1) = 0$, it is necessary to test to determine whether (x_1,y_1) really yields a relative extremum. The usual test for extrema in E^3 is a second derivative test, called the M-test, and it will be stated without proof.

The M-Test for Relative Extrema

Let $z = f(x,y)$ be continuous and have continuous first- and second-order partial derivatives in a circular neighborhood of $(x_1, y_1, 0)$. Also, let $f_x(x_1, y_1) = f_y(x_1, y_1) = 0$.

By definition

$$M(x,y) = \frac{\partial^2 z}{\partial x^2} \cdot \frac{\partial^2 z}{\partial y^2} - \left(\frac{\partial^2 z}{\partial y \partial x} \right)^2.$$

Then:

(A) If $M(x_1, y_1) > 0$ and $f_{xx}(x_1, y_1) < 0$, there exists a relative maximum at (x_1, y_1).

(B) If $M(x_1, y_1) > 0$ and $f_{xx}(x_1, y_1) > 0$, there exists a relative minimum at (x_1, y_1).

(C) If $M(x_1, y_1) < 0$, there exists a saddle point at (x_1, y_1).

(D) If $M(x_1, y_1) = 0$, no information is given by this test.

This test gives us a method for finding relative extrema.

Method of Testing for Relative Extrema

Given $z = f(x,y)$.

1. Find $\dfrac{\partial z}{\partial x}$ and $\dfrac{\partial z}{\partial y}$.

2. Find the critical point(s) where $\dfrac{\partial z}{\partial x}$ and $\dfrac{\partial z}{\partial y}$ are equal to zero. Call (x_1, y_1) such a point.

3. Find the values of M and $\dfrac{\partial^2 z}{\partial x^2}$ at point (x_1, y_1).

M at point (x_1, y_1)	$\dfrac{\partial^2 z}{\partial x^2}\Big]_{(x_1, y_1)}$	Conclusion
$M > 0$	$f_{xx}(x_1 y_1) < 0$	Relative maximum at (x_1, y_1)
$M > 0$	$f_{xx}(x_1 y_1) > 0$	Relative minimum at (x_1, y_1)
$M < 0$		Saddle point at (x_1, y_1)
$M = 0$		No conclusion possible

It can also be shown that the search for absolute extrema in E^3 is comparable to our search for absolute extrema in E^2. For in E^3, it can be shown that absolute extrema for $z = f(x,y)$ either occur at a point of relative extrema or at a point on the boundary of the domain of $z = f(x,y)$.

Example 1

Purpose To use the M-test to find relative extrema:

Problem Find relative extrema for $z = 2xy + x^3 - y^3$.

Solution
$$f_x(x,y) = 2y + 3x^2 \qquad f_{xx}(x,y) = 6x$$
$$f_y(x,y) = 2x - 3y^2 \qquad f_{yy}(x,y) = -6y$$
$$f_{xy}(x,y) = 2$$

Solve: $f_x(x,y) = 0 \qquad 2y + 3x^2 = 0 \qquad (1)$
$\qquad\quad f_y(x,y) = 0 \qquad 2x - 3y^2 = 0 \qquad (2)$

From (2): $2x - 3y^2 = 0$

$$x = \tfrac{3}{2}y^2$$

Substituting this result in (1):

$$2y + 3(\tfrac{3}{2}y^2)^2 = 0$$
$$2y + \tfrac{27}{4}y^4 = 0$$
$$y(2 + \tfrac{27}{4}y^3) = 0$$
$$y = 0 \qquad \text{or} \qquad 2 + \tfrac{27}{4}y^3 = 0$$
$$y^3 = \tfrac{-8}{27}$$
$$y = \tfrac{-2}{3}$$

From $x = \tfrac{3}{2}y^2$;

$$\text{if } y = 0, \ x = 0$$
$$\text{if } y = \tfrac{-2}{3}, \ x = \tfrac{3}{2}(\tfrac{-2}{3})^2 = \tfrac{2}{3}$$

The critical points are $(0,0)$ and $(\tfrac{2}{3}, -\tfrac{2}{3})$.
Test each of these by the M-test.

1. At $(0,0)$, $\qquad\qquad f_{xx}(0,0) = 6x]_{(0,0)} = 0$
$$f_{yy}(0,0) = -6y]_{(0,0)} = 0$$
$$f_{xy}(0,0) = 2]_{(0,0)} = 2$$

Therefore, at $(0,0)$, $M = 0 \cdot 0 - 2^2 = -4$.
Conclusion: $(0,0)$ is a saddle point.

2. At $(\tfrac{2}{3}, -\tfrac{2}{3})$, $\qquad f_{xx}(\tfrac{2}{3}, -\tfrac{2}{3}) = 6x]_{(2/3, -2/3)} = 4$
$$f_{yy}(\tfrac{2}{3}, -\tfrac{2}{3}) = -6y]_{(2/3, -2/3)} = 4$$
$$f_{xy}(\tfrac{2}{3}, -\tfrac{2}{3}) = 2]_{(2/3, -2/3)} = 2.$$

Therefore, at $(\frac{2}{3}, -\frac{2}{3})$, $M = 4 \cdot 4 - 2^2 = 12$.

Since at $(\frac{2}{3}, -\frac{2}{3})$, $M > 0$ and $f_{xx}(\frac{2}{3}, -\frac{2}{3}) = 4 > 0$, it follows that $(\frac{2}{3}, -\frac{2}{3})$ will yield a relative minimum.

In summary, $z = 2xy + x^3 - y^3$ has a relative minimum at $(\frac{2}{3}, -\frac{2}{3})$ where the value of z is $\frac{-8}{27}$.

Example 2

Purpose To use the M-test in an applied business problem:

Problem The Bodwell Company finds its profit function to be $P = f(x,y) = 30x + 140y + 5xy - 3x^2 - 3y^2 - 2000$ where x and y represent the numbers of two different models produced and sold per week. Determine the number of each model that should be produced each week to maximize profit. What is the maximum weekly profit?

Solution $f(x,y) = 30x + 140y + 5xy - 3x^2 - 3y^2 - 2000$

$\quad f_x(x,y) = 30 + 5y - 6x \qquad f_{xx}(x,y) = -6$

$\quad f_y(x,y) = 140 + 5x - 6y \qquad f_{yy}(x,y) = -6$

$\qquad\qquad\qquad\qquad\qquad\quad f_{xy}(x,y) = 5$

Find the critical point where $f_x(x,y)$ and $f_y(x,y)$ are equal to zero

$$f_x(x,y) = 30 + 5y - 6x = 0 \qquad 30x - 25y = 150$$
$$f_y(x,y) = 140 + 5x - 6y = 0 \qquad \underline{-30x + 36y = 840}$$
$$11y = 990$$
$$y = 90$$

By substitution into $30 + 5y - 6x = 0$, we obtain $x = 80$. The critical point is $(80,90)$.

$$M = 11 > 0 \qquad \text{and} \qquad f_{xx}(80,90) = -6 < 0.$$

Therefore, by the M-test, when $x = 80$ units and $y = 90$ units, the company will realize a maximum profit. The maximum weekly profit is

$$P = f(80,90) = 30(80) + 140(90) + 5(80)(90) - 3(80)^2 - 3(90)^2 - 2000$$
$$= \$5500.$$

Exercise 7.7

In Problems 1 to 6, examine the function for relative maxima and minima. Verify all conclusions by the M-test.

1. $f(x,y) = x^2 + y^2 - 3x$
2. $f(x,y) = 4x^3 + \frac{1}{2}y^3 - xy$
3. $f(x,y) = 6xy - x^3 - y^3$

4. $f(x,y) = 2x - 4y - x^2 - y^2 - 2$

5. $f(x,y) = xy^2 - 6x^2 - 3y^2$

6. $f(x,y) = x^3 + 2y^3 + \frac{3}{2}x^2 - 3y^2 - 60x - 36y$

7. If the demand functions n and m for two products (quantity x of one product, y of the other produced a day) are $n = 110 - x$, $m = 170 - \frac{1}{2}y$ and the joint cost function is $C = \$(x^2 + xy + 2y^2)$ then the profit $P = \$(nx + my - C)$. Determine the quantities x and y of the two products, and the prices n and m that will maximize the profit P. Also, find the maximum profit P.

8. A store's profit function P was determined to be $P = f(x,y) = \$[1400 - (12 - x)^2 - (40 - y)^2]$ (see Exercise 7.5–7.6, Problem 18). What values of x and y will maximize the profit? Find the maximum profit.

9. The labor cost to manufacture a precision instrument is $C = f(x,y) = \$(x^2 + y^3 - 2x - 3y - 2xy + 320)$ where x is the number of days required by a skilled craftsman and y is the number of days required by a semiskilled craftsman to produce an instrument. Find the x and y values that will minimize the cost.

10. A data processing firm pays senior programmers \$20 per hour and junior programmers \$12 per hour. A specific project will cost $z = 6000 + 6x^3 - 36xy + 3y^2$ where z is in dollars, x represents the number of junior programmers, and y the number of senior programmers used. How many employees of each kind should be assigned to the project in order to minimize the cost z?

11. A freeze-dried instant coffee manufacturer finds that its monthly profit is determined by $P(x,y) = 2400 + 18x + 76y - x^2 - \frac{1}{2}y^2$, where x is the cost of labor in dollars per hour and y the cost of coffee beans in dollars per 100 pound bags. Find the values of x and y that will yield a maximum profit, and what is the maximum profit?

12. The Backpack Company used past experiences to establish the following data on their Model I and Model II backpacks:

Model	Number Produced and Sold (daily)	Selling Price
I	x	$80 - x$
II	y	$50 - y$

The cost of manufacturing x units of Model I and y units of Model II is $C(x,y) = \$(x^2 + 2y^2 + 2xy)$.
(a) Show that the profit function is $P = f(x,y) = -2x^2 - 3y^2 + 80x + 50y - 2xy$.
(b) How many of each kind should be scheduled in order to maximize profit?

13. The Thermogram Company advertises in the newspapers, via radio and on television. They estimate that when the advertising budget is allocated by:

Medium	Advertising Dollars Spent
Newspapers	x
Radio	y
Television	z

sales of $x^{.3}y^{.5}z^{.6}$ are generated. Find the relationship between x, y, and z that will maximize net sales. (Assume that an extension of the previous method holds—If net sales $N = f(x,y,z) = x^{.3}y^{.5}z^{.6} - x - y - z$, solve simultaneously $f_x(x,y,z) = f_y(x,y,z) = f_z(x,y,z) = 0$.)

14. A closed rectangular box is to be constructed at a cost of $500. The material for the bottom of the box costs $1 per square foot and the material for the sides and the top costs $4 per square foot. Find the dimensions of the box of greatest volume that can be made.

15. What values of x and y will yield a maximum for the laboratory learning equation $L = f(x,y) = -x^2 + 6x - y^2 + 8y$?

16. Looking at a set of data $\{(x_i,y_i), i = 1, \ldots, n\}$, it is decided that a straight line $y = a + bx$ is the best fit to this data. To determine a and b the imposed criterion is that $\sum\limits_{i=1}^{n} (y_{actual} - y_{predicted})^2$ is to be minimized.

Therefore, a and b should be selected in such a way that $D = \sum\limits_{i=1}^{n} [y_i - (a + bx_i)]^2 = \sum\limits_{i=1}^{n} [y_i - a - bx_i]^2$ is a minimum. Minimize D by first finding $\dfrac{\partial D}{\partial a}$ and $\dfrac{\partial D}{\partial b}$. Then set these partial derivatives equal to zero and obtain the equations:

$$a = \frac{\sum\limits_{i=1}^{n} x_i^2 \sum\limits_{i=1}^{n} y_i - \sum\limits_{i=1}^{n} x_i y_i \sum\limits_{i=1}^{n} x_i}{n \sum\limits_{i=1}^{n} x_i^2 - \left(\sum\limits_{i=1}^{n} x_i \right)^2}$$

and

$$b = \frac{n \sum\limits_{i=1}^{n} x_i y_i - \sum\limits_{i=1}^{n} x_i \sum\limits_{i=1}^{n} y_i}{n \sum\limits_{i=1}^{n} x_i^2 - \left(\sum\limits_{i=1}^{n} x_i \right)^2}$$

$$\left(\text{Remember } \sum\limits_{i=1}^{n} a = na. \right)$$

These values of a and b give the best fitting straight line, in the least-squares sense, to a set of data. The technique described in this problem is called the *least-squares technique*.

7.8 RELATIVE EXTREMA WITH CONSTRAINTS, LAGRANGIAN MULTIPLIERS

When we wish to maximize or minimize $z = f(x,y)$ subject to another condition or constraint that $g(x,y) = 0$, there exists a method usually simpler than the preceding M-test. This is known as the method of Lagrangian multipliers. The method is stated without proof.

Method of Lagrangian Multipliers

If we wish to maximize or minimize $z = f(x,y)$ subject to the constraint $g(x,y) = 0$:

1. Form the Lagrangian function F where λ is called a multiplier:

$$F(x,y,\lambda) = f(x,y) + \lambda g(x,y)$$

2. Then the simultaneous solution of the three equations, $F_x(x,y,\lambda) = 0$, $F_y(x,y,\lambda) = 0$, $F_\lambda(x,y,\lambda) = 0$, contain those points (x_1,y_1) that will yield relative extrema. Notice the order in which the Lagrangian function is formed.

$$F(x,y,\lambda) = \underset{\substack{\text{function whose} \\ \text{extrema are to} \\ \text{be found}}}{f(x,y)} + \underset{\substack{\text{constraint} \\ \text{function}}}{\lambda g(x,y)}$$

The Lagrangian function must be formed in that order.

Although stated here for only two independent variables, the method of Lagrange is applicable when there are more than two independent variables present. See Problems B and C that follow.

Example 1

Purpose To use the method of Lagrange to find relative extrema with constraints:

Problem A Maximize the product of two numbers given their sum is 24.

Solution Let x and y be the two numbers. $P = f(x,y) = xy$ is to be maximized subject to the constraint $g(x,y) = x + y - 24 = 0$.

Form the Lagrangian function

$$F(x,y,\lambda) = xy + \lambda(x + y - 24)$$
$$F_x(x,y,\lambda) = y + \lambda$$
$$F_y(x,y,\lambda) = x + \lambda$$
$$F_\lambda(x,y,\lambda) = x + y - 24$$

Set $F_x(x,y,\lambda)$ and $F_y(x,y,\lambda)$ each equal to zero and solve each for λ.

$$y + \lambda = 0 \quad \text{or} \quad y = -\lambda$$
$$x + \lambda = 0 \quad \text{or} \quad x = -\lambda$$

Now set $F_\lambda(x,y,\lambda) = 0$, obtaining $x + y - 24 = 0$, and replace x and y by their equivalent expressions in terms of λ.

$$(-\lambda) + (-\lambda) - 24 = 0$$
$$-2\lambda = 24$$
$$\lambda = -12$$

Therefore $y = -\lambda = 12$ and $x = -\lambda = 12$.
The maximum product is $12 \cdot 12 = 144$.

Problem B   The Conglomerate Company has three plants, A, B, and C. The yearly production levels of the plants are: Plant A, x units per year; Plant B, y units per year; and Plant C, z units per year. A computer has established that Conglomerate's yearly revenue is given by $R = f(x,y,z) = \$[15x^2yz - 10^6x - 10^6y - 10^6z]$. Conglomerate has a government contract to produce 200 units per year. Each plant, by adding more turns, is capable of producing all 200 units per year, if desirable. How should Conglomerate allocate its production to meets its contract obligation and also maximize revenue? What is the maximum yearly revenue?

Solution   Since the total output is to be 200 units per year, we have the constraint: total output is $x + y + z = 200$. Therefore, we must maximize the revenue R subject to the above constraint.
Form the Lagrangian function.

$$F(x,y,z,\lambda) = 15x^2yz - 10^6x - 10^6y - 10^6z + \lambda(x + y + z - 200)$$
$$F_x(x,y,z,\lambda) = 30xyz - 10^6 + \lambda$$
$$F_y(x,y,z,\lambda) = 15x^2z - 10^6 + \lambda$$
$$F_z(x,y,z,\lambda) = 15x^2y - 10^6 + \lambda$$
$$F_\lambda(x,y,z,\lambda) = x + y + z - 200$$

$F_x(x,y,z,\lambda) = F_y(x,y,z,\lambda) = F_z(x,y,z,\lambda) = 0$
yields

$$10^6 - \lambda = 30xyz$$
$$10^6 - \lambda = 15x^2z$$
$$10^6 - \lambda = 15x^2y$$

or

$$30xyz = 15x^2z \quad \text{and} \quad 15x^2z = 15x^2y$$
$$2y = x \qquad\qquad\qquad z = y$$

$F_\lambda(x,y,z,\lambda) = 0$ yields $x + y + z - 200 = 0$.
But, $x = 2y$ and $z = y$. Therefore

$$2y + y + y - 200 = 0$$
$$y = 50$$
$$z = 50$$
$$x = 100$$

The maximum yearly revenue is
$R = \$[15(100)^2(50)(50) - 10^6(100) - 10^6(50) - 10^6(50)] = \$175,000,000.$

Problem C Maximize the product of three numbers given their sum is 24.

Solution Let x, y, and z be the three numbers. $P = f(x,y,z) = xyz$ is to be maximized subject to the constraint $g(x,y,z) = x + y + z - 24 = 0$.

$$F(x,y,z,\lambda) = xyz + \lambda(x + y + z - 24)$$
$$F_x(x,y,z,\lambda) = yz + \lambda$$
$$F_y(x,y,z,\lambda) = xz + \lambda$$
$$F_z(x,y,z,\lambda) = xy + \lambda$$
$$F_\lambda(x,y,z,\lambda) = x + y + z - 24$$

$F_x(x,y,z,\lambda) = F_y(x,y,z,\lambda) = F_z(x,y,z,\lambda) = 0$
yields

$$\lambda = -yz$$
$$\lambda = -xz$$
$$\lambda = -xy$$

From $\lambda = -yz = -xz = -xy$

$$-y = -x \quad \text{and} \quad -y = -z$$

So,

$$-x = -y = -z \quad \text{or} \quad x = y = z$$

Substituting this result into $F_\lambda(x,y,z,\lambda) = 0$ gives

$$x + y + z - 24 = 0$$
$$x + x + x - 24 = 0$$
$$x = 8$$
$$y = 8$$
$$z = 8$$

Maximum product is $8 \cdot 8 \cdot 8 = 512.$

Exercise 7.8

In Problems 1 to 3, find the maxima and minima (if any) of each of the functions subject to the stated constraint.

1. $f(x,y) = 5x^2 - 3y^2 + xy$ subject to the constraint $2x - y = 20$.

2. $f(x,y) = x^2 - 4xy + 4y^2$ subject to the constraint $x + y = 1$.

3. $f(x,y) = 2xy$ subject to the constraint $x^2 + y^2 = 32$.

Work the following problems by the method of Lagrangian multipliers.

4. Find the maximum and minimum values of the function $z = f(x,y) = x^2 - 4xy + y^2$ with the constraint that (x,y) lie on the circle $x^2 + y^2 = 1$.

5. Find the values K and L, the unit inputs of capital and labor, that will yield the maximum production Q where $Q = f(K,L) = K^{.4}L^{.6}$, $K \geqslant 0$, $L \geqslant 0$; subject to the budget constraints, $10K + 8L = 1000$.

6. For shipping purposes the Dunder Manufacturing Company wishes to produce a daily total of 70 units of their two models of electric tow motors. The joint daily cost function for producing x units of Model Deluxe and y units of Model Standard is given by $C = f(x,y) = 2x^2 + y^2 - \frac{1}{2}xy$. How should the production be allocated to minimize cost but to produce a total of 70 units?

7. The ABC Company produces two models of washing machines. The joint cost function C is given by $C = f(x,y) = \$[2x^2 + y^2 + xy]$ where x represents the number of machines of Model I built and y represents the number of machines of Model II built. If a total of 12 machines must be built, how should production be allocated in order to minimize the cost?

8. The production quantity, Q, as a function of inputs of x units of Type I and y units of Type II is $Q = f(x,y) = x^2 + 3xy - 6y$. Find the amounts of x and y that maximize production if $x + y = 50$.

9. The surface area of an open rectangular box is equal to 8 square inches. Find the dimensions of the box that has maximum volume.

10. The base of an open rectangular box costs \$3 per square foot to construct, while the sides cost only \$1 per square foot. Find the dimensions of the box of largest volume that can be constructed for \$36.

11. A warehouse is needed to contain a million cubic feet. It is estimated that the floor and ceiling of the building will cost \$3 per square foot to construct and the side walls will cost \$7 per square foot. Find the cost of the most economical rectangular building. Assume that the roof is flat.

12. A closed rectangular box of constant volume is to be constructed. What would be the dimensions of such a box if the surface area is to be a minimum?

13. Maximize the sum of three numbers x, y, and z if $x^2 + y^2 + z = 0$.

7.9 MULTIPLE INTEGRALS

Just as an area may be found by integration, it is possible to find a volume under a surface by double integration of a function over a region. For example, the volume under $z = f(x,y) = x + y$ and over the rectangular region R where $0 \leqslant x \leqslant 1$, $0 \leqslant y \leqslant 2$ is given by the double integral

$$\int_0^1 \int_0^2 z \, dy \, dx = \int_0^1 \int_0^2 (x + y) \, dy \, dx.$$

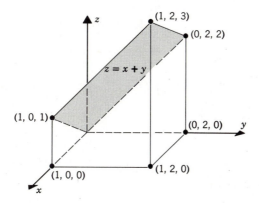

A theorem in advanced calculus states that under certain conditions a double integral may be evaluated by repeated or iterated integration. Thus to evaluate $\int_0^1 \left[\int_0^2 (x + y) \, dy \right] dx$, we first evaluate the inner integral $\int_0^2 (x + y) \, dy$. Notice that in this first integration y is treated as the independent variable and x is treated as a constant.

$$\int_0^2 (x + y) \, dy = xy + \frac{y^2}{2} \bigg]_0^2 = 2x + 2$$

This result is then integrated with respect to x in the outer integral.

$$\int_0^1 (2x + 2) \, dx = x^2 + 2x \bigg]_0^1 = 1 + 2 = 3$$

Therefore, $\int_0^1 \int_0^2 (x + y) \, dy \, dx = 3$.

In general, a function $f(x,y)$ integrated over a region R in the xy plane is expressed as

$$\iint_R f(x,y) \, dx \, dy.$$

The relationship between double integrals and iterated integrals over a rectangular region R in the xy plane where $a \leqslant x \leqslant b$ and $c \leqslant y \leqslant d$ is

$$\iint_R f(x,y)\, dx\, dy = \int_c^d \left[\int_a^b f(x,y)\, dx \right] dy = \int_a^b \left[\int_c^d f(x,y)\, dy \right] dx.$$

Example 1

Purpose   To set up and solve a double integral:

Problem   Evaluate $\displaystyle\iint_R xy^2\, dx\, dy$ over a rectangular region, R, in the xy plane where $3 \leqslant x \leqslant 4$ and $1 \leqslant y \leqslant 2$.

Solution
$$\int_1^2 \left[\int_3^4 xy^2\, dx \right] dy$$

Evaluate the inner integral

$$\int_3^4 xy^2\, dx = \frac{1}{2} x^2 y^2 \Big]_3^4 = \frac{1}{2}(16y^2 - 9y^2) = \frac{7}{2} y^2$$

Now integrate

$$\int_1^2 \frac{7}{2} y^2\, dy = \frac{7}{6} y^3 \Big]_1^2 = \frac{7}{6}(8 - 1) = \frac{49}{6}$$

An alternate setup is

$$\int_3^4 \left[\int_1^2 xy^2\, dy \right] dx$$

Evaluate the inner integral

$$\int_1^2 xy^2\, dy = \frac{1}{3} xy^3 \Big]_1^2 = \frac{1}{3}(8x - x) = \frac{7}{3} x$$

Now integrate

$$\int_3^4 \frac{7}{3} x\, dx = \frac{7}{6} x^2 \Big]_3^4 = \frac{7}{6}(16 - 9) = \frac{49}{6}$$

The next problem to consider is that of setting up a double integral over a region R that is not a rectangle. Given the function f and the region R, the region R must be sketched to determine the proper limits of integration. This is illustrated by the next example.

Example 2

Purpose   To illustrate the technique of setting up limits of integration in a double integral:

Problem   Find the volume under $z = x^2 + y^2$ and above the first quadrant region bounded by $y = 0$, $y = 2x$, $x = 1$. Do the integration first with respect to y.

Solution   Graph the region R either in E^2 or E^3.

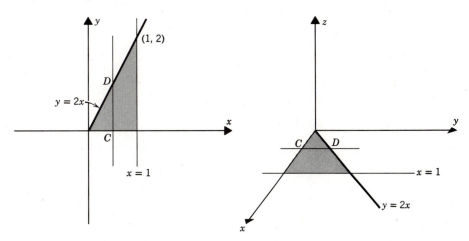

Since the first integration is with respect to y draw a line across R parallel to the y axis. y goes from its value at C to its value at D. Therefore, y goes from 0 to $2x$. Meanwhile, to sweep over R, x will go from its minimum value attained on R at (0,0) to its maximum value attained on R along line $x = 1$. Therefore, x goes from 0 to 1. The correct double integral is then $\displaystyle\int_0^1 \int_0^{2x} (x^2 + y^2)\, dy\, dx$. Do the inner integration treating x as a constant.

$$\int_0^{2x} (x^2 + y^2)\, dy = x^2y + \frac{1}{3}y^3 \Big]_0^{2x} = x^2(2x) + \frac{1}{3}(2x)^3$$

$$= 2x^3 + \frac{8}{3}x^3$$

$$= \frac{14}{3}x^3$$

Do the outer integration.

$$\int_0^1 \frac{14}{3}x^3\, dx = \frac{14}{12}x^4 \Big]_0^1 = \frac{14}{12} = \frac{7}{6}$$

Therefore, $\displaystyle\int_0^1 \int_0^{2x} (x^2 + y^2)\, dy\, dx = \frac{7}{6}$.

The concept of an iterated integral may be extended to that of a triple integral. This is illustrated by the example that follows.

Example 3

Purpose   To evaluate a triple integral:

Problem   Evaluate $\int_0^1 \int_0^x \int_0^y (x + y + z)\, dz\, dy\, dx$.

Solution   Evaluate first inner integral. Treat y and x as constants.

$$\int_0^y (x + y + z)\, dz = xz + yz + \frac{z^2}{2} \Big]_0^y$$

$$= xy + y^2 + \frac{y^2}{2} = xy + \frac{3y^2}{2}$$

Thus, $\int_0^1 \int_0^x \int_0^y (x + y + z)\, dz\, dy\, dx = \int_0^1 \int_0^x \left(xy + \frac{3y^2}{2}\right) dy\, dx$.   Evaluate the second inner integral. Treat x as a constant.

$$\int_0^x \left(xy + \frac{3y^2}{2}\right) dy = \frac{xy^2}{2} + \frac{y^3}{2} \Big]_0^x$$

$$= \frac{x^3}{2} + \frac{x^3}{2} = x^3$$

Thus, $\int_0^1 \int_0^x \int_0^y (x + y + z)\, dz\, dy\, dx = \int_0^1 x^3\, dx$. Evaluate the outer integral.

$$\int_0^1 x^3\, dx = \frac{x^4}{4} \Big]_0^1 = \frac{1}{4}$$

$$\int_0^1 \int_0^x \int_0^y (x + y + z)\, dz\, dy\, dx = \frac{1}{4}.$$

This section is not meant to be a complete treatment of the multiple integral. It has been concerned with the evaluation of multiple integrals and also with one of the many applications of the multiple integral, that of a volume. However, a multiple integral does not always represent a volume. It has many other interpretations in probability theory, physics, psychology, and in the social studies.

Exercise 7.9

In Problems 1 to 18, evaluate the integrals.

1. $\int_2^4 \int_0^3 dy\, dx$

2. $\int_0^3 \int_2^4 dx\, dy$

3. $\int_{-1}^3 \int_0^4 dx\, dy$

4. $\int_{-1}^3 \int_0^4 dy\, dx$

5. $\int_0^{a^2} \int_0^a dx\, dy$

6. $\int_0^2 \int_0^{x\sqrt{4-x^2}} dy\, dx$

7. $\displaystyle\int_{-1}^{1}\int_{0}^{3} (4 - x^2)\, dy\, dx$

13. $\displaystyle\int_{0}^{2}\int_{x}^{2x} (x^2 + y^3)\, dy\, dx$

8. $\displaystyle\int_{0}^{1}\int_{0}^{4} (xy)\, dx\, dy$

14. $\displaystyle\int_{4}^{8}\int_{0}^{y^{3/2}} \left(\frac{x}{y^2}\right) dx\, dy$

9. $\displaystyle\int_{0}^{2}\int_{0}^{y} (x + y)^2\, dx\, dy$

15. $\displaystyle\int_{-a}^{a}\int_{0}^{x} (x)\, dy\, dx$

10. $\displaystyle\int_{0}^{2}\int_{x}^{2x} (xy)\, dy\, dx$

16. $\displaystyle\int_{0}^{1}\int_{0}^{x} (e^{x^2} + x)\, dy\, dx$

11. $\displaystyle\int_{1}^{3}\int_{0}^{1/x} (x^2)\, dy\, dx$

17. $\displaystyle\int_{0}^{2}\int_{\ln(x)}^{\ln(x^2)} (e^y)\, dy\, dx$

12. $\displaystyle\int_{0}^{4}\int_{\sqrt{x}}^{\sqrt{x+1}} (y)\, dy\, dx$

18. $\displaystyle\int_{0}^{\ln 5}\int_{0}^{y} (e^x)\, dx\, dy$

19. Determine the volume under the surface $z = y^2$ and above the region where $0 \le x \le 3$ and $0 \le y \le 2$.

20. Determine the volume under the surface $z = 4 - \sqrt{x}$ and above the region where $0 \le x \le 4$ and $0 \le y \le 6$.

21. Determine the volume under the plane $z = 6 - x - y$ and above the region where $0 \le x \le 6 - y$ and $0 \le y \le 6$.

22. Determine the volume under the plane $z = 4 - 2x - 2y$ and above the region where $0 \le y \le 2 - x$ and $0 \le x \le 2$.

23. Find the volume of the solid in the first octant bounded by $x + y + z = 4$.

24. Find the area bounded by the following two curves in the xy-plane: $y = x^3$, $y = x^2$. $\left(A = \displaystyle\iint_{R} dy\, dx.\right)$

25. Find the area of the region bounded by $y^2 = 4x$ and $x^2 = 4y$. $\left(A = \displaystyle\iint_{R} dy\, dx.\right)$

In Problems 26 to 33, evaluate the integrals.

26. $\displaystyle\int_{0}^{3a}\int_{0}^{2a}\int_{0}^{a} dz\, dy\, dx$

30. $\displaystyle\int_{0}^{2}\int_{0}^{x}\int_{-x^2y}^{xy^2} dz\, dy\, dx$

27. $\displaystyle\int_{1}^{2}\int_{0}^{2x}\int_{0}^{3y} dz\, dy\, dx$

31. $\displaystyle\int_{0}^{4}\int_{0}^{(4-x)/2}\int_{0}^{(4-x-2y)/2} dz\, dy\, dx$

28. $\displaystyle\int_{0}^{6}\int_{2}^{x}\int_{4y}^{3y^2} dz\, dy\, dx$

32. $\displaystyle\int_{-3}^{0}\int_{0}^{6}\int_{y}^{x} (xy)\, dz\, dy\, dx$

29. $\displaystyle\int_{1}^{2}\int_{1}^{x}\int_{x+y}^{6y^2} dz\, dy\, dx$

33. $\displaystyle\int_{0}^{1}\int_{0}^{1}\int_{-xy}^{xy} e^{x^2+y^2}\, dz\, dx\, dy$

■ 7.10 CHAPTER REVIEW

Important Ideas

three-dimensional coordinate system
xy-, yz-, and xz-planes
function
limit of a function
continuous function at (a,b)
traces in the xy-, yz-, xz-planes
plane
cylinder
hyperbolic paraboloid
saddle point
first partial derivative of f with respect to x
first partial derivative of f with respect to y
second-order partial derivatives
relative maximum
relative minimum
M-test for relative extrema
method of Lagrangian multipliers
multiple integrals

■ REVIEW EXERCISE (optional)

1. Define a function in E^3.

2. Plot, in E^3, the points whose coordinates are: $(0,0,-2)$, $(3,2,5)$, $(-2,0,4)$, $(1,-2,0)$, $(5,4,-2)$.

3. Given $f(x,y) = x^2 + \dfrac{x}{y} - \ln(xy)$

   Find $f(3,1)$, $f(4,2)$.

4. Let $P_1(x_1,y_1,z_1)$ and $P_2(x_2,y_2,z_2)$ be two points in E^3. Let P_3 be (x_2,y_2,z_1).

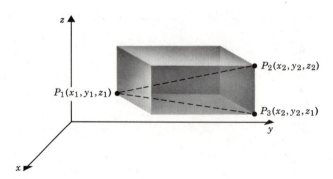

Since P_1 and P_3 lie in the same plane, you may use the formula $d = \sqrt{(x_2 - x_1)^2 + (y_2 - y_1)^2}$ for the distance P_1P_3. Also, $P_3P_2 = (z_2 - z_1)$. Use the Pythagorean theorem on right triangle $P_1P_2P_3$ to find the distance from P_1 to P_2.

In Problems 5 to 8, sketch the surface represented by the equation in E^3 space. Use traces in the xy-, xz-, and yz-planes to aid your graphing.

5. $4x + 3y + 6z = 12$ **7.** $y^2 + z^2 = 4x$

6. $3x + z = 9$ **8.** $x - 2y + 3z = 6$

In Problems 9 to 12, find $f_x(x,y)$, $f_y(x,y)$, $f_{xx}(x,y)$, $f_{yy}(x,y)$, $f_{xy}(x,y)$, $f_{yx}(x,y)$ for each function.

9. $f(x,y) = x^2y - y^2x + 10x$ **11.** $f(x,y) = xe^y - \ln(xy)$

10. $f(x,y) = \dfrac{x}{y} + \dfrac{y}{x}$ **12.** $f(x,y) = (2x + 3y)^2$

13. If $f(x,y) = \dfrac{e^{xy}}{x + y}$, find $f_x(x,y)$.

14. If $f(x,y) = \dfrac{1}{x^2 + y^2}$, find $f_x(-1,1)$ and $f_y(-1,1)$.

15. If $z = f(x,y) = \sqrt{9 - x^2 - y^2}$ show that $\dfrac{\partial z}{\partial x} = \dfrac{-x}{z}$ and $\dfrac{\partial z}{\partial y} = \dfrac{-y}{z}$.

16. Given $f(x,y) = \dfrac{x}{x + y}$.

(a) Show that $f_{xy}(x,y) - f_{xx}(x,y) + f_y(x,y) = \dfrac{1 - x}{(x + y)^2}$.

(b) Show that $f_{xy}(x,y) = f_{yx}(x,y)$.

17. $Q = x^2y + y$ is a production function that gives the amount of output Q as a function of the inputs x and y. $\dfrac{\partial Q}{\partial x}$ is called the marginal productivity of x and represents the rate of change of Q as input x is increased, if input y remains constant. Find and interpret $\dfrac{\partial Q}{\partial x}$ and $\dfrac{\partial Q}{\partial y}$.

18. Assume the productivity function for the Lisbon Company is $Q = f(x,y) = \dfrac{e^{xy}}{x^2 + y^2}$.

(a) Find and interpret $\dfrac{\partial Q}{\partial x}$, $\dfrac{\partial Q}{\partial y}$.

(b) Find and interpret $\dfrac{\partial Q}{\partial x}\bigg]_{(100,5)}$

(c) Find and interpret $\dfrac{\partial Q}{\partial y}\bigg]_{(3,50)}$

In Exercises 19 to 21, examine the functions for relative maxima and minima. Verify all conclusions by the M-test.

19. $f(x,y) = x^3 + y^3 - 6xy$

20. $f(x,y) = 4x + 2y - x^2 - y^2$

21. $f(x,y) = xy + (x + y)(60 - x - y)$

22. A daily cost function is determined to be $C = f(x,y) = 36x + 25y + \dfrac{49x + 64y}{xy}$. Here, x represents the number of units in hundreds of product A and y represents the number of units in hundreds of product B produced per day. How many units of products A and B should be produced in a 60-day time period to minimize the daily cost?

23. The Crihfield Company finds that the weekly profit when x units of the standard model and y units of the deluxe model are produced is given by $P = f(x,y) = 2xy + 500x + 40y$. If the union contract specifies that a total of 1000 units per week must be built, how should the company allocate production to maximize the profit? Use the method of Lagrangian multipliers.

24. What are the three numbers whose sum is 36 and whose product is maximum?

25. A closed rectangular box of constant volume is to be constructed. What would be the dimensions of such a box if the surface area is to be a minimum? Use the M-test. (Let the dimensions of the box be x, y, and z. State z in terms of x and y; or $z = \dfrac{V}{xy}$. Then the surface area can be expressed as a function of x and y.)

In Problems 26 to 31, evaluate the integrals.

26. $\displaystyle\int_1^3 \int_3^5 x \, dy \, dx$

27. $\displaystyle\int_1^3 \int_3^5 x \, dx \, dy$

28. $\displaystyle\int_0^2 \int_x^{x^2} xy \, dy \, dx$

29. $\displaystyle\int_0^1 \int_0^{\sqrt{x}} e^{x^2} y \, dy \, dx$

30. $\displaystyle\int_0^1 \int_y^{\sqrt{y}} xy \, dx \, dy$

31. $\displaystyle\int_1^2 \int_0^x x^2 e^{xy} \, dy \, dx$

32. Determine the volume under the surface $z = f(x,y) = x^2 + y^2$ and above the region where $1 \leq x \leq 2$ and $1 \leq y \leq 3$.

33. Find the volume under $z = f(x,y) = 4$ and above the triangle $0 \leqslant x \leqslant 2$, $0 \leqslant y \leqslant x$.

34. Find the volume of the solid in the first octant bounded by $3x + 4y + 6z = 12$.

In Problems 35 and 36, evaluate the integrals.

35. $\displaystyle\int_0^9 \int_0^x \int_{-\sqrt{y}}^{\sqrt{x}} dz\, dy\, dx$

36. $\displaystyle\int_0^1 \int_0^x \int_{2y}^{1+y^2} x\, dz\, dy\, dx$

Appendix

I. REVIEW OF BASIC MATHEMATICAL SKILLS

Product

1. Distributive property

$$a(x + y) = ax + ay$$

2. (Sum of two terms) times (difference of same two terms)

$$(x + y)(x - y) = x^2 - y^2$$

3. (a) Sum of two terms squared

$$(x + y)^2 = x^2 + 2xy + y^2$$

3. (b) Difference of two terms squared

$$(x - y)^2 = x^2 - 2xy + y^2$$

4. $(x + b)(x + d) =$
$$x^2 + (b + d)x + bd$$

often called FOIL rule, where F = first, O = outside, I = inside, and L = last.

$$
\begin{array}{cc}
F & L \\
(3x - 4)(2x + 1) = \\
I \\
O
\end{array}
$$

$$
\begin{array}{cccc}
F & O & I & L \\
\end{array}
$$
$$6x^2 + 3x - 8x - 4 =$$
$$6x^2 - 5x - 4$$

Factoring

1. Removing highest common factor

$$ax + ay = a(x + y)$$

2. Difference of two squares

$$x^2 - y^2 = (x + y)(x - y)$$

3. (a) Perfect square

$$x^2 + 2xy + y^2 = (x + y)^2$$

Perfect Squares

$2xy$ is twice the product of the square roots of x^2 and y^2.

3. (b) Perfect square

$$x^2 - 2xy + y^2 = (x - y)^2$$

4. Three-term quadratic

$$6x^2 - 5x - 4 = (3x - 4)(2x + 1)$$

Try different combinations and check. This is usually called factoring by inspection or by trial and error.

Examples

(A) $6x + 9y = 3(2x + 3y)$

(B) $4x^2 - 25 = (2x + 5)(2x - 5)$

(C) $x^2 + 6xy + 9y^2 = (x + 3y)^2$

(D) $4x^2 - 4x + 1 = (2x - 1)^2$

(E) $x^2 - x - 6 = (x - 3)(x + 2)$

(F) $6x^2 - 13xy - 5y^2 = (3x + y)(2x - 5y)$

Exercise A.I.1

Factor each of the following.

1. $4xy + 8xz$
2. $2x^2 - 10x + 16$
3. $-6x^2 + 3xy$
4. $x^2 - 9$
5. $16x^2 - w^2$
6. $75 - 27x^2$
7. $x^4 - y^4$
8. $x^2 + 2x + 1$
9. $x^2 + 4x + 4$
10. $12x^2 - 12x + 3$
11. $x^2 - 2x - 3$
12. $x^2 + 2x - 3$
13. $x^3 - 16x$
14. $6x^2 - 11x - 10$
15. $3z(x + y) - (x + y)$
16. $2x^4 + 8x^3 + 6x^2$
17. $8x^2 + 18y^2$
18. $ac - ad + c^2 - cd$
19. $x^2y^2 - 36y^2$
20. $2x^2 + 7xy + 6y^2$

21. $x^2 - 6x + 8$
22. $3x + 4xy - 7x^2$
23. $15x^2 - 60$
24. $w^4 + w^2 - 2$
25. $t^2 - 4t + 4$
26. $x^5 + 5x^4 - 6x^3$
27. $x^4 - 10x^2 + 9$
28. $x^2z^2 - 25z^4$
29. $3x^2 + 6x + 12$
30. $2 + 3x + x^2$
31. $x^3y - y^3x$
32. $u^2 - 8u + 16$
33. $x^2y^3 - x^3y^2$
34. $4x^2 - 10x + 4$
35. $(x + 2)^2 + 3(x + 2) + 2$
36. $25x^2 + 60xy + 36y^2$
37. $9t^2 + 12tv + 4v^2$
38. $5a^2 + 14ab + 9b^2$
39. $2x^2 + 7xy + 6y^2$
40. $6x^2 + 3x - 9$

Algebraic Fractions

	Method	*Example*
1.	$\dfrac{a}{b} = \dfrac{ak}{bk}$	$\dfrac{9x}{6xy} = \dfrac{3 \cdot 3x}{2y \cdot 3x} = \dfrac{3}{2y}$
2.	$\dfrac{a}{b} + \dfrac{c}{d} = \dfrac{ad + bc}{bd}$	$\dfrac{2x}{x - 2} + \dfrac{3}{x + 2} = \dfrac{2x(x + 2) + 3(x - 2)}{(x - 2)(x + 2)}$
		$= \dfrac{2x^2 + 7x - 6}{x^2 - 4}$
3.	$\dfrac{a}{b} - \dfrac{c}{d} = \dfrac{ad - bc}{bd}$	$\dfrac{2x}{3} - \dfrac{5y}{2} = \dfrac{4x - 15y}{6}$
4.	$\dfrac{a}{b} \cdot \dfrac{c}{d} = \dfrac{ac}{bd}$	$\dfrac{x^2 - 4}{xy^2} \cdot \dfrac{2xy}{x - 2} = \dfrac{(x + 2)(x - 2)(2xy)}{xy^2(x - 2)}$
		$= \dfrac{2(x + 2)}{y}$

5. $\dfrac{\dfrac{a}{b}}{\dfrac{c}{d}} = \dfrac{ad}{bc}$

$\dfrac{\dfrac{x + 2xy}{3x^2}}{\dfrac{2y + 1}{6x}} = \dfrac{x(1 + 2y)(6x)}{3x^2(2y + 1)} = 2$

Examples

(A) $\dfrac{x^2 - 7x + 12}{2x - 8} = \dfrac{(x - 3)(x - 4)}{2(x - 4)} = \dfrac{x - 3}{2}$

(B) $\dfrac{x}{x^2 - 9} + \dfrac{3x}{x + 3} = \dfrac{x}{(x + 3)(x - 3)} + \dfrac{3x(x - 3)}{(x + 3)(x - 3)} =$

$\dfrac{x + 3x(x - 3)}{(x + 3)(x - 3)} = \dfrac{x + 3x^2 - 9x}{(x + 3)(x - 3)} = \dfrac{3x^2 - 8x}{(x + 3)(x - 3)}$

(C) $\dfrac{x - 1}{2x^2 - 7x + 3} - \dfrac{4}{2x - 1} = \dfrac{x - 1}{(2x - 1)(x - 3)} - \dfrac{4}{2x - 1}$

$= \dfrac{x - 1}{(2x - 1)(x - 3)} - \dfrac{4(x - 3)}{(2x - 1)(x - 3)} = \dfrac{x - 1 - 4(x - 3)}{(2x - 1)(x - 3)}$

$= \dfrac{x - 1 - 4x + 12}{(2x - 1)(x - 3)} = \dfrac{-3x + 11}{(2x - 1)(x - 3)}$

(D) $\dfrac{2x - y}{x^2 + 7x + 10} \cdot \dfrac{2x^2 + 9x - 5}{4x^2 - y^2} = \dfrac{(2x - y)(2x - 1)(x + 5)}{(x + 5)(x + 2)(2x + y)(2x - y)} =$

$\dfrac{2x - 1}{(x + 2)(2x + y)}$

(E) $\dfrac{x^2 - 16}{2y^2} \div \dfrac{xy - 4y}{6y^2 - 4xy} = \dfrac{x^2 - 16}{2y^2} \cdot \dfrac{6y^2 - 4xy}{xy - 4y}$

$= \dfrac{(x + 4)(x - 4)(2y)(3y - 2x)}{(2y^2)(y)(x - 4)} = \dfrac{(x + 4)(3y - 2x)}{y^2}$

(F) $\dfrac{\dfrac{y + 2}{6x^2}}{\dfrac{y + 2}{x}} = \dfrac{y + 2}{6x^2} \cdot \dfrac{x}{y + 2} = \dfrac{(y + 2)(x)}{(6x^2)(y + 2)} = \dfrac{1}{6x}.$

Exercise A.I.2

Perform the indicated operations and simplify.

1. $\dfrac{2xy^3}{16y}$

2. $\dfrac{2 - x}{3x^2y} - \dfrac{4 - 3y}{2xy^2}$

3. $\dfrac{4x^2 - 16}{x^2 - 2x}$

4. $\dfrac{2x}{3} \cdot \dfrac{9x}{4x^2y}$

5. $\dfrac{2x}{x-2} + \dfrac{3}{x-2}$

6. $\dfrac{2x}{x-2} - \dfrac{3}{x+2}$

7. $\dfrac{3}{2x-4y} - \dfrac{5}{x^2-4y^2}$

8. $\dfrac{4x}{x^2-4} - \dfrac{2}{x+2} + \dfrac{2}{x-2}$

9. $\dfrac{4x^2 - 20xy + 25y^2}{6x^2 - 15xy}$

10. $\dfrac{4x^2 - 20x}{x^2 - 4x - 5}$

11. $\dfrac{x}{x-5} + \dfrac{3x}{4x-20}$

12. $\dfrac{3}{2x} + \dfrac{2}{3x^2} - \dfrac{5}{6x^2}$

13. $\dfrac{6}{x-2} + \dfrac{3}{x} \cdot \dfrac{1-2x}{x-2}$

14. $\dfrac{2a^2 - 5a + 2}{\dfrac{2a-1}{3}}$

15. $(b^2 - a^2) \div \dfrac{a^2 + 2ab + b^2}{2a - 3b}$

16. $\dfrac{\dfrac{3}{2x}}{\dfrac{6}{4y}}$

17. $\dfrac{x^2 + 2x}{2x}$

18. $\dfrac{x^2 - 3x - 4}{\dfrac{x+1}{2}}$

19. $\dfrac{\dfrac{1}{x} + \dfrac{1}{y}}{x^2 - y^2}$

20. $\dfrac{\dfrac{x^2 - 9}{3y}}{y(x+3)}$

Solution of Equations

General linear equation. $ax + b = 0, \quad a \neq 0$

Solution of equation for x. $\quad x = -\dfrac{b}{a}$

Example: $\quad 5x + 9 = 0; \quad x = -\dfrac{9}{5}$

General quadratic equation. $ax^2 + bx + c = 0, \quad a \neq 0.$

Solution of equation for x.

(a) Factor (if possible)

Example: $2x^2 + 5x - 12 = 0$

$(2x - 3)(x + 4) = 0$

$2x - 3 = 0 \qquad x - 4 = 0$

$x = \dfrac{3}{2} \qquad x = 4$

(b) Use quadratic formula which is

$$x = \frac{-b \pm \sqrt{b^2 - 4ac}}{2a}$$

Example: $2x^2 + 7x - 6 = 0$

$$x = \frac{-7 \pm \sqrt{49 - 4(2)(-6)}}{2 \cdot 2} = \frac{-7 \pm \sqrt{97}}{4}.$$

Examples

(A) $x - 1 + 4x = 6 + 3x - 10$

$5x - 1 = 3x - 4$

$2x = -3$

$x = -\frac{3}{2}$

(B) $6x^2 + 5x = 4$

$6x^2 + 5x - 4 = 0$

$(3x + 4)(2x - 1) = 0$

$3x + 4 = 0 \qquad 2x - 1 = 0$

$x = -\frac{4}{3} \qquad x = \frac{1}{2}$

(C) $\frac{1}{2}x^2 + 4x + 8 = 0$

$x^2 + 8x + 16 = 0$

$(x + 4)^2 = 0$

$x + 4 = 0$

$x = -4$

(D) $3x^2 - 7x - 3 = 0$

$$x = \frac{-(-7) \pm \sqrt{(-7)^2 - 4(3)(-3)}}{2 \cdot 3}$$

$$= \frac{7 \pm \sqrt{49 + 36}}{6}$$

$$= \frac{7 \pm \sqrt{85}}{6}$$

Exercise A.I.3

Solve the following equations for x.

1. $3x + 7 = 0$
2. $3 - 2x = 7 - x$
3. $\dfrac{x + 7}{4} = \dfrac{2}{3}$
4. $3(x - 1) = x + 8$
5. $\frac{2}{3}(x + 7) = -(x - 7)$
6. $x^2 - 2x - 24 = 0$
7. $2x^2 - 7x - 4 = 0$
8. $x(x - 3) = 10$
9. $x^2 + 3x = 0$
10. $3x^2 - 27 = 0$
11. $x^2 - 6x + 9 = 0$
12. $x^2 + 6x + 9 = 0$
13. $x^2 - 10x + 9 = 0$
14. $x^2 + 8x - 9 = 0$
15. $9x^2 - 16 = 0$

16. $9x^2 + 12x + 4 = 0$
17. $x^2 - x - 1 = 0$
18. $6x^2 - 5x - 4 = 0$
19. $x^2 - 4x + 2 = 0$
20. $3x^2 + 2x = 2$
21. $12x^3 - 27x = 0$
22. $x^2 - 10x + 20 = 0$
23. $2x^2 - 10x + 11 = 0$
24. $3x^2 + 3x - 1 = 0$
25. $x^2 + 6x - 8 = 0$
26. $36x^2 + 24x - 1 = 0$
27. $2x^2 - 7x - 9 = 0$
28. $x^2 + 2x - 1 = 0$
29. $3x^2 - 2x - 1 = 0$
30. $x^2 + 4x + 3 = 2$

Exponents, Radicals, and Logarithms

(See Chapter 6 for the restrictions on b, N, L, m, and n.)

| *Definition* | *Example* |

$$\log_b N = L \text{ if } b^L = N$$

$$b^0 = 1, \; b \neq 0$$

$$b^n = \underbrace{b \cdot b \cdot b \cdots b}_{n \text{ times}}$$

$$b^{-n} = \frac{1}{b^n}$$

$$\sqrt[n]{b} = a \text{ if } b = a^n$$

$$\sqrt[n]{b} = b^{1/n}$$

$$\log_3 81 = 4 \text{ if } 3^4 = 81$$

$$5^0 = 1$$

$$x^4 = x \cdot x \cdot x \cdot x$$

$$y^{-2} = \frac{1}{y^2}; \qquad \frac{1}{3} = 3^{-1}$$

$$\sqrt[3]{64} = 4 \text{ if } 64 = 4^3$$

$$\sqrt{x} = x^{1/2}; \qquad \frac{1}{\sqrt[3]{x}} = x^{-1/3}$$

Properties

$$b^m \cdot b^n = b^{m+n}$$

$$\frac{b^m}{b^n} = b^{m-n}$$

$$(b^m)^n = b^{m \cdot n}$$

$$(ab)^m = a^m \cdot b^m$$

$$b^{m/n} = \sqrt[n]{b^m} = (\sqrt[n]{b})^m$$

$$\log_b (m^n) = n \log_b m$$

$$\log_b (mn) = \log_b m + \log_b n$$

$$\log_b \left(\frac{m}{n}\right) = \log_b m - \log_b n$$

Examples

$$x^2 \cdot x^3 = x^5; \qquad x^{1/2} \cdot x^{1/3} = x^{5/6}$$

$$\frac{7^8}{7^3} = 7^5; \qquad \frac{y^4}{y^5} = y^{-1}$$

$$(x^2)^3 = x^6; \qquad (x^{1/2})^4 = x^2$$

$$(2x^4)^2 = 4x^8; \qquad \sqrt{9x} = 3\sqrt{x}$$

$$8^{2/3} = \sqrt[3]{8^2} = (\sqrt[3]{8})^2 = 4$$

$$\log_b x^3 = 3 \cdot \log_b x$$

$$\log_b (4 \cdot 3) = \log_b 4 + \log_b 3$$

$$\log_b \left(\frac{x}{y}\right) = \log_b x - \log_b y$$

Examples

(A) $\quad 3^0 \cdot 3^4 = 3^{0+4} = 3^4 \quad$ or

$\qquad 3^0 \cdot 3^4 = 1 \cdot 3^4 = 3^4$

(B) $\quad (xy^2)^3 = x^3(y^2)^3 = x^3 y^6$

(C) $\quad b^5 \cdot b^{-3} = b^{5+(-3)} = b^2 \quad$ or

$\qquad b^5 \cdot b^{-3} = \dfrac{b^5}{b^3} = b^{5-3} = b^2$

(D) $\quad (2x^2) + (2x)^2 = 2x^2 + 4x^2 = 6x^2$

(E) $\quad \left(\dfrac{x^5}{x^2}\right)^3 = (x^{5-2})^3 = (x^3)^3 = x^9 \quad$ or

$\qquad \left(\dfrac{x^5}{x^2}\right)^3 = \dfrac{x^{15}}{x^6} = x^{15-6} = x^9$

(F) $\quad \dfrac{10^2 \cdot 10^3}{10^{-2}} = \dfrac{10^5}{10^{-2}} = 10^{5-(-2)} = 10^7 \quad$ or

$\qquad \dfrac{10^2 \cdot 10^3}{10^{-2}} = \dfrac{10^2 \cdot 10^3}{\dfrac{1}{10^2}} = 10^2 \cdot 10^3 \cdot 10^2 = 10^7.$

Exercise A.I.4.

Simplify each of the following expressions.

1. $(2a^2)^3$

2. $\dfrac{2^0 - 2^{-2}}{2 - 2(2)^{-2}}$

3. $(xy)^2 \cdot x$

4. $4^2 + (\frac{1}{4})^{-2}$

5. $(5^2 \cdot 5^3) \div 5$

6. $8^{4/3}$

7. $2^3 \cdot 2^5$

8. $\dfrac{2^3}{2^5}$

9. $(2^3)^5$

10. $125^{2/3}$

11. $9^{3/2}$

12. $(6x)^2 - 6x^2$

13. $\left(\dfrac{a}{b}\right)^{-2}$

14. $(2^2)^3$

15. $2^3 \cdot 2^2$

16. $2^3 + 2^2$

17. $\dfrac{1}{(\frac{1}{2})^{-3}}$

18. $\sqrt[3]{64x^7y^{-6}}$

19. $\sqrt[4]{9} \cdot \sqrt[4]{9}$

20. $5^0 \cdot 2^5 - 2^0 \cdot 5^2$

21. $\dfrac{x^3}{x^{-3}}$

22. $\dfrac{x^{-3}}{x^3}$

23. $\dfrac{1}{(\frac{1}{2})^3}$

24. $\dfrac{x^{-1} + y^{-1}}{(xy)^{-1}}$

25. $\frac{1}{2} \log_3 9$

26. $\dfrac{3^2 - 3^0}{3^3 - 3^2}$

27. $\dfrac{4^{3n}}{2^n}$

28. $(\sqrt{16} + 3)(\sqrt{16} - 3)$

29. $\log_{12} 9 + \log_{12} 16$

30. $\left(\dfrac{x}{y}\right)^{-1}$

31. $\sqrt{16 + 9}$

32. $\sqrt{16} + \sqrt{9}$

33. $\log_2 32 - \log_2 4$

34. $\log_e 1$

35. $\log_4 64$

36. $\sqrt{3x} + (3x)^{-1/2}$

37. $\log_{10} \sqrt{x} + \log_{10} \dfrac{1}{\sqrt{x}}$

38. $\log_e e$

39. State whether the following are true or false.

    (a) $x^3 + x^3 = 2x^3$ (d) $2^2 \cdot 3^2 = 6^4$

    (b) $x^3 \cdot x^3 = x^9$ (e) $2^2 \cdot 3^2 = 36$

    (c) $2 \cdot 3^2 = 6^2$ (f) $4^2 + 4 = 4^3$

40. State whether the following are true or false.

    (a) $\dfrac{1}{\sqrt{5} + \sqrt{3}} = \dfrac{\sqrt{5} - \sqrt{3}}{2}$ (e) $4^2 \cdot 4 = 4^3$

    (b) $\dfrac{1}{\sqrt{x}} = x^{1/2}$ (f) $\dfrac{1}{x^3} = \sqrt[3]{x}$

    (c) $\sqrt[3]{x^2} = x^{2/3}$ (g) $\dfrac{1}{4} \log_b 16 = \log_b 2$

    (d) $\dfrac{1}{x^4} = x^{1/4}$ (h) $\log_2 1 - \log_2 8 = -3$

II.   LIMITS

Limit of a Function, $f(x)$

What is meant by $\lim_{x \to a} f(x) = L$?

As we mentioned earlier, the concept of limit was used in mathematics long before Cauchy gave a suitable definition of limit. We now examine more closely the idea of limit, the "glue" that holds all of the calculus together.

Let us begin by asking: What do we mean when we say $\lim_{x \to 3} (2x + 1) = 7$?

1.   The first attempt to explain $\lim_{x \to 3} (2x + 1) = 7$.

This means that when x is very close to 3 in value, then $2x + 1$ is very close to 7.

2.   The second attempt to explain $\lim_{x \to 3} (2x + 1) = 7$.

Since "close to" is not a definite mathematical concept, we have trouble knowing what we mean by these words. Is Boston close to New York? Is the earth close to its moon? Answers to such questions depend on the definition of "close to." Therefore, our first attempt to explain $\lim_{x \to 3} (2x + 1) = 7$ is unsatisfactory. For our second attempt to explain $\lim_{x \to 3} (2x + 1) = 7$, we will state: By $\lim_{x \to 3} (2x + 1) = 7$ we mean that the value of $2x + 1$ can be made as near 7 as you would ever wish to have it, by making x near enough to 3. That is, if you wish $2x + 1$ to differ from 7 by less than .0000000001 then I can tell you how near to 3 you must take x to insure this condition.

And, moreover, if you wish $2x + 1$ to differ from 7 by less than any amount, however small, I can always tell you how near to 3 in value x must be taken in order to guarantee your wish.

3.   The third attempt to explain $\lim_{x \to 3} (2x + 1) = 7$.

Let us say that you select a positive number, however small, and insist that the absolute value by which $2x + 1$ differs from 7 is to be less than your selected positive number. Call your selected number ϵ. That is, you are insisting:

$$|(2x + 1) - 7| < \epsilon \leftrightarrow -\epsilon < (2x + 1) - 7 < \epsilon$$
$$\leftrightarrow 7 - \epsilon < (2x + 1) < 7 + \epsilon.$$

Now, if I can always tell you what deleted neighborhood about 3, $DN_\delta(3)$, you should use, so that when x belongs to this deleted neighborhood, your statement $|(2x + 1) - 7| < \epsilon$ is true, then by Cauchy's definition the $\lim_{x \to 3} (2x + 1) = 7$.

A *deleted neighborhood* of a, $DN_\delta(a)$, is an open interval with the point a removed from the center of the open interval.

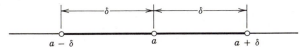

Deleted neighborhood of a.

To represent this deleted neighborhood of a, $DN_\delta(a)$, by an inequality, we write: $0 < |x - a| < \delta$. This is equivalent to $a - \delta < x < a$ and $a < x < a + \delta$ (from properties of absolute inequalities).

Example 1

Purpose To illustrate the use of inequalities to represent a deleted neighborhood:

Problem Write an inequality for a deleted neighborhood of 3 with half width of .5.

Solution

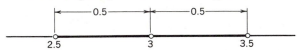

$$DN_\delta(3) \text{ where } \delta = .5 \leftrightarrow 0 < |x - 3| < .5.$$

To review, you picked an $\epsilon > 0$. You want $|(2x + 1) - 7| < \epsilon$. To prove that $\lim_{x \to 3} (2x + 1) = 7$, for every $\epsilon > 0$, I must be able to tell how to select a deleted neighborhood about 3 so that when x takes on any value in that $DN_\delta(3)$, it follows that $|(2x + 1) - 7| < \epsilon$. That is, I must exhibit a deleted neighborhood

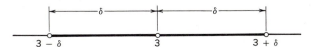

about 3, $DN_\delta(3)$, $0 < |x - 3| < \delta$, such that when x takes on any value in this deleted neighborhood, $|(2x + 1) - 7| < \epsilon$ is true.

This leads us to the Cauchy Definition for limit of a function.

DEFINITION

Limit of a Function $f(x)$. The $\lim_{x \to a} f(x) = L$ if for every $\epsilon > 0$ there can be found a $\delta > 0$ such that $|f(x) - L| < \epsilon$ whenever x belongs to the deleted neighborhood, $0 < |x - a| < \delta$.

Example 2

Purpose To use the definition of limit in order to prove the correctness of an answer:

Problem Prove: $\lim_{x \to 3} (2x + 1) = 7$.

Solution I must show that for every $\epsilon > 0$, I can find a δ such that:

$$|f(x) - L| < \epsilon \qquad \text{whenever} \qquad 0 < |x - a| < \delta$$

or $\qquad |(2x + 1) - 7| < \epsilon \qquad$ whenever $\qquad 0 < |x - 3| < \delta$

or $\qquad\qquad |2x - 6| < \epsilon \qquad$ whenever $\qquad 0 < |x - 3| < \delta$

or $\qquad\qquad 2|x - 3| < \epsilon \qquad$ whenever $\qquad 0 < |x - 3| < \delta$

or $\qquad\qquad |x - 3| < \dfrac{\epsilon}{2} \qquad$ whenever $\qquad 0 < |x - 3| < \delta$

Compare the inequalities $|x - 3| < \dfrac{\epsilon}{2}$ and $0 < |x - 3| < \delta$, and note a suitable choice for δ.

Let $\delta = \dfrac{\epsilon}{2}$. Then $0 < |x - 3| < \dfrac{\epsilon}{2}$ or $|x - 3| < \dfrac{\epsilon}{2}$ and, by reversing the steps, if $|x - 3| < \dfrac{\epsilon}{2}$ then $|(2x + 1) - 71 < \epsilon$. The important idea is: I have a δ that will work for any positive ϵ that you give me. For example, if you select $\epsilon = .00001$ we have

ϵ	Condition desired	$\delta = \dfrac{\epsilon}{2}$
.00001	$|(2x + 1) - .7| < .00001$	.000005

As long as x is in the following deleted neighborhood,

|———o—————————o—————————o———|
2.999995   3   3.000005

$|(2x + 1) - 7| < .00001$. And, moreover, no matter what positive ϵ you give to me, I can establish δ for this problem by letting $\delta = \dfrac{\epsilon}{2}$. Graphical interpretation of $\lim\limits_{x \to 3} (2x + 1) = 7$.

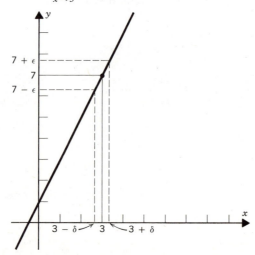

As long as x is in the interval between $3 - \delta$ and $3 + \delta$, $x \neq 3$, and $\delta = \dfrac{\epsilon}{2}$, then $|(2x + 1) - 7| < \epsilon$.

Example 3

Purpose To prove that a function has a certain limit:

Problem Prove: $\lim\limits_{x \to 2} (\tfrac{1}{2}x - 4) = -3$.

Solution

$$
\begin{array}{lll}
|(\tfrac{1}{2}x - 4) - (-3)| < \epsilon & \text{whenever} & 0 < |x - 2| < \delta \\
|\tfrac{1}{2}x - 1| < \epsilon & \text{whenever} & 0 < |x - 2| < \delta \\
\tfrac{1}{2}|x - 2| < \epsilon & \text{whenever} & 0 < |x - 2| < \delta \\
|x - 2| < 2\epsilon & \text{whenever} & 0 < |x - 2| < \delta
\end{array}
$$

Let $\delta = 2\epsilon$. Then $|x - 2| < 2\epsilon$ and $|(\tfrac{1}{2}x - 4) - (-3)| < \epsilon$.

Exercise A.II

In Problems 1 and 2 use inequalities and absolute values to represent the deleted neighborhood.

1. A deleted neighborhood of 1 with half width $\tfrac{1}{2}$.
2. A deleted neighborhood of -2 with half width 2.
3. Show that:
    (a) $|x - 2| < 1$ is equivalent to $|(3x - 4) - 2| < 3$.
    (b) $|x - 2| < \dfrac{\epsilon}{3}$ is equivalent to $|(3x - 4) - 2| < \epsilon$ (ϵ, epsilon, is an arbitrary small positive number).
    (c) Find a value for δ, such that if $|x - 2| < \delta$, then $|(3x - 4) - 2| < \epsilon$.
4. Show that:
    (a) $|x - 3| < 1$ is equivalent to $|(2x + 1) - 7| < 2$.
    (b) $|x - 3| < \dfrac{\epsilon}{2}$ is equivalent to $|(2x + 1) - 7| < \epsilon$.
    (c) Find a value for δ, such that if $|x - 3| < \delta$, then $|(2x + 1) - 7| < \epsilon$.
5. Find a value for δ such that if $|x - 1| < \delta$, then $|(4x - 1) - 3| < \epsilon$.
6. Find a value for δ such that $|(\tfrac{1}{2}x + 1) - 3| < \epsilon$ whenever $|x - 4| < \delta$.

In Problems 7 to 12, find $\lim\limits_{x \to a} f(x)$ and verify your answer by the ϵ, δ definition of limit.

7. $f(x) = x + 2$, $a = 5$
8. $f(x) = 2x - 3$, $a = 1$
9. $f(x) = \tfrac{1}{3}x + 1$, $a = 3$
10. $f(x) = 4x + 7$, $a = -1$
11. $f(x) = \tfrac{1}{2}x - 2$, $a = 4$
12. $f(x) = 3x + 2$, $a = -2$

Additional Limit Definitions

> **DEFINITION**
>
> $\lim_{x \to a^+} f(x) = \infty$. $\lim_{x \to a^+} f(x) = \infty$ if for every arbitrarily large number, $N > 0$, there exists a $\delta > 0$ such that $f(x) > N$ whenever x belongs to the deleted neighborhood $0 < x - a < \delta$. (See Figure A.II.1.)

Example 4

Problem Prove $\lim_{x \to 0^+} \dfrac{1}{x} = \infty$.

Solution Let $\delta = \dfrac{1}{N}$, then $0 < x - 0 < \delta$ implies $x < \dfrac{1}{N}$ and $\dfrac{1}{x} > N$.

There are similar definitions for: $\lim_{x \to a^-} f(x) = \infty$, $\lim_{x \to a^+} f(x) = -\infty$, and $\lim_{x \to a^-} f(x) = -\infty$.

> **DEFINITION**
>
> $\lim_{x \to \infty} f(x) = L$. $\lim_{x \to \infty} f(x) = L$ if for every $\epsilon > 0$, there exists a sufficiently large number M, such that $|f(x) - L| < \epsilon$ whenever $x > M$. (See Figure A.II.2.)

Example 5

Problem Prove $\lim_{x \to \infty} \dfrac{1}{x} = 0$.

Solution Let $M = \dfrac{1}{\epsilon}$, then $x > M$ implies $x > \dfrac{1}{\epsilon}$ and $\dfrac{1}{x} < \epsilon$. Thus, $\left| \dfrac{1}{x} - 0 \right| < \epsilon$ since $x > 0$.

There is a similar definition for $\lim\limits_{x \to -\infty} f(x) = L$.

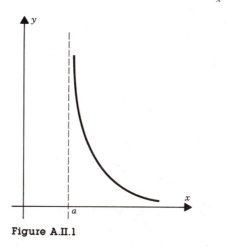

Figure A.II.1

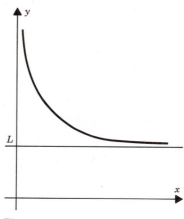

Figure A.II.2

Limit of a Riemann Sum

$\int_a^b f(x)\ dx$, when it exists, represents a number. What number? The number that

is $\lim\limits_{n \to \infty} \sum\limits_{i=1}^{n} f(x_i)\ \Delta x$. This limit is a new type of limit and is defined as follows.

DEFINITION

Limit of a Riemann Sum. $\int_a^b f(x)\ dx = \lim\limits_{n \to \infty} \sum\limits_{i=1}^{n} f(x_i)\ \Delta x = L$ if for every

$\epsilon > 0$ there exists a $\delta > 0$ such that $\left| \sum\limits_{i=1}^{n} f(x_i)\ \Delta x - L \right| < \epsilon$ for

every choice of subinterval, Δx, where $\Delta x < \delta$.

Limit of a Function, $f(x,y)$

DEFINITION

Limit of a Function, $f(x,y)$. The limit of $f(x,y)$ as $(x,y) \to (a,b)$ is said to
be L, $\lim\limits_{(x,y) \to (a,b)} f(x,y) = L$, if for every $\epsilon > 0$ there exists a circular dele-

ted neighborhood about (a,b), $0 < [(x - a)^2 + (y - b)^2] < \delta^2$, such that for all points belonging to this circular deleted neighborhood it follows that $|f(x,y) - L| < \epsilon$.

III.   PROOFS OF DIFFERENTIATION THEOREMS

Power Rule

THEOREM 3.1α

If n is a positive integer, then $\dfrac{d}{dx}(x^n) = nx^{n-1}$.

GIVEN. n is a positive integer, and the function f is defined by $y = f(x) = x^n$.

TO PROVE. $\dfrac{d}{dx}(x^n) = nx^{n-1}$.

Proof

1. $y = f(x) = x^n$
2. $y + \Delta y = f(x + \Delta x) = (x + \Delta x)^n$
3. $\Delta y = f(x + \Delta x) - f(x) = (x + \Delta x)^n - x^n$
4.   By the Binomial theorem expand $(x + \Delta x)^n$ obtaining:

$$\Delta y = \left[x^n + \frac{n}{1!}x^{n-1}\,\Delta x + \frac{n(n-1)}{2!}x^{n-2}(\Delta x)^2 + \cdots \right.$$
$$\left. + nx(\Delta x)^{n-1} + (\Delta x)^n \right] - x^n$$

5.   Divide by Δx to form a difference quotient

$$\frac{\Delta y}{\Delta x} = nx^{n-1} + \frac{n(n-1)}{2!}x^{n-2}(\Delta x) + \cdots + nx(\Delta x)^{n-2} + (\Delta x)^{n-1}$$

6.   Take $\lim\limits_{\Delta x \to 0} \dfrac{\Delta y}{\Delta x}$ and use the theorem: the limit of a sum is equal to the sum of the limits. Every term except the first term has Δx as a factor. As $\Delta x \to 0$, all terms except the first term will approach zero. Therefore,

$$\lim_{\Delta x \to 0} \frac{\Delta y}{\Delta x} = \lim_{\Delta x \to 0} \left[nx^{n-1} + \frac{n(n-1)}{2!} x^{n-2}(\Delta x) + \cdots \right.$$
$$\left. + nx(\Delta x)^{n-2} + (\Delta x)^{n-1} \right]$$
$$= nx^{n-1}$$

or, $y' = \dfrac{dy}{dx} = \dfrac{d}{dx}(x^n) = nx^{n-1}$ when n is a positive integer.

We accept the fact that the theorem does hold true if n is any real number, although we have only proved it true for n being a positive integer. The corollary to Theorem 3.1d proves this theorem if n is a negative integer. To prove the theorem when n is a rational number, let $n = \dfrac{p}{q}$ where p and q are integers, $q \neq 0$, then $y = x^{p/q}$ or $y^q = x^p$, and use implicit differentiation. Also, $\dfrac{d}{dx}(x^n) = nx^{n-1}$ is true when n is an irrational number. However, this portion of the proof will be omitted.

Product Rule

THEOREM 3.1c

For any value of x where the functions g and h are differentiable, the function $[g \cdot h]$ is also differentiable with its derivative given by

$$\frac{d}{dx}[g(x) \cdot h(x)] = g(x) \cdot h'(x) + h(x) \cdot g'(x).$$

GIVEN. Let x be any value where the functions g and h are differentiable, that is, $g'(x)$ and $h'(x)$ exist.

TO PROVE. $\dfrac{d}{dx}[g(x) \cdot h(x)] = g(x) \cdot h'(x) + h(x) \cdot g'(x).$

Proof
1. $y = f(x) = g(x) \cdot h(x)$
2. $y + \Delta y = f(x + \Delta x) = g(x + \Delta x) \cdot h(x + \Delta x)$
3. $\Delta y = g(x + \Delta x) \cdot h(x + \Delta x) - g(x) \cdot h(x)$

4. Form the difference quotient:

$$\frac{\Delta y}{\Delta x} = \frac{g(x + \Delta x) \cdot h(x + \Delta x) - g(x) \cdot h(x)}{\Delta x}$$

5. Take the limit of both sides of the equation as $\Delta x \to 0$:

$$\lim_{\Delta x \to 0} \frac{\Delta y}{\Delta x} = \lim_{\Delta x \to 0} \frac{g(x + \Delta x) \cdot h(x + \Delta x) - g(x) \cdot h(x)}{\Delta x}$$

If this limit exists, it is $\frac{d}{dx}[g(x) \cdot h(x)]$. However, to take the limit, the right-hand side of the equation must be rewritten. This is done by adding zero to the numerator in the form of

$$-g(x + \Delta x) \cdot h(x) + g(x + \Delta x) \cdot h(x).$$

6. $$\lim_{\Delta x \to 0} \frac{\Delta y}{\Delta x} = \lim_{\Delta x \to 0} \left\{ \frac{[g(x + \Delta x) \cdot h(x + \Delta x) - g(x + \Delta x) \cdot h(x)]}{\Delta x} \right.$$
$$\left. + \frac{[g(x + \Delta x) \cdot h(x) - g(x) \cdot h(x)]}{\Delta x} \right\}$$

7. Regroup and factor:

$$\lim_{\Delta x \to 0} \frac{\Delta y}{\Delta x} = \lim_{\Delta x \to 0} \left\{ g(x + \Delta x) \cdot \frac{[h(x + \Delta x) - h(x)]}{\Delta x} \right.$$
$$\left. + h(x) \cdot \frac{[g(x + \Delta x) - g(x)]}{\Delta x} \right\}$$

8. Use the limit theorems to rewrite:

$$\lim_{\Delta x \to 0} \frac{\Delta y}{\Delta x} = \lim_{\Delta x \to 0} g(x + \Delta x) \cdot \lim_{\Delta x \to 0} \frac{h(x + \Delta x) - h(x)}{\Delta x}$$
$$+ \lim_{\Delta x \to 0} h(x) \cdot \lim_{\Delta x \to 0} \frac{g(x + \Delta x) - g(x)}{\Delta x}$$

9. $$\frac{dy}{dx} = f'(x) = g(x) \cdot h'(x) + h(x) \cdot g'(x)$$

or, $\frac{d}{dx}[g(x) \cdot h(x)] = g(x) \cdot h'(x) + h(x) \cdot g'(x).$

It is hoped that the student examines the limits involved in this proof.

$$\lim_{\Delta x \to 0} \frac{h(x + \Delta x) - h(x)}{\Delta x} \quad \text{and} \quad \lim_{\Delta x \to 0} \frac{g(x + \Delta x) - g(x)}{\Delta x}$$

are merely definitions for $h'(x)$ and $g'(x)$, respectively, and these limits do exist, since x represents a value where the functions g and h were given to be differentiable. For a specific value of x, $h(x)$ will be a constant and $\lim\limits_{\Delta x \to 0} h(x) = h(x)$, since the limit of a constant is that constant or $\lim\limits_{\Delta x \to 0} c = c$. However, $\lim\limits_{\Delta x \to 0} g(x + \Delta x)$ becomes a little "sticky." For $\lim\limits_{\Delta x \to 0} g(x + \Delta x)$ does not always equal $g(x)$. Continuity of the function g is the "key" to evaluating this limit. We know that g is continuous at x, since g is differentiable at x. Once this continuity is established, the alternate definition of continuity may be used to obtain the desired result, $\lim\limits_{\Delta x \to 0} g(x + \Delta x) = g(x)$.

Quotient Rule

Before proving Theorem 3.1d, which gives us a shortcut method for finding the derivative of $\dfrac{g(x)}{h(x)}$, $h(x) \neq 0$, let us prove this theorem for the special case where $g(x) = 1$. This result in conjunction with Theorem 3.1c will then be used to prove the generalized quotient rule.

THEOREM

For any value of x where the function h is differentiable, the function $\dfrac{1}{h}$ is also differentiable with its derivative given by

$$\frac{d}{dx}\left[\frac{1}{h(x)}\right] = -\frac{h'(x)}{[h(x)]^2}, \text{ provided that } h(x) \neq 0.$$

GIVEN. Let x be any value where the function h is differentiable, that is, $h'(x)$ exists. Also, $h(x) \neq 0$.

TO PROVE. $\dfrac{d}{dx}\left[\dfrac{1}{h(x)}\right] = -\dfrac{h'(x)}{[h(x)]^2}$.

Proof

1. $y = f(x) = \dfrac{1}{h(x)}$

2. $y + \Delta y = f(x + \Delta x) = \dfrac{1}{h(x + \Delta x)}$

3. $\Delta y = f(x + \Delta x) - f(x) = \dfrac{1}{h(x + \Delta x)} - \dfrac{1}{h(x)} = \dfrac{h(x) - h(x + \Delta x)}{h(x) \cdot h(x + \Delta x)}$

4. $\dfrac{\Delta y}{\Delta x} = \dfrac{h(x) - h(x + \Delta x)}{\Delta x} \cdot \dfrac{1}{h(x) \cdot h(x + \Delta x)}$

5. $\displaystyle\lim_{\Delta x \to 0} \dfrac{\Delta y}{\Delta x} = -\lim_{\Delta x \to 0} \dfrac{h(x + \Delta x) - h(x)}{\Delta x} \cdot \lim_{\Delta x \to 0} \dfrac{1}{h(x) \cdot h(x + \Delta x)}$

6. $f'(x) = -h'(x) \cdot \dfrac{1}{h(x) \cdot h(x)} = -\dfrac{h'(x)}{[h(x)]^2}$

or, $\dfrac{d}{dx}\left[\dfrac{1}{h(x)}\right] = -\dfrac{h'(x)}{[h(x)]^2} .$

Note the limits in step 5. The $\displaystyle\lim_{\Delta x \to 0} \dfrac{[h(x + \Delta x) - h(x)]}{\Delta x}$ is the definition for $h'(x)$. Also, a value of x where the function h is differentiable implies h is continuous for that value of x and $\displaystyle\lim_{\Delta x \to 0} h(x + \Delta x) = h(x)$.

The theorem for differentiating the quotient of two functions is now easily proved.

THEOREM 3.1d

For any value of x where the functions g and h are differentiable and $h(x) \neq 0$, the function $\left[\dfrac{g}{h}\right]$ is also differentiable with its derivative given by:

$$\dfrac{d}{dx}\left[\dfrac{g(x)}{h(x)}\right] = \dfrac{h(x) \cdot g'(x) - g(x) \cdot h'(x)}{[h(x)]^2} .$$

GIVEN. The functions g and h are differentiable. Also, $h(x) \neq 0$.

TO PROVE. $\dfrac{d}{dx}\left[\dfrac{g(x)}{h(x)}\right] = \dfrac{h(x) \cdot g'(x) - g(x) \cdot h'(x)}{[h(x)]^2} .$

Proof

$y = \dfrac{g(x)}{h(x)} = g(x) \cdot \dfrac{1}{h(x)}$

1. By Theorem 3.1c,

$\dfrac{dy}{dx} = g(x) \cdot \dfrac{d}{dx}\left[\dfrac{1}{h(x)}\right] + \dfrac{1}{h(x)} \cdot \dfrac{d}{dx}[g(x)]$

2. $\dfrac{dy}{dx} = g(x) \cdot -\dfrac{h'(x)}{[h(x)]^2} + \dfrac{1}{h(x)} \cdot g'(x)$

3. Combining the terms,

$$\dfrac{d}{dx}\left[\dfrac{g(x)}{h(x)}\right] = \dfrac{h(x) \cdot g'(x) - g(x) \cdot h'(x)}{[h(x)]^2}$$

COROLLARY TO THEOREM 3.1d

If $y = x^n$ (n is a negative integer), then $y' = nx^{n-1}$.

Proof

Let $n = -m$ (since n is negative, $-m$ is negative, and m is positive). By substitution: $y = x^n = x^{-m} = \dfrac{1}{x^m}$, m represents a positive integer. Using the quotient rule (Theorem 3.1d):

$$y' = \dfrac{d}{dx}\left(\dfrac{1}{x^m}\right) = \dfrac{(x^m)\dfrac{d}{dx}(1) - (1)\dfrac{d}{dx}(x^m)}{(x^m)^2}$$

$$= \dfrac{(x^m) \cdot (0) - (1) \cdot m \cdot x^{m-1}}{(x^m)^2}$$

$$= \dfrac{-m \cdot x^{m-1}}{x^{2m}} = -m \cdot \dfrac{x^{m-1}}{x^{2m}} = -m \cdot x^{-m-1}$$

$y' = nx^{n-1}$, since $n = -m$.

IV. THE TRIGONOMETRIC FUNCTIONS

This section contains information on a special type of function, the trigonometric functions, which are used to describe phenomena that are cyclical in nature.

Radian Measure

In the calculus it is desirable to measure angles in radians. Let us review the definition of one radian. Begin with a circle of radius equal to one unit. Place a central angle α at the center of this unit circle. If angle α intercepts an arc of length one on the circumference of the circle, then angle α has measure of one radian.

DEFINITION

Radian. An angle of *one radian* is the measure of a central angle that intercepts on the circumference of the unit circle an arc of length equal to one.

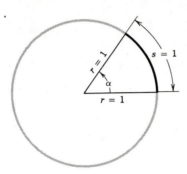

Observe that if $s = 2$, then the central angle α would have a measure of 2 radians, etc. Also, if s were equal to the circumference of the circle, $2\pi r = 2\pi(1) = 2\pi$, then the central angle would have a measure of 2π radians. But for s to be equal to the circumference of the circle, the central angle α would have to be 360°. Therefore, an angle of 360° is equivalent to an angle of 2π radians.

$$2\pi \text{ radians} = 360°$$
$$\pi \text{ radians} = 180°$$

Our conclusion is that π radians = 180° or, since $\pi \approx 3.14$, 3.14 radians $\approx$ 180°. Indeed, it can be shown that an angle of one radian is equal to

$$\frac{180°}{\pi} \approx \frac{180°}{3.14} \approx 57°18'.$$

Example 1

Purpose   To convert from degree measure to radian measure and from radian measure to degree measure:

Problem   Convert (A) 90° and (B) 60° into radian measure.

Convert (C) $\frac{3}{2}\pi$ and (D) $\frac{1}{6}\pi$ into degree measure.

Solution   Since π radians = 180°, it follows that

$$1 \text{ radian} = \frac{180°}{\pi}$$

$$1° = \frac{\pi}{180}.$$

The conversions are:

(A) $90° = \dfrac{1}{2}(180°) = \dfrac{1}{2}\pi$ \qquad (C) $\dfrac{3}{2}\pi = \dfrac{3}{2}\pi \cdot \dfrac{180°}{\pi} = 270°$

(B) $60° = \dfrac{1}{3}(180°) = \dfrac{1}{3}\pi$ \qquad (D) $\dfrac{1}{6}\pi = \dfrac{1}{6}\pi \cdot \dfrac{180°}{\pi} = 30°$

The Trigonometric Functions

Let θ be a central angle measured in radians. Place θ at the center of a unit circle with the vertex of θ at the origin and the initial side of θ on the x axis. See the sketch below.

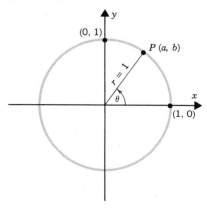

Let (a,b) be the coordinates of point P, the point where the terminal side of θ intercepts the unit circle.

DEFINITIONS

Trigonometric Functions. We will define the following trigonometric functions:

$$\text{sine } \theta = \sin \theta = b$$
$$\text{cosine } \theta = \cos \theta = a$$
$$\text{tangent } \theta = \tan \theta = \frac{b}{a}, \qquad a \neq 0$$
$$\text{cosecant } \theta = \csc \theta = \frac{1}{b}, \qquad b \neq 0$$
$$\text{secant } \theta = \sec \theta = \frac{1}{a}, \qquad a \neq 0$$
$$\text{cotangent } \theta = \cot \theta = \frac{a}{b}, \qquad b \neq 0$$

Example 2

Purpose   To find the sine, cosine, and tangent of some frequently used angles:

Problem A   Find the sin θ, cos θ, and tan θ if $\theta = \frac{1}{4}\pi$ or 45°.

Solution

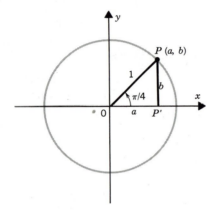

By the Pythagorean theorem $a^2 + b^2 = 1$; $a = b$, and since $\Delta OPP'$ is an isosceles right triangle,

$$a^2 + a^2 = 1$$

$$2a^2 = 1$$

$$a^2 = \frac{1}{2}$$

$$a = \frac{1}{\sqrt{2}} = \frac{\sqrt{2}}{2}, \qquad b = \frac{\sqrt{2}}{2}$$

Therefore,

$$\sin \frac{1}{4}\pi = b = \frac{\sqrt{2}}{2}$$

$$\cos \frac{1}{4}\pi = a = \frac{\sqrt{2}}{2}$$

$$\tan \frac{1}{4}\pi = \frac{b}{a} = 1$$

Problem B   Find: sin $\frac{1}{2}\pi$; cos π; tan 0.

Solution $\sin \frac{1}{2}\pi = 1$

$\cos \pi = -1$

$\tan 0 = 0$

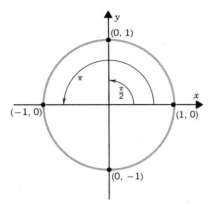

Graphs of the Sine, Cosine, and Tangent Functions

It could be established that the previous definitions enable us to determine three functions represented by $y = \sin x$, $y = \cos x$, and $y = \tan x$. The values of the sine, cosine, and tangent found in the last section along with other selected values enable us to graph these functions.

Table of values for $y = \sin x$:

x	0	$\frac{1}{6}\pi$	$\frac{1}{2}\pi$	π	$\frac{3}{2}\pi$	2π
$y = \sin x$	0	$\frac{1}{2}$	1	0	-1	0

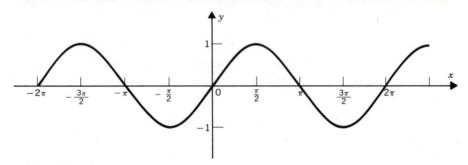

Graph of $y = \sin x$.

Table of values for $y = \cos x$:

x	0	$\frac{1}{3}\pi$	$\frac{1}{2}\pi$	π	$\frac{3}{2}\pi$	2π
$y = \cos x$	1	$\frac{1}{2}$	0	-1	0	1

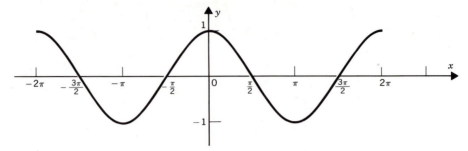

Graph of $y = \cos x$.

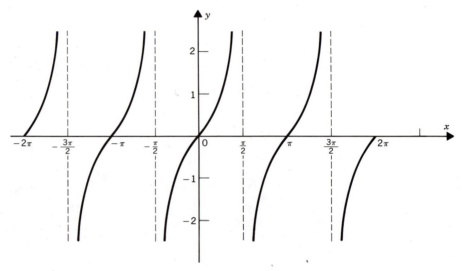

Graph of $y = \tan x$.

From inspecting the graphs it is observed that the period of the sine and cosine functions is 2π. That is, every 2π radians the graphs of the sine and cosine functions repeat the same set of y values. However, the tangent function has a period of only π radians. Do you see why?

Fundamental Identities

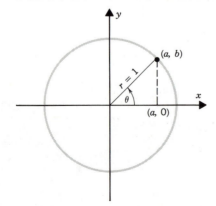

In our unit circle $r = 1$ and $\sin \theta = b$, $\cos \theta = a$, $\tan \theta = \dfrac{b}{a}$, and $\sec \theta = \dfrac{1}{a}$.

From these definitions we may establish certain fundamental identities. Note from the Pythagorean right triangle theorem that $a^2 + b^2 = 1^2 = 1$. Therefore, we have:

$$\tan \theta = \frac{b}{a} = \frac{\sin \theta}{\cos \theta}; \qquad \cot \theta = \frac{a}{b} = \frac{\cos \theta}{\sin \theta};$$

$$\sec \theta = \frac{1}{a} = \frac{1}{\cos \theta}; \qquad \csc \theta = \frac{1}{b} = \frac{1}{\sin \theta};$$

$$(\sin \theta)^2 + (\cos \theta)^2 = \sin^2 \theta + \cos^2 \theta = b^2 + a^2 = 1$$

$$(\tan \theta)^2 + 1 = \tan^2 \theta + 1 = \left(\frac{b}{a}\right)^2 + 1 = \frac{a^2 + b^2}{a^2} = \frac{1}{a^2}$$

$$= \left(\frac{1}{a}\right)^2 = \sec^2 \theta$$

We will use the following identities throughout this appendix.

$$\tan \theta = \frac{\sin \theta}{\cos \theta} \qquad \cot \theta = \frac{\cos \theta}{\sin \theta}$$

$$\sec \theta = \frac{1}{\cos \theta} \qquad \csc \theta = \frac{1}{\sin \theta}$$

$$\sin^2 \theta + \cos^2 \theta = 1 \qquad 1 + \tan^2 \theta = \sec^2 \theta$$

Exercise A.IV.1

1. Convert the following measures from degree measure to radian measure.
   (a) 270°   (b) 135°   (c) 30°   (d) 210°

2. Convert the following measures from radian measure to degree measure.
   (a) $\frac{1}{2}\pi$ (b) $\frac{1}{4}\pi$ (c) $\frac{1}{3}\pi$ (d) $\frac{5}{3}\pi$

3.

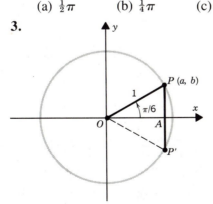

Figure A.1

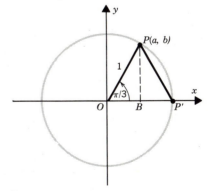

Figure A.2

OPP' is an isosceles triangle.

In Figure A.1, $P'P = 1$, $AP = \frac{1}{2} = b$, and

$$(OA)^2 + (AP)^2 = 1$$

$$a^2 + \left(\frac{1}{2}\right)^2 = 1$$

$$a = \frac{\sqrt{3}}{2}$$

In Figure A.2, $OP' = 1$, $OB = \frac{1}{2} = a$, and

$$(OB)^2 + (BP)^2 = 1$$

$$\left(\frac{1}{2}\right)^2 + b^2 = 1$$

$$b = \frac{\sqrt{3}}{2}$$

Find the value of the $\sin \theta$, $\cos \theta$, and $\tan \theta$ when

(a) $\theta = \frac{1}{6}\pi$ (c) $\theta = \frac{1}{3}\pi$

(b) $\theta = \frac{5}{6}\pi$ (d) $\theta = \frac{4}{3}\pi$

4. Given angle θ in standard position and the point $P(\sqrt{2}, \sqrt{2})$ on the terminal side of the angle, find $\sin \theta$, $\cos \theta$, $\tan \theta$, $\csc \theta$, $\sec \theta$, and $\cot \theta$.

5. Find x in the interval $\left[0, \dfrac{\pi}{2}\right]$ such that $\sin x = \frac{1}{2}$.

6. Find x in the interval $\left[0, \dfrac{\pi}{2}\right]$ such that $\sin x = \cos x$.

7. Prove the following identities:

(a) $\cos \theta \cdot \tan \theta \cdot \csc \theta = 1$

(b) $\dfrac{1}{\tan \theta + \cot \theta} = \sin \theta \cdot \cos \theta$

(c) $\sin \theta + \cos \theta \cdot \cot \theta = \csc \theta$

(d) $\dfrac{\cos x}{1 - \sin x} - \dfrac{1 - \sin x}{\cos x} = 2 \tan x$

(e) $\dfrac{\tan x}{\tan^2 x + 1} = \sin x \cdot \cos x$

Derivatives of the Trigonometric Functions

Before finding the derivative of the sine function it is necessary to establish the value of the $\lim\limits_{x \to 0} \dfrac{\sin x}{x}$. A proof that this limit does exist and is equal to one may be found in more advanced calculus texts. That the $\lim\limits_{x \to 0} \dfrac{\sin x}{x} = 1$ seems plausible is illustrated by the following table.

x (radians)	$\sin x$	$\dfrac{\sin x}{x}$
$-\dfrac{\pi}{3}$	$-\dfrac{\sqrt{3}}{2}$	.83
$-\dfrac{\pi}{6}$	$-\dfrac{1}{2}$	.95
$-\dfrac{\pi}{180}$	$-.0175$	.99
$\dfrac{\pi}{180}$	$.0175$	.99
$\dfrac{\pi}{6}$	$\dfrac{1}{2}$	.95
$\dfrac{\pi}{3}$	$\dfrac{\sqrt{3}}{2}$	.83

Another limit that will be utilized in our work is $\lim\limits_{x \to 0} \dfrac{1 - \cos x}{x} = 0$. This limit is now established. $\left(\text{Assume that } \lim\limits_{x \to 0} \dfrac{\sin x}{x} = 1.\right)$

$$\lim_{x \to 0} \frac{1 - \cos x}{x} = \lim_{x \to 0} \frac{(1 - \cos x)(1 + \cos x)}{x(1 + \cos x)}$$

$$= \lim_{x \to 0} \frac{1 - \cos^2 x}{x(1 + \cos x)}$$

$$= \lim_{x \to 0} \frac{\sin^2 x}{x(1 + \cos x)}$$

$$= \lim_{x \to 0} \frac{\sin x}{x} \cdot \lim_{x \to 0} \frac{\sin x}{1 + \cos x}$$

$$\lim_{x \to 0} \frac{1 - \cos x}{x} = 1 \cdot \frac{0}{2} = 1 \cdot 0 = 0$$

By the use of these two limits, $\lim\limits_{x \to 0} \dfrac{\sin x}{x} = 1$ and $\lim\limits_{x \to 0} \dfrac{1 - \cos x}{x} = 0$, along with your recall of the identity $\sin(\alpha + \beta) = \sin\alpha \cos\beta + \cos\alpha \sin\beta$, we proceed to find $\dfrac{d}{dx}(\sin x) = \cos x$.

THEOREM A.IVα

If f is a function defined by $f(x) = \sin x$, then

$$\frac{d}{dx}(\sin x) = \cos x.$$

GIVEN. $y = f(x) = \sin x$.

TO PROVE. $\dfrac{d}{dx}(\sin x) = \cos x.$

Proof

$$y = f(x) = \sin x \quad \text{and} \quad y + \Delta y = f(x + \Delta x) = \sin(x + \Delta x)$$
$$\Delta y = \sin(x + \Delta x) - \sin x$$
$$\Delta y = (\sin x \cos \Delta x + \cos x \sin \Delta x) - \sin x$$
$$\Delta y = -\sin x(1 - \cos \Delta x) + \cos x \sin \Delta x$$
$$\frac{\Delta y}{\Delta x} = -\sin x \frac{1 - \cos \Delta x}{\Delta x} + \cos x \frac{\sin \Delta x}{\Delta x}$$
$$\lim_{\Delta x \to 0} \frac{\Delta y}{\Delta x} = \lim_{\Delta x \to 0} (-\sin x) \cdot \lim_{\Delta x \to 0} \frac{1 - \cos \Delta x}{\Delta x} + \lim_{\Delta x \to 0} (\cos x) \cdot \lim_{\Delta x \to 0} \frac{\sin \Delta x}{\Delta x}$$
$$\frac{dy}{dx} = (-\sin x) \cdot 0 + \cos x \cdot 1 = \cos x$$

or, $\dfrac{d}{dx}(\sin x) = \cos x$

By the chain rule we obtain:

THEOREM A.IVb

If $y = \sin u$ and u represents a differentiable function of x which causes y to be a function of x, then

$$\frac{d}{dx}(\sin u) = \frac{d}{du}(\sin u) \cdot \frac{du}{dx} = (\cos u) \cdot \frac{du}{dx}.$$

In obtaining the derivative of the function defined by $y = \cos x$, we wish to avoid the lengthy development used in the preceding proof. To do this we recall from trigonometry that $\cos x = \sin\left(x + \dfrac{\pi}{2}\right)$ and $\cos\left(x + \dfrac{\pi}{2}\right) = -\sin x$. The fact that $\cos x = \sin\left(x + \dfrac{\pi}{2}\right)$ can be seen from the graphs of $y = \sin x$ and $y = \cos x$. For if we were to move the graph of $y = \sin x$ to the left $\dfrac{\pi}{2}$ units, the displaced graph $y = \sin\left(x + \dfrac{\pi}{2}\right)$ would coincide with the graph of $y = \cos x$. In a similar manner $\cos\left(x + \dfrac{\pi}{2}\right) = -\sin x$ could be established.

THEOREM A.IVc

If f is a function defined by $f(x) = \cos x$, then

$$\frac{d}{dx}(\cos x) = -\sin x.$$

GIVEN. $f(x) = \cos x$.

TO PROVE. $\dfrac{d}{dx}(\cos x) = -\sin x$.

Proof

$$\frac{d}{dx}(\cos x) = \frac{d}{dx}\sin\left(x + \frac{\pi}{2}\right)$$

$$= \cos\left(x + \frac{\pi}{2}\right)\frac{d}{dx}\left(x + \frac{\pi}{2}\right) = \cos\left(x + \frac{\pi}{2}\right)$$

$$\frac{d}{dx}(\cos x) = -\sin x$$

By applying the chain rule we obtain:

THEOREM A.IVd

If $y = \cos u$ and u represents a differentiable function of x which causes y to be a function of x, then

$$\frac{d}{dx}(\cos u) = \frac{d}{du}(\cos u) \cdot \frac{du}{dx} = (-\sin u) \cdot \frac{du}{dx}.$$

The derivative of the tangent function is easy to obtain.

> **THEOREM A.IVe**
>
> If f is a function defined by $f(x) = \tan x$, then
>
> $$\frac{d}{dx}(\tan x) = \sec^2 x.$$

GIVEN. $f(x) = \tan x$.

TO PROVE. $\dfrac{d}{dx}(\tan x) = \sec^2 x$.

Proof

(Use quotient rule.)

$$\frac{d}{dx}(\tan x) = \frac{d}{dx}\left(\frac{\sin x}{\cos x}\right)$$

$$= \frac{\cos x\,(\cos x) - \sin x\,(-\sin x)}{(\cos x)^2}$$

$$= \frac{\cos^2 x + \sin^2 x}{\cos^2 x}$$

$$= \frac{1}{\cos^2 x}$$

$$= \left(\frac{1}{\cos x}\right)^2$$

$$\frac{d}{dx}(\tan x) = (\sec x)^2 = \sec^2 x$$

The derivatives of the other trigonometric functions can be derived in a similar manner (see Problem 1 of the following exercise). The formulas for differentiation of the trigonometric functions are now summarized.

Formulas for Differentiation of the Trigonometric Functions

$$\frac{d}{dx}(\sin u) = \cos u\,\frac{du}{dx} \qquad\qquad \frac{d}{dx}(\csc u) = -\csc u \cot u\,\frac{du}{dx}$$

$$\frac{d}{dx}(\cos u) = -\sin u\,\frac{du}{dx} \qquad\qquad \frac{d}{dx}(\sec u) = \sec u \tan u\,\frac{du}{dx}$$

$$\frac{d}{dx}(\tan u) = \sec^2 u \, \frac{du}{dx} \qquad\qquad \frac{d}{dx}(\cot u) = -\csc^2 u \, \frac{du}{dx}$$

Example 3

Purpose To use the above formulas to find the derivatives of trigonometric functions:

Problem Differentiate the following functions.

(A) $y = \sin x \cdot \tan 2x$ (C) $y = \dfrac{\tan^2 x}{\sin \sqrt{x}}$

(B) $y = \sqrt{\cos x}$

Solution (A) $y = \sin x \cdot \tan 2x$

$\qquad y' = \sin x \cdot \sec^2 2x \cdot 2 + \tan 2x \cdot \cos x$

$\qquad y' = 2 \sin x \sec^2 2x + \tan 2x \cos x$

(B) $y = \sqrt{\cos x}$

$\qquad y' = \dfrac{1}{2}(\cos x)^{-(1/2)}(-\sin x) = \dfrac{-\sin x}{2\sqrt{\cos x}}$

(C) $y = \dfrac{\tan^2 x}{\sin \sqrt{x}}$

$\qquad y' = \dfrac{\sin \sqrt{x} \cdot 2 \tan x \cdot \sec^2 x - \tan^2 x \cdot \cos \sqrt{x} \cdot \frac{1}{2}x^{-(1/2)}}{\sin^2 \sqrt{x}}$

$\qquad y' = \dfrac{4\sqrt{x} \sin \sqrt{x} \cdot \tan x \sec^2 x - \tan^2 x \cdot \cos \sqrt{x}}{2\sqrt{x} \sin^2 \sqrt{x}}$

Exercise A.IV.2

1. Verify the following.

(a) $\dfrac{d}{dx}(\csc x) = -\csc x \cot x$ *Hint.* Use $\csc x = \dfrac{1}{\sin x}$.

(b) $\dfrac{d}{dx}(\csc u) = -\csc u \cot u \, \dfrac{du}{dx}$; u is a differentiable function of x.

(c) $\dfrac{d}{dx}(\sec x) = \sec x \tan x$

(d) $\dfrac{d}{dx}(\sec u) = \sec u \tan u \, \dfrac{du}{dx}$; u is a differentiable function of x.

(e) $\dfrac{d}{dx}(\cot x) = -\csc^2 x$

(f) $\dfrac{d}{dx}(\cot u) = -\csc^2 u \, \dfrac{du}{dx}$; u is a differentiable function of x.

2. Using the identity $\cos (\alpha + \beta) = \cos \alpha \cos \beta - \sin \alpha \sin \beta$, develop an alternate proof for $\dfrac{d}{dx}(\cos x) = -\sin x$ by forming the difference quotient and taking the limit as $\Delta x \rightarrow 0$.

3. Differentiate the following functions.

   (a) $y = \sin 2x$ (f) $y = \csc (x^2)$

   (b) $y = \cos^2 x$ (g) $y = e^x \cot x$

   (c) $y = \sin 3x \cdot \cos 2x$ (h) $y = \ln (\sec x)$

   (d) $y = \tan \frac{1}{2}x$ (i) $y = \tan^3 x$

   (e) $y = \dfrac{\cot x}{\sin x}$ (j) $y = \dfrac{\sec x}{\sqrt{x}}$

4. Differentiate the following functions.

   (a) $y = \sin 5x$ (f) $y = \sqrt[3]{\tan x}$

   (b) $y = 6 \cos 2x$ (g) $y = 4 \sin^3 x$

   (c) $y = \tan (\frac{1}{2}x^2)$ (h) $y = \frac{1}{3} \sec^6(2x)$

   (d) $y = \tan^2 \frac{1}{2}x$ (i) $y = \sec (x^3)$

   (e) $y = \ln (\cot 2x)$ (j) $y = \cot \frac{2}{3}x$

5. A company finds that its profits are cyclic in nature, and its profit function is given by $P(t) = 6 + 5 \sin \left(\dfrac{\pi}{180}t\right)$ where t is in days and $P(t)$ is in hundreds. For simplicity, assume that there are 360 business days in a year, and January 1 is day 0. Find the first and second derivatives, determine relative extrema, where the function is increasing or decreasing, where the curve is concave upward or concave downward, and graph the function for a one-year period ($0 \leqslant t \leqslant 360$).

6. A politician finds that the support she receives from a small community is cyclic in nature. She finds that the number of individuals who would vote for her in a given month, $N(t)$, may be calculated using the function $N(t) = 1000 - 500 \sin \left(\dfrac{\pi}{4}t\right)$ where t is in months (January $= 1$). Using information provided by the first and second derivatives of the function, graph $N(t)$ for a year ($1 \leqslant t \leqslant 12$). Find all relative extrema in this interval.

7. Find the first and second derivatives, determine relative extrema, where the function is increasing or decreasing, where the curve is concave upward or concave downward, and graph the function.

   (a) $y = \sin x$ (c) $y = \sin x + \cos x$

   (b) $y = \cos 2x$ (d) $y = \sin x - \cos x$

Integration of Trigonometric Functions

We may prove $\int \cos \, du = \sin u + c$ by showing that $\dfrac{d}{du}(\sin u + c) = \cos u$. This equation is true, since $\dfrac{d}{du}(\sin u + c) = \cos u + 0 = \cos u$. In this manner, we can establish the following table of integrals by making use of the table of derivatives of the trigonometric functions.

Integration Formulas for some Trigonometric Functions

$\int \sin u \, du = -\cos u + c$ $\qquad$ $\int \csc u \cot u \, du = -\csc u + c$

$\int \cos u \, du = \sin u + c$ $\qquad$ $\int \sec u \tan u \, du = \sec u + c$

$\int \sec^2 u \, du = \tan u + c$ $\qquad$ $\int \csc^2 u \, du = -\cot u + c$

$\int \tan u \, du = -\ln |\cos u| + c$ $\qquad$ $\int \cot u \, du = \ln |\sin u| + c$

Example 4

Purpose To illustrate a method for establishing

$$\int \tan u \, du = -\ln |\cos u| + c:$$

Problem Establish $\displaystyle\int \tan u \, du = -\ln |\cos u| + c$ by using the integration technique of substitution.

Solution Let $w = \cos u$ $\qquad$ $dw = -\sin u \, du$
Then

$$\int \tan u \, du = \int \frac{\sin u}{\cos u} \, du$$

$$= -\int \frac{(-\sin u) \, du}{\cos u}$$

$$= -\int \frac{dw}{w}$$

$$= -\ln |w| + c = -\ln |\cos u| + c$$

Example 5

Purpose To use the above integration formulas to do integration problems involving trigonometric functions:

Problem Find

$$(A) \int \sec 3x \cdot \tan 3x \, dx$$

(B) $\displaystyle\int_{\pi/6}^{\pi/2} \sin x \cdot \cos x\, dx$

(C) $\displaystyle\int x \cos x\, dx$

Solution

(A) $\displaystyle\int \sec 3x \cdot \tan 3x\, dx$

Let $u = 3x \qquad du = 3\, dx$
Then

$$\int \sec 3x \tan 3x\, dx = \tfrac{1}{3} \int \sec 3x \tan 3x(3\, dx)$$

$$= \tfrac{1}{3} \int \sec u \tan u\, du$$

$$= \tfrac{1}{3} \sec u + c$$

Therefore, $\displaystyle\int \sec 3x \tan 3x\, dx = \tfrac{1}{3} \sec 3x + c$

(B) $\displaystyle\int_{\pi/6}^{\pi/2} \sin x \cos x\, dx$

Let $u = \sin x \qquad du = \cos x\, dx$
Integrate

$$\int \sin x \cos x\, dx = \int u\, du$$

$$= \tfrac{1}{2}u^2 + c$$

$$= \tfrac{1}{2} \sin^2 x + c$$

Evaluate

$$\int_{\pi/6}^{\pi/2} \sin x \cos x\, dx = \left[\tfrac{1}{2} \sin^2 x \right]_{\pi/6}^{\pi/2}$$

$$= (\tfrac{1}{2} \cdot 1) - (\tfrac{1}{2} \cdot \tfrac{1}{4}) = \tfrac{3}{8}$$

(C) $\displaystyle\int x \cos x\, dx$

Use integration by parts, $\displaystyle\int u\, dv = uv - \int v\, du$

$$u = x \qquad dv = \cos x\, dx$$

$$du = dx \qquad v = \sin x$$

$$\int x \cos x\, dx = x \sin x - \int \sin x\, dx$$

$$= x \sin x + \cos x + c$$

Example 6

Purpose   To find the area of a region involving trigonometric functions:

Problem   By calculus, show that the area of a circle of radius r is πr^2.

Solution   The equation of a circle is $x^2 + y^2 = r^2$. Consider the first quadrant and the corresponding function $y = \sqrt{r^2 - x^2}$. Find the first quadrant area and multiply the answer by 4.

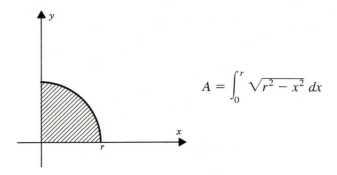

$$A = \int_0^r \sqrt{r^2 - x^2}\, dx$$

$\int_0^r \sqrt{r^2 - x^2}\, dx$ cannot be evaluated by any of our previous techniques. We will show the solution by introducing a new technique called trigonometric substitution.

Let $x = r \sin \theta$
$\qquad dx = r \cos \theta$

$r^2 - x^2 = r^2 - r^2 \sin^2 \theta = r^2(1 - \sin^2 \theta) = r^2 \cos^2 \theta$

$\sqrt{r^2 - x^2} = r \cos \theta$

Upon substitution in terms of θ and putting limits of integration in terms of θ (if $x = 0$, $\theta = 0$, and if $x = r$, $\theta = \frac{1}{2}\pi$), we have

$$\int_0^r \sqrt{r^2 - x^2}\, dx = \int_0^{\pi/2} r \cos \theta \cdot r \cos \theta\, d\theta$$

$$= \int_0^{\pi/2} r^2 \cos^2 \theta\, d\theta \qquad \left(\cos^2 \theta = \frac{1 + \cos 2\theta}{2}\right)$$

$$= r^2 \int_0^{\pi/2} \frac{1}{2}(1 + \cos 2\theta)\, d\theta$$

$$= \frac{1}{2} r^2 (\theta + \tfrac{1}{2} \sin 2\theta) \Big]_0^{\pi/2}$$

$$= \tfrac{1}{2} r^2 [(\tfrac{1}{2}\pi + 0) - (0)]$$

$$= \tfrac{1}{4} \pi r^2$$

The area of the circle is $4(\tfrac{1}{4}\pi r^2) = \pi r^2$.

Example 7

Purpose   To use integration in an applied problem:

Problem   A company finds its profit function to be $P = 2 + \dfrac{\pi}{90} \sin\left(\dfrac{\pi}{180}t\right)$ where t is in days and P is the daily profit in hundreds of dollars. Find the company's total profit during one year. (Assume a 360-day business year.)

Solution   To find the company's total profit for one year integrate $P = 2 + \dfrac{\pi}{90} \sin\left(\dfrac{\pi}{180}t\right)$ from $t = 0$ to $t = 360$.

$$\int_0^{360}\left(2 + \frac{\pi}{90}\sin\frac{\pi}{180}t\right)dt = 2t - \frac{\pi}{90}\cdot\frac{180}{\pi}\cos\frac{\pi}{180}t\,\Big]_0^{360}$$

$$= 2t - 2\cos\frac{\pi}{180}t\,\Big]_0^{360}$$

$$= (720 - 2) - (0 - 2) = 720$$

The profit is approximately \$72,000.

Exercise A.IV.3

1.   Verify $\int \cot u \, du = \ln|\sin u| + c$ by using the integration technique of substitution.

In Problems 2 to 15 perform the indicated integration.

2. $\displaystyle\int \cos 5x \, dx$

3. $\displaystyle\int \tan 3x \, dx$

4. $\displaystyle\int 6x^2 \cdot \tan x^3 \, dx$

5. $\displaystyle\int x \cos(x^2) \, dx$

6. $\displaystyle\int \csc(x + \tfrac{1}{4}\pi) \cot(x + \tfrac{1}{4}\pi)dx$

7. $\displaystyle\int \sec(x + \pi) \tan(x + \pi)dx$

8. $\displaystyle\int (\sec^2 4x + \csc^2 4x)dx$

9. $\displaystyle\int \frac{\cos x}{\sin^2 x} \, dx$

10. $\displaystyle\int \frac{\sin 2x}{\cos^2 2x} \, dx$

11. $\displaystyle\int \frac{\csc x}{\tan x}\, dx$

12. $\displaystyle\int 7(\csc^2 x + x)dx$

13. $\displaystyle\int x \sin x\, dx$

14. $\displaystyle\int \sin x \sqrt{1 - \cos x}\, dx$

15. $\displaystyle\int \sin^2 x\, dx \left(\text{Use the identity } \sin^2 x = \frac{1 - \cos 2x}{2}\right)$

In Problems 16 to 21, evaluate the definite integral.

16. $\displaystyle\int_0^{\pi/4} \tan x\, dx$

17. $\displaystyle\int_0^{\pi/2} (1 + \cos x)\, dx$

18. $\displaystyle\int_0^{\pi/4} \sin^2 x \cos x\, dx$

19. $\displaystyle\int_{\pi/2}^{\pi} \sin x \cos^2 x\, dx$

20. $\displaystyle\int_0^2 \sqrt{4 - x^2}\, dx$

21. $\displaystyle\int_0^2 x\sqrt{4 - x^2}\, dx$

22. Find the area of the region bounded by the curve $y = \sin x$ and the x axis from $x = 0$ to $x = \pi$.

23. Find the area of the region bounded by the curves $y = \sin x$ and $y = \cos x$ from $x = \frac{1}{4}\pi$ to $x = \frac{5}{4}\pi$.

24. A company notices that its sales fluctuate according to the equation $S = 10 + 2\pi \cos\left(\frac{\pi}{90}t\right)$ where t is in days and S represents the number of sales on that day. Find the total number of sales made during one year. (For convenience, assume a 360-day work year.)

25. A company finds its profit function to be $P = 4 + \pi \sin\left(\frac{\pi}{30}t\right)$ where t is in days and P is in hundreds of dollars. Find the company's total profit during one year. (Assume a 360-day business year.)

V. SEQUENCES AND SERIES

A sequence is a function with a special set for its domain.

> **DEFINITION**
>
> *Sequence.* A *sequence* is a function whose domain is the set of natural numbers.

Let us illustrate with an example. If $f_n = \dfrac{1}{n}$, write the first five terms of this sequence. Let $n = 1, 2, 3, 4, 5$ and obtain the first five terms of the sequence: $\frac{1}{1}, \frac{1}{2}, \frac{1}{3}, \frac{1}{4}, \frac{1}{5}$.

When the domain of f is the set of all natural numbers, the corresponding functional values are called an infinite sequence or a sequence. The sequence is specified by giving its general term. These ideas will be clarified in the following examples.

Sequence	*General Term*	*Sequence*
$\left\{\dfrac{1}{n}\right\}$	$\dfrac{1}{n}$	$1, \dfrac{1}{2}, \dfrac{1}{3}, \dfrac{1}{4}, \cdots$

Graph:

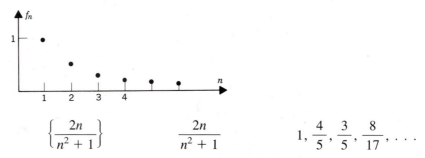

$\left\{\dfrac{2n}{n^2 + 1}\right\}$	$\dfrac{2n}{n^2 + 1}$	$1, \dfrac{4}{5}, \dfrac{3}{5}, \dfrac{8}{17}, \cdots$

Graph:

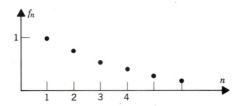

$\{2n\}$ $2n$ 2, 4, 6, 8, 10, . . .

Graph:

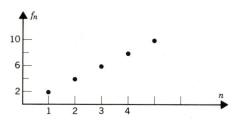

In the sequences $\left\{\dfrac{1}{n}\right\}$ and $\left\{\dfrac{2n}{n^2 + 1}\right\}$ it appears that as you consider subsequent terms of these sequences, their values approach zero. This is not true for the sequence $\{2n\}$. To distinguish between the behavior of these sequences we make these statements:

$$\lim_{n \to \infty} \frac{1}{n} = 0; \quad \lim_{n \to \infty} \frac{2n}{n^2 + 1} = \lim_{n \to \infty} \frac{\dfrac{2}{n}}{1 + \dfrac{1}{n^2}} = \frac{0}{1} = 0;$$

and

$$\lim_{n \to \infty} 2n = \infty,$$

where ∞ is a symbol, not a real number, which represents the phenomenon of increasing without bound.

The technical definition of limit of a sequence is given as:

DEFINITION

Limit of Sequence. The *limit of sequence* $\{x_n\}$ is said to be the real number L, $\lim_{n \to \infty} x_n = L$, if for every $\epsilon > 0$ there is an integer N such that $|x_n - L| < \epsilon$ for all $n \geq N$.

If a sequence has as a limit the real number L, the sequence is said to converge. Otherwise, the sequence is said to diverge. Fortunately, limits of sequences are not usually proved by an ϵ technique but are obtained by using the limit theorems found in Chapter 2. This technique will be illustrated by the following example.

Example 1

Purpose To find the limits of certain sequences:

Problem Determine if the following sequences converge to a limit.

(A) $\left\{\dfrac{1}{n+1}\right\}$;   (B) $\left\{\dfrac{n}{2n+3}\right\}$;   (C) $\left\{\dfrac{4n^2+1}{3n-2}\right\}$.

Solution (A) $\left\{\dfrac{1}{n+1}\right\}$

$$\lim_{n\to\infty}\frac{1}{n+1}=0$$

The sequence $\left\{\dfrac{1}{n+1}\right\}$ converges to a limit 0 or $\left\{\dfrac{1}{n+1}\right\}\to 0$.

(B) $\left\{\dfrac{n}{2n+3}\right\}$

$$\lim_{n\to\infty}\frac{n}{2n+3}=\lim_{n\to\infty}\frac{1}{2+\dfrac{3}{n}}=\frac{1}{2}$$

The sequence $\left\{\dfrac{n}{2n+3}\right\}$ converges to a limit $\dfrac{1}{2}$ or $\left\{\dfrac{n}{2n+3}\right\}\to\dfrac{1}{2}$.

(C) $\left\{\dfrac{4n^2+1}{3n-2}\right\}$

$$\lim_{n\to\infty}\frac{4n^2+1}{3n-2}=\infty$$

The sequence $\left\{\dfrac{4n^2+1}{3n-2}\right\}$ diverges.

Applications of Sequences

Consider the sequences $\{a+(n-1)d\}$ and $\{ar^{n-1}\}$, called the arithmetic and geometric sequences, respectively.

$\{a+(n-1)d\}$ or $a, a+d, a+2d, a+3d, a+4d, \ldots$
$\{ar^{n-1}\}$ or $a, ar, ar^2, ar^3, ar^4, \ldots$

Example 2

Purpose To illustrate an arithmetic and geometric sequence:

Problem Form the first four terms of $\{3+(n-1)2\}$ and $\{3\cdot 2^{n-1}\}$.

Solution 3, 5, 7, 9 are the first four terms of the arithmetic sequence $\{3+(n-1)2\}$.
3, 6, 12, 24 are the first four terms of the geometric sequence $\{3\cdot 2^{n-1}\}$.

From the arithmetic and geometric sequences we will construct *new* functions called the *arithmetic series* and the *geometric series*, respectively. They will be defined by:

Arithmetic series $\qquad S_A = a + (a + d) + (a + 2d) + (a + 3d) + \cdots$
Geometric series $\qquad S_G = a + ar + ar^2 + ar^3 + \cdots$

We emphasize that S_A and S_G are no longer sequences. They were constructed from infinite sequences by addition. For this reason S_A and S_G are often called the infinite arithmetic series and the infinite geometric series, respectively.

Let us take an example of S_G. Let $a = 1$ and $r = \frac{1}{2}$, then

$$S = 1 + \tfrac{1}{2} + \tfrac{1}{4} + \tfrac{1}{8} + \tfrac{1}{16} + \cdots.$$

The question we now consider is: Can we assign a real number to the series S and call this number the sum of the series? In other words, as you add more and more terms, does the sum get close to some number and stay close to that number? Let us calculate a few so-called partial sums of S.

$$S = 1 + \tfrac{1}{2} + \tfrac{1}{4} + \tfrac{1}{8} + \tfrac{1}{16} + \cdots.$$

Partial Sum	*Meaning*	*Value*
s_1	1	$s_1 = 1$
s_2	$1 + \frac{1}{2}$	$s_2 = \frac{3}{2}$
s_3	$1 + \frac{1}{2} + \frac{1}{4}$	$s_3 = 1\frac{3}{4} = \frac{7}{4}$
s_4	$1 + \frac{1}{2} + \frac{1}{4} + \frac{1}{8}$	$s_4 = 1\frac{7}{8} = \frac{15}{8}$
s_5	$1 + \frac{1}{2} + \frac{1}{4} + \frac{1}{8} + \frac{1}{16}$	$s_5 = 1\frac{15}{16} = \frac{31}{16}$
$\vdots$	$\vdots$	$\vdots$

Graphically, we have the following:

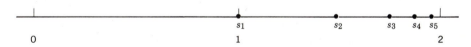

It appears, from our examination of the first five partial sums, that the partial sum $\rightarrow 2$. That is, the sequence of partial sums, $1, \frac{3}{2}, \frac{7}{4}, \frac{15}{8}, \frac{31}{16}, \ldots$, appears to be approaching 2. This motivates our definition for the sum of the terms of any infinite series.

DEFINITION

Convergent Series. The infinite series, $a_1 + a_2 + a_3 + a_4 + a_5 + \cdots$ is said to converge to or have a sum S if the infinite sequence of partial sums, $s_1, s_2, s_3, \ldots, s_n, \ldots$, converges to S. Here $s_1 = a_1$; $s_2 = a_1 + a_2, \ldots, s_n = a_1 + a_2 + \cdots + a_n$. If the sequence of partial sums diverges, then no sum will be assigned to the infinite series.

Let us return to our problem and see if $S = 1 + \frac{1}{2} + \frac{1}{4} + \frac{1}{8} + \cdots$ will have a sum assigned to it. This is a specific case of the problem: When will the geometric series $S_G = a + ar + ar^2 + \cdots + ar^n + \cdots$ have a sum assigned to it?

To answer this we inspect $\{s_n\}$, where

$$s_n = a + ar + ar^2 + \cdots + ar^{n-1}.$$

Multiply by r to obtain

$$rs_n = ar + ar^2 + ar^3 + \cdots + ar^n.$$

Subtract the second equation from the first equation to obtain

$$s_n - rs_n = a - ar^n$$

or

$$s_n(1 - r) = a - ar^n$$

Thus,

$$s_n = \frac{a - ar^n}{1 - r}, \; r \neq 1.$$

Now $S_G = a + ar + ar^2 + \cdots + ar^n + \cdots$ will converge if the corresponding sequence of partial sums, $\{s_n\} = \left\{\dfrac{a - ar^n}{1 - r}\right\}$, converges to a real number L. We investigate $\lim\limits_{n \to \infty} \dfrac{a - ar^n}{1 - r}$ and notice that $r^n \to 0$ if r is a fraction between -1 and 1, $-1 < r < 1$. If $r > 1$ or $r < -1$, the sequence diverges. If $r = \pm 1$, the sequence diverges unless $a = 0$. However, if $-1 < r < 1$, then $\lim\limits_{n \to \infty} \dfrac{a - ar^n}{1 - r} = \lim\limits_{n \to \infty} \dfrac{a - a \cdot 0}{1 - r} = \dfrac{a}{1 - r}$. This establishes the following theorem.

THEOREM A.V

Let S_G be the geometric series $S_G = a + ar + ar^2 + \cdots + ar^n + \cdots$.

This series converges to the sum $\dfrac{a}{1 - r}$ if $-1 < r < 1$. If $r \geq 1$ or $r \leq -1$, S_G diverges unless $a = 0$ and no sum will be assigned to it. Also, s_n, the sum of the first n terms of geometric series, is given by

$$s_n = \frac{a - ar^n}{1 - r}.$$

Exercise A.V

1. Given the following sequences, write the first four terms of the sequence, and determine whether the sequence converges.

    (a) $\left\{\dfrac{n+1}{n^2}\right\}$

    (d) $\left\{\dfrac{n^2+1}{n^2-2n+1}\right\}$, $n = 2,3,4,5$

    (b) $\left\{\dfrac{n(n+1)}{2}\right\}$

    (e) $\left\{\dfrac{1}{2^n}\right\}$

    (c) $\left\{\dfrac{3}{1-n^3}\right\}$, $n = 2,3,4,5$

    (f) $\left\{\dfrac{n^2-1}{n-1}\right\}$, $n = 2,3,4,5$

2. Find the simplest "obvious" expression for the nth term of the following sequences and determine whether the sequence converges.

    (a) $\{\frac{1}{2}, \frac{2}{3}, \frac{3}{4}, \frac{4}{5}, \ldots\}$

    (b) $\{1, \frac{1}{3}, \frac{1}{5}, \frac{1}{7}, \ldots\}$

    (c) $\left\{1, \dfrac{1}{2-\frac{3}{2}}, \dfrac{1}{2-\frac{5}{3}}, \dfrac{1}{2-\frac{7}{4}}, \ldots\right\}$

3. A rubber ball is dropped from a height of 5 feet and always rebounds $\frac{1}{2}$ of the height it falls. Assuming that the ball bounces infinitely often, how far does it travel?

4. At the end of each year a computer depreciates 10 percent of the value it has at the start of the year. If it was worth $100,000 new, what is its worth at the end of six years? $(.9)^6 = .531441$.

5. A house costing $20,000 appreciates 10 percent in value each year. What is the house worth at the end of 7 years?
    $(1.1)^7 \approx 1.948717$

6. As an inducement for signing his contract, a man is offered the following terms: a salary of $10,000 the first year and a 5 percent raise every year for 5 years. What is his salary during the fifth year, and what is the total amount of salary paid him?
    $(1.05)^4 \approx 1.215506$
    $(1.05)^5 \approx 1.276281$

7. Find the sum of each of the following infinite geometric series.
    (a) $1 + \frac{1}{3} + \frac{1}{9} + \frac{1}{27} + \cdots$
    (b) $2 - .2 + .02 - .002 + \cdots$

8. Find a rational number that is equal to each of the following decimals by first expressing the decimal as a geometric series.
    (a) $.666 \ldots = \frac{6}{10} + \frac{6}{10} \cdot \frac{1}{10} + \frac{6}{10} \cdot (\frac{1}{10})^2 + \cdots$
    (b) $.999 \ldots$
    (c) $.1414 \ldots$

9. Jones and Smith flip a coin. The first person to flip a head wins. Let:

   JT represent Jones flips a tail
   JH represent Jones flips a head
   ST represent Smith flips a tail
   SH represent Smith flips a head.

   Jones can win on the first toss if the sequence of plays is JH. If Jones loses on his first toss, he may win on his next turn at play if the sequence of play is JT, ST, JH. In general, Jones can win on the 1st, 3rd, 5th, . . . $(2n - 1)$, . . . toss if Smith loses on the intervening tosses.

Toss	Jones' Winning Sequence	Probability
1	JH	$\frac{1}{2}$
3	JT, ST, JH	$\frac{1}{2} \cdot \frac{1}{2} \cdot \frac{1}{2} = \frac{1}{8}$
5	JT, ST, JT, ST, JH	$(\frac{1}{2})^5 = \frac{1}{32}$
$\vdots$	$\vdots$	$\vdots$

   In probability theory it is shown that the probability that Jones wins this game is equal to the sum of the probabilities $\frac{1}{2} + \frac{1}{8} + \frac{1}{32} + \cdots$. Find the probability that Jones wins the game. Also, calculate the probability that Smith will win the game.

10. For every dollar the government pumps into the economy the individual receiving the dollar takes 10 cents out of circulation and spends or passes along 90 cents to another individual. Assuming an infinite number of transactions of this type, determine the total number of dollars spent if the government initially pumped 1 million dollars into the economy.

11. If P dollars is invested at a yearly rate of interest r, for t years, and the interest is compounded k times per year, the amount A present at the end of t years is given by $A = P\left(1 + \dfrac{r}{k}\right)^{kt}$.

    If $1000 is invested on January 1 each year at a yearly rate of 8 percent compounded quarterly, what amount will be in the account at the end of 10 years?
    $(1.02)^{40} \approx 2.20805$
    $(1.02)^{-40} \approx .45289$
    $(1.02)^{-4} \approx .92385$

VI. TAYLOR'S THEOREM

One of the immediate uses of higher order derivatives is in Taylor's theorem. Taylor's theorem gives a method for expanding a function in terms of powers of $(x - a)$. This expansion is then valid for certain values of x and is useful in many problems, such as integration, differentiation, and moment generating functions in statistics. The proof of Taylor's theorem is complex, and the determination of the values of x for which the expansion is valid requires work with convergent infinite series. Rather than digress on the topic of convergent infinite series, we will state the theorem without proof and caution the student that, before he blindly applies the theorem, he should become familiar with the method for determining the range of x values for which the expansion is valid.

TAYLOR'S THEOREM

If f possesses derivatives of all order at $x = a$ and if f can be represented by an expansion in terms of powers of $x - a$, then

$$f(x) = f(a) + \frac{f'(a)}{1!}(x - a) + \frac{f''(a)}{2!}(x - a)^2$$

$$+ \cdots + \frac{f^{(n-1)}(a)}{(n - 1)!}(x - a)^{n-1} + \cdots$$

($f^{(n-1)}$ is the $n - 1$st derivative of f.)

Example 1

Purpose To illustrate the use of Taylor's theorem:

Problem Expand $x^{-(1/3)}$ in powers of $(x - 1)$.

Solution $(x - a) = (x - 1)$. Therefore, $a = 1$.

$$f(x) = x^{-(1/3)} \qquad\qquad f(a) = f(1) = 1^{-(1/3)} = 1$$

$$f'(x) = -\frac{1}{3}x^{-(4/3)} \qquad\qquad f'(1) = -\frac{1}{3}1^{-(4/3)} = -\frac{1}{3}$$

$$f''(x) = \frac{1 \cdot 4}{3^2}x^{-(7/3)} \qquad\qquad f''(1) = \frac{1 \cdot 4}{3^2}1^{-(7/3)} = \frac{1 \cdot 4}{3^2}$$

$$f'''(x) = -\frac{1 \cdot 4 \cdot 7}{3^3}x^{-(10/3)} \qquad\qquad f'''(1) = -\frac{1 \cdot 4 \cdot 7}{3^3}$$

$$f^{iv}(x) = \frac{1 \cdot 4 \cdot 7 \cdot 10}{3^4}x^{-(13/3)} \qquad\qquad f^{iv}(1) = \frac{1 \cdot 4 \cdot 7 \cdot 10}{3^4}$$

$$\vdots \qquad\qquad\qquad \vdots$$

$$f(x) = x^{-(1/3)} = 1 - \frac{\dfrac{1}{3}}{1!}(x - 1) + \frac{\dfrac{1 \cdot 4}{3^2}}{2!}(x - 1)^2 - \frac{\dfrac{1 \cdot 4 \cdot 7}{3^3}}{3!}(x - 1)^3$$

$$+ \frac{\dfrac{1 \cdot 4 \cdot 7 \cdot 10}{3^4}}{4!}(x - 1)^4 + \cdots.$$

Simplifying, we obtain the first five terms of the expansion as:

$$x^{-(1/3)} = 1 - \frac{1}{3}(x-1) + \frac{2}{9}(x-1)^2 - \frac{14}{81}(x-1)^3 + \frac{35}{243}(x-1)^4 + \cdots.$$

Note. The theory of convergent infinite series shows that this expansion is valid for only those values of x, $0 < x < 2$.

Example 2

Purpose To illustrate the use of Taylor's theorem when $a = 0$. A Taylor's series with $a = 0$ is called a *Maclaurin series*.

Problem Expand $\dfrac{1}{1-x}$ in a Maclaurin series, that is, a Taylor's series with $a = 0$.

Solution

$$f(x) = \frac{1}{1-x} = (1-x)^{-1} \qquad\qquad f(a) = f(0) = 1$$

$$f'(x) = -(1-x)^{-2}(-1) \qquad\qquad f'(0) = 1$$
$$f''(x) = -2(1-x)^{-3}(-1) \qquad\qquad f''(0) = +2$$
$$f'''(x) = -6(1-x)^{-4}(-1) \qquad\qquad f'''(0) = +6$$
$$f^{iv}(x) = -24(1-x)^{-5}(-1) \qquad\qquad f^{iv}(0) = +24$$

$$\vdots \qquad\qquad\qquad\qquad\qquad \vdots$$

$$f(x) = \frac{1}{1-x} = 1 + \frac{1}{1!}x + \frac{2}{2!}x^2 + \frac{6}{3!}x^3 + \frac{24}{4!}x^4 + \cdots,$$

or

$$f(x) = \frac{1}{1-x} = 1 + x + x^2 + x^3 + x^4 + \cdots.$$

Again, the theory of infinite series shows us that this expansion is valid for $-1 < x < 1$.

Exercise A.VI

1. Use Taylor's theorem and expand through five terms $x^{1/2}$ in powers of $(x - 2)$.

2. Use a Maclaurin series and expand through five terms $f(x) = (1 + x)^m$ (m is any real number). (Notice this is the binomial expansion!)

3. Using Taylor's series with powers of x (Maclaurin series), state the first four terms of $\ln (1 + x)$. This series converges on the interval $-1 < x \leqslant 1$.

4. Find the first four terms of a Maclaurin series for $y = e^{\theta x}$. This series converges for every value of x (θ may be treated as a constant).

5. Expand by Taylor's series $e^{x/2}$ in powers of $(x - 2)$ through four terms. The series converges for every value of x.

TABLE OF EXPONENTIALS AND NATURAL LOGARITHMS

x	e^x	$\ln x$	x	e^x	$\ln x$
0.0	1.0000		2.5	12.182	0.91629
0.1	1.1052	-2.3026	2.6	13.464	0.95551
0.2	1.2214	-1.6094	2.7	14.880	0.99325
0.3	1.3499	-1.2040	2.8	16.445	1.02962
0.4	1.4918	-0.91629	2.9	18.174	1.06471
0.5	1.6487	-0.69315	3.0	20.086	1.09861
0.6	1.8221	-0.51083	3.1	22.198	1.13140
0.7	2.0138	-0.35667	3.2	24.533	1.16315
0.8	2.2255	-0.22314	3.3	27.113	1.19392
0.9	2.4596	-0.10536	3.4	29.964	1.22378
1.0	2.7183	0.00000	3.5	33.115	1.25276
1.1	3.0042	0.09531	3.6	36.598	1.28093
1.2	3.3201	0.18232	3.7	40.447	1.30833
1.3	3.6693	0.26236	3.8	44.701	1.33500
1.4	4.0552	0.33647	3.9	49.402	1.36098
1.5	4.4817	0.40547	4.0	54.598	1.38629
1.6	4.9530	0.47000	4.1	60.340	1.41099
1.7	5.4739	0.53063	4.2	66.686	1.43508
1.8	6.0496	0.58779	4.3	73.700	1.45862
1.9	6.6859	0.64185	4.4	81.451	1.48160
2.0	7.3891	0.69315	4.5	90.017	1.50408
2.1	8.1662	0.74194	4.6	99.484	1.52606
2.2	9.0250	0.78846	4.7	109.95	1.54756
2.3	9.9742	0.83291	4.8	121.51	1.56862
2.4	11.023	0.87547	4.9	134.29	1.58924

x	e^x	$\ln x$	x	e^x	$\ln x$
5.0	148.41	1.60944	7.5	1808.0	2.14090
5.1	164.02	1.62924	7.6	1998.2	2.02815
5.2	181.27	1.64866	7.7	2208.3	2.04122
5.3	200.34	1.66771	7.8	2440.6	2.05412
5.4	221.41	1.68640	7.9	2697.3	2.06686
5.5	244.69	1.70475	8.0	2981.0	2.07944
5.6	270.43	1.72275	8.1	3294.5	2.09186
5.7	298.87	1.74047	8.2	3641.0	2.10413
5.8	330.30	1.75786	8.3	4023.9	2.11626
5.9	365.04	1.77495	8.4	4447.1	2.12823
6.0	403.43	1.79176	8.5	4914.8	2.14007
6.1	445.86	1.80829	8.6	5431.7	2.15176
6.2	492.75	1.82455	8.7	6002.9	2.16332
6.3	544.57	1.84055	8.8	6634.2	2.17475
6.4	601.85	1.85630	8.9	7332.0	2.18605
6.5	665.14	1.87180	9.0	8103.1	2.19722
6.6	735.10	1.88707	9.1	8955.3	2.20827
6.7	812.41	1.90211	9.2	9897.1	2.21920
6.8	897.85	1.91692	9.3	10938.	2.23001
6.9	992.27	1.93152	9.4	12088.	2.24071
7.0	1096.6	1.94591	9.5	13360.	2.25129
7.1	1212.0	1.96009	9.6	14765.	2.26176
7.2	1339.4	1.97408	9.7	16318.	2.27213
7.3	1480.3	1.98787	9.8	18034.	2.28238
7.4	1636.0	2.00148	9.9	19930.	2.29253
			10.0	22026.	2.30259

ANSWERS TO SELECTED PROBLEMS

Chapter 1

Exercise 1.4–1.5

1. (a) natural, integer, rational;   (c) irrational;

   (e) irrational;   (g) integer, rational;

   (i) undefined;   (k) natural, integer, rational.

2. Definition of division: $\dfrac{a}{b} = c$ if $a = b \cdot c$ (c, a unique number).

   Case I: $a \neq 0,\ b = 0$

$$\frac{a}{0} = 0 \quad \text{if } a = 0 \cdot c$$

     $a = 0$ is impossible, since $a \neq 0$ by hypothesis.

   Case II: $a = 0,\ b = 0$

$$\frac{0}{0} = c \quad \text{if } 0 = 0 \cdot c$$

     $0 = 0$ is indeterminate, since c can be any number.

3.

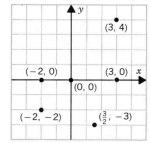

7.

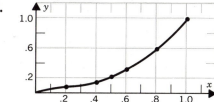

9.

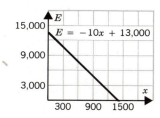

11. 400

13. (b) Vertex at $(-1, 3)$;

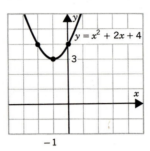

(c) Vertex at $(-1, -3)$.

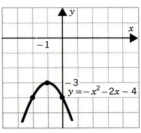

Estimate of maximum
profit is $25,000

14. (b)

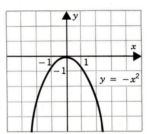

(d)

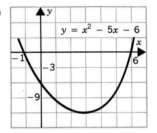

15.

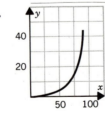

Estimate of maximum profit is $25,000

17. The maximum y value is 25.

19.

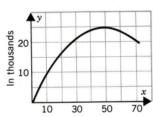

$x\%$	y
0	0
25	$\frac{5}{3}$
50	5
75	15
90	45

21. (b) The ideal temperature is $60°$ F.

23. 5. **25.** $4\sqrt{2}$.

27. $2c$.

29. $AB = \sqrt{41}$, $AC = 2\sqrt{10}$, $BC = \sqrt{5}$, neither.

31. $AB = 2\sqrt{17}$, $AC = \sqrt{34}$, $BC = \sqrt{34}$, isosceles and right triangle.

33. $x^2 + y^2 = 4$.

35. $(x - 1)^2 + (y + 2)^2 = 25$.

37. $(x - a)^2 + y^2 = a^2$.

39. (23) $(\frac{1}{2}, -3)$.

(25) $(-2, -1)$.

(27) $(0, d)$.

41. $x = -1$.

Exercise 1.6

1. $-\frac{4}{9}$.

3. $\frac{4}{3}$.

5. Undefined.

9. $-\frac{25}{9}$.

11. $-\frac{1}{5}$.

13. $\frac{3}{5}$.

15. 3.

17. $L_1 \parallel L_5$, $L_2 \parallel L_4$, $L_1 \perp L_2$, $L_1 \perp L_4$, $L_2 \perp L_5$, $L_4 \perp L_5$.

19. (a) $2x + 3y = -11$;

(b) $5x - 3y = 1$;

(c) $x = 3$;

(d) $y = 3$;

(e) $x - 3y = 6$.

21. $y = 0$.

22. $x = 0$.

25. $4x - 3y = -18$.

27. (a) $5x - 4y - 17 = 0$;

(b) $5x - 4y + 31 = 0$;

(c) $8x + 10y - 19 = 0$;

(d) $4x + 5y - 8 = 0$;

(e) $3x + 4y - 7 = 0$.

31. (c), (d), (e). These are of the form $Ax + By = C$.

33. $y = .1x$, 11 grams per liter.

35. (b) $y = \frac{13}{4}x + \frac{23}{4}$;

(c) $y = 28\frac{1}{2}$.

37. (b) $y = \frac{1}{5}x$, 6 responses;

(c) $y = \frac{1}{3}x - 2$; 8 responses.

39. (a)

(b) $x + 23y = 96$;

(c) 6150, 400.

41. $(-1, 10)$; $(3, -2)$.

43. Equilibrium quantity $(\frac{7}{2}, 3)$, 3500 bacteria at $300.

45. (a) $y = 25 - x$;

(b) left;

(c) $(12, 13)$, 1200 calculators at $13.

47. 30 radios at $40.

49. (a) $TC = 15000 + 10x$;

(b) $TR = 15x$;

(d) 3000 items.

51. (a) $TR = 4x$;

(b) $TC = 2800 + 1.2x$;

(c) 1000 items;

(d) 700 items.

52. (a) $x = 7$;   (b) $y = 77$;
(c) $x = 5$, $y = 32$.

Exercise 1.7

1. (a) 4, 11, 0;   (b) $7 - x_1$, $7 - x_1 - \Delta x$,
$-\Delta x$, -1.

3. (a) 19, 3, 1;
(b) $2x_1^2 + 1$, $2x_1^2 + 4x_1 \Delta x + 2(\Delta x)^2 + 1$, $4x_1 \Delta x + 2(\Delta x)^2$, $4x_1 + 2 \Delta x$.

5. (a) 0, 3, 7;
(b) $7 + 9 \Delta x + 2(\Delta x)^2$, $9 \Delta x + 2(\Delta x)^2$, $9 + 2 \Delta x$.

7. (a) 1, -1, undefined;
(b) $\dfrac{2}{t_1 - 2}$; $\dfrac{2}{t_1 + \Delta t - 2}$; $\dfrac{-2 \Delta t}{(t_1 - 2)(t_1 + \Delta t - 2)}$; $\dfrac{-2}{(t_1 - 2)(t_1 + \Delta t - 2)}$.

9. (a) $\sqrt{2}$, 0, 3;
(b) $\sqrt{2 - t_1}$, $\sqrt{2 - t_1 - \Delta t}$, $\sqrt{2 - t_1 - \Delta t} - \sqrt{2 - t_1}$
$\dfrac{\sqrt{2 - t_1 - \Delta t} - \sqrt{2 - t_1}}{\Delta t} = \dfrac{-1}{\sqrt{2 - t_1 - \Delta t} + \sqrt{2 - t_1}}$.

11. (a) \$0.06;   (b) \$0.60;
(c) \$6.00.

13. (b) \$10,000, \$5,000.

15. (a) 44 minutes;   (b) 121.

17. (b) Multiply x by 5 and subtract this quantity from 2 to obtain a value for y.

19. Function, all real numbers.

21. Not a function.

23. Function, $x \geq 3$.

25. Not a function.

27. Function, $x \neq \pm 1$.

29. Function, all real numbers.

31.

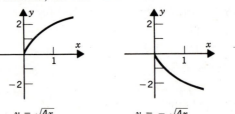

$y = \sqrt{4x}$ $y = -\sqrt{4x}$ $y^2 = 4x$

$y = \sqrt{4x}$ and $y = -\sqrt{4x}$ are functions.

33. $y = 185 + 5x$.

35. (c) $P(x) = -300x^2 + 1200x - 18000$, maximum profit is \$102,000.

36. (b) 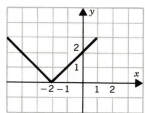 (c)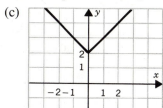

Exercise 1.8

1. $x < 8$.

3. $x > \frac{8}{3}$.

5. $x \geqslant -\frac{1}{2}$.

7. $1 < x < 3$.

9. $-3 < x < 3$.

11. $x > 8$ or $x < -1$.

13. $x > 1$ or $x < -1$.

15. $-2 < x < 3$.

17. $-1 < x < \frac{1}{2}$.

19. $-2 < x < -\frac{2}{3}$.

21. $-2 < x < \frac{1}{2}$.

23. $x < -1$.

25. $x > 1$.

26. (a) $x \geqslant 5$ or $x \leqslant -5$;
(b) $x \geqslant 3$ or $x \leqslant 2$.

27., 28., 29.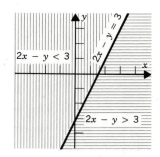

37. $x < 30$.

39. $45x + 1500 \geqslant 3670$; 49.

41. $x \geqslant 62$; $y \geqslant 4275$; \$375.

43. 10 of Model A, 12 of Model B.

45. 32 cans of A, 10 cans of B.

47. 6 of Line A, 0 of Line B.

Exercise 1.9

1. $M = 300 + .30x$ where x is number of cards punched.

3. (a) $S = 75x - 575$ where x is number of cars sold and $x > 15$;
(b) \$550; (c) \$550.

5. $y = -\frac{1}{2}x + 35$.   **7.** $N = 10 + t$, $0 \leqslant t \leqslant 14$.   **9.** $R = 4t - t^2$.

11. $R(x) = x[2 - \frac{1}{12}(x - 50)] = \frac{37}{6}x - \frac{1}{12}x^2$ where x is number of trees and
$x > 50$.

13.   (a) $R(x) = 500x - x^2$;   (c) $P(x) = -\frac{3}{2}x^2 + 500x - 150$.
(b) $C(x) = \frac{1}{2}x^2 + 150$;

15.   (a) $3w$;   (b) $V = 6w^3$.

17.   (a) πr^2; (b) $2r$; (c) $4r^2$; (d) $r^2 - \dfrac{\pi}{4}r^2$.

19. $\frac{1}{2}h \sqrt{625 - h^2}$ where h is the height of the triangle.

21. $(v + 2)$ by $(v + 5)$.   **23.** $x^2(96 - x)^2$; $[x(96 - x)]^2$; yes.   **25.** $.16\pi r^2$.

Review Exercise

3.   (a) A set of lines intersecting at $(0, b)$ with various slopes;
(b) A set of parallel lines.

5.   (a) $y = -\frac{5}{4}x + 3$;   (b) $x - 3y + 5 = 0$;
(c) $y = 4$;   (d) $y = -4$;
(e) $y = \frac{3}{4}x - 1$;   (f) $\dfrac{x}{2} + \dfrac{y}{-5} = 1$.

7. $3x + 2y = -1$.   **8.** $2x - 3y = -5$.

9.   (a) $\frac{1}{2}$;   (b) $y = \frac{1}{2}x + \frac{9}{2}$;
(c) $y = \frac{1}{2}x - 3$;   (d) $(1, 5)$;
(e) $2x + y = 7$;   (f) $2x + y = 7$;
(g) $2x + y = 7$;   (h) $5\sqrt{2}$;
(i) $5\sqrt{2}$;   (j) yes, length of AB = length
of BC.

11. $C = 2000 + 720x$.

13. $(3, 4)$; $(-3, 4)$.   **15.**   2500 units.

17. $(x - 3)^2 + y^2 = 25$.   **19.**   (a) 3; (b) no; (c) $x \geqslant 3$.

21.   (a) $0, -6, -4$;
(b) $x_1^2 - 3x_1 - 4$, $x_1^2 + 2x_1 \Delta x + (\Delta x)^2 - 3x_1 - 3 \Delta x - 4$, $2x_1 \Delta x + (\Delta x)^2 - 3 \Delta x$, $2x_1 + \Delta x - 3$.

23.   (b), (c), (e), (f), and (g) are functions.
(f)

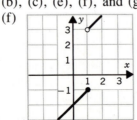

25. (a) Remains the same;   (b) remains the same;
(c) changes the sense of an inequality.

27. $x < \frac{2}{3}$.   **29.** $x > \frac{2}{3}$.

31. $-1 < x < 0$.   **33.** $-2 \leqslant x \leqslant 2$.

35. $x > \frac{8}{3}$ or $x < -1$.   **37.** $-1 < x < 2$.

39. 20 modern, 40 traditional.

41. $T(x) = 1.50 + .10x$.   **43.** $P(x) = -1.2x^2 + 150x - 80$.

44. (a) $3r$;   (b) $\frac{2}{3}\pi r^2$;
(c) $3\pi r^3$.   (d) $\frac{11}{3}\pi r^3$.

Chapter 2

Exercise 2.2

1. (a) 1.5;   (b) -1.5;
(c) 3;   (d) -1.5;
(e) 3;   (f) 3;
(g) for part (b) 3, -0.5, 2.5; for part (c) -2, 1, -1;
(h) 3.

3. (a) 10;   (b) 5;
(c) 5;   (d) they are the same;
(e) -2;   (f) -1.

7. $\frac{1}{2}$.   **9.** -4.

11. $2x_1 + \Delta x + 6$.   **13.** $\dfrac{-3}{x_1(x_1 + \Delta x)}$.

15. $\dfrac{-1}{(x_1 + 4)(x_1 + \Delta x + 4)}$.

17. $\dfrac{\sqrt{2x_1 + 2\Delta x + 5} - \sqrt{2x_1 + 5}}{\Delta x} = \dfrac{2}{\sqrt{2x_1 + 2\Delta x + 5} + \sqrt{2x_1 + 5}}$.

19. (a) -2.19;   (b) 0.34.

21. a.   **22.** $2ax_1 + a\Delta x + b$.

23. 2 cm^2/hr.   **25.** 32 mph.

27. (a)

t	0	1	2	3	4	5
	0	55	105	150	190	225

(b) 55, 50, 45, 40, 35;   (c) 45 mph.

28. (a) $2\pi r_1 \Delta r + \pi(\Delta r)^2$;   (b) $2\pi \Delta r$;
(c) $2\pi r_1 + \pi \Delta r$;   (d) 2π.

29. (a) $\frac{9}{4}\pi$; (b) π; (c) $\frac{9}{2}\pi$; (d) 2π.

Exercise 2.3

1. 6.9, 6.99, 7.01, 7.1, f(x) is very close to 7.

3. 10.24, 9.61, 8.41, 7.84, f(x) is very close to 9.

5. 2, 2.59374, 2.70481, 2.71693, 2.71814, 2.71828, 2.71828.

7. (a) 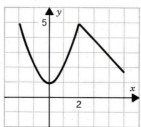 (b) 5; (c) yes.

9. (a) (b) 35, 39, does not exist.

11. 7.	**13.** 4.
15. -1.	**17.** -44.
19. $-\frac{1}{2}$.	**21.** 5.
23. $-\frac{1}{3}$.	**25.** -1.
27. $\frac{7}{5}$.	**29.** -6.
31. 8.	**33.** 8.
35. 0.	**37.** 1.
39. $\frac{1}{32}$.	**41.** -4.

43. (a) 0; (b) 0; (c) 0; (d) 2; (e) 1; (f) does not exist.

Exercise 2.4–2.5

1. $\frac{13}{5}$.	**3.** 6.
5. $2x_1$.	**7.** 4.
9. $-\infty$.	**11.** 0.
13. 2.	**15.** 0.
17. 0.	**19.** 0.
21. $\frac{1}{2}$.	**23.** $-\frac{1}{4}$.

25. $\dfrac{1}{2\sqrt{5}}$.

27. (a) $\frac{3}{2}$; (b) $\frac{4}{3}$; (c) $\frac{5}{4}$; (d) 0; (e) ∞; (f) $-\infty$, (g) 1.

29. (a) $\frac{49}{6}$; (b) $\frac{1}{8}$; (c) ∞; (d) $-\infty$; (e) ∞; (f) 0; (g) $\frac{289}{14}$; (h) $\frac{16}{7}$.

31. (a) 4; (b) does not exist.

33. (a) 8, 6, does not exist;   (b) 5, 5, 5.

35. 3 minutes.   **36.** (a) $\frac{3}{7}$; (b) 0.

37. Discontinuous nowhere.   **39.** 2, -2.

41. Discontinuous nowhere.   **43.** 0.

45. 0.

47. (a) 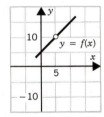   (b) $W < 10$, 15, 18, 22.

49. All x except $x = 2$.

51. (a) No. $f(x)$ does not exist at $x = 5$.   (b) No. $\lim\limits_{x \to 5} g(x) \neq g(5)$.

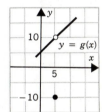

51. (c) Yes.

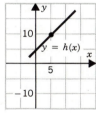

53. $y = [x]$ is discontinuous at each integer x; $\lim\limits_{x \to a^+} [x] \neq \lim\limits_{x \to a^-} [x]$ when a is an integer.

55. (a) $\lim f(x)$ exists for all x.
     (b) $f(x)$ is continuous for all x.

Exercise 2.6

1. (a) $2x - 2$;   (b) 4;
     (c) 4;   (d) $4x - y = 12$;

(e) -3, $12x + 4y = -13$; (f) $4x - 12y = 19$;
(g) 0; (h) $y = -4$, horizontal line.

3. $1; f'(c) = 1$ (c, any constant); slope of the tangent line at a point on the graph of this function is always 1.

4. 0; the value of f$'$ is always 0; slope of the tangent line at a point on the graph of this function is always 0.

5. 3. **7.** $\frac{1}{2}$.

9. -4. **11.** $2x + 6$.

13. $\dfrac{-3}{x^2}$. **15.** $\dfrac{-1}{(x + 4)^2}$.

17. $\dfrac{1}{\sqrt{2x + 5}}$. **19.** 7.

21. $-\frac{1}{4}$. **23.** $-2x$.

25. $\dfrac{1}{2\sqrt{x}}$. **27.** $6x + 3y = 13$.

29. $2x + y = 5$ toward $(\frac{5}{2}, 0)$. **31.** 25.

33. $x + 1$. **35.** 32.

37. $2t$. **38.** $at + v_0$.

Review Exercise

3. 2. **5.** $\frac{7}{3}$.

7. $-\frac{3}{2}$. **9.** 0.

11. 0. **13.** $4x^2 - 7x$.

15. $\dfrac{3}{x^2}$. **16.** -2.

17. $f(3)$ is undefined because you cannot divide by zero; $\lim_{x \to 3} f(x) = 6$.

19. (b) -2; (c) 2; (d) no.

20. (a) $f(x_1 + \Delta x) - f(x_1)$; (b) $\dfrac{\Delta y}{\Delta x} = \dfrac{f(x_1 + \Delta x) - f(x_1)}{\Delta x}$;

(c) $\lim\limits_{\Delta x \to 0} \dfrac{\Delta y}{\Delta x} = \lim\limits_{\Delta x \to 0} \dfrac{f(x_1 + \Delta x) - f(x_1)}{\Delta x}$;

(d) limiting position of the secant line $P_1 P_2$ as $P_2 \to P_1$ or $\Delta x \to 0$.

24. (a) $6x - 10$; (b) -10;
(c) -10; (d) $10x + y = -8$;
(e) 14, $y = 14x - 56$; (f) $x + 14y = 4$;
(g) 0; (h) horizontal line.

25. $3x + y = 12$, $3x + y = 0$. **27.** $\frac{3}{4}$.

29. -5.   **31.** $\dfrac{4}{(x+2)^2}$.   **33.** $9 - 2x$.   **35.** $.4x + 4$.

37. (a) All x; (b) $x \neq \pm 2$; (c) $x \neq 1$; (d) $x \geq -3$.

38. Yes. $f(0) = 1$, $\lim\limits_{x \to 0} f(x) = 1$, and $\lim\limits_{x \to 0} f(x) = f(0) = 1$;

Yes. $f(-1) = 0$, $\lim\limits_{x \to -1} f(x) = 0$, and $\lim\limits_{x \to -1} f(x) = f(-1) = 0$;

No. $f(1)$ does not exist.

41. (a) $\lim\limits_{\Delta x \to 0} (5 - 2 - 2\,\Delta x) = 3$, $g(1) = 6$;

(b) the function is discontinuous at $x = 1$;

(c) the function must be continuous.

Chapter 3

Exercise 3.1

1. $6x$.

3. $x^2 - 1$.

5. $-x^4 + 6x^2 - 72$.

7. $\frac{1}{3}(1 - x)$.

9. $3t^3 - 4t^2 + t$.

11. $x^2 - 4x$.

13. $v + v^{-3}$.

15. $\dfrac{-6}{x^3}$.

17. $x^4 + x^{-3}$.

19. $\dfrac{1}{2\sqrt{x}} - \dfrac{1}{2\sqrt{x^3}}$.

21. $x^2 - \dfrac{3}{x^4}$.

23. $\dfrac{3}{2\sqrt{u}} - \dfrac{1}{\sqrt{u^3}}$.

25. $-\dfrac{81}{t^{10}}$.

27. $\dfrac{1}{2\sqrt{t}} - \dfrac{32}{t^3}$.

29. $-5x^4 - 6x^2 + 4x$.

31. $-\dfrac{133}{4}\, x^3 \sqrt[4]{x^3}$.

33. $\dfrac{1}{\sqrt[3]{v^2}} - 4\sqrt[3]{v}$.

34. $(16t^2 - t)[\frac{2}{3}t^{-(1/3)}] + (2 + t^{2/3})(32t - 1) = \frac{128}{3}t\sqrt[3]{t^2} + 64t - \frac{5}{3}\sqrt[3]{t^2} - 2$.

35. $-24t^3 - 21t^2 - 2t + 1$.

36. $\dfrac{3\sqrt{x}}{2} + \dfrac{2}{3\sqrt[3]{x}}$.

37. $\dfrac{4(3u^2 + 2)}{3u^2}$.

39. $\dfrac{-6}{(x-3)^2}$.

41. $\dfrac{(t-1)[\frac{1}{2}t^{-(1/2)}] - (t^{1/2})(1)}{(t-1)^2} = \dfrac{-(t+1)}{2\sqrt{t}(t-1)^2}$.

43. $\dfrac{18x^2 - 1}{(6x^3 - x + 12)^2}.$

45. $\dfrac{12(3x^6 - 40x^3 + 64)}{(4x - x^4)^2}.$

47. $\dfrac{-4v^4 - 23v^2 - 8v + 10}{(v^3 - 2v - 1)^2}.$

49. $\dfrac{x^4 - 2x^3 - 12x^2 - 6x + 3}{(3 + x^3)^2}.$

52. (a) $2x - 6$;   (b) $-6, 0$;

   (c) $6x + y = 9,\ y = 0.$

53. (a) $3x^2 + 6x - 4$;   (b) $-7, 41$;

   (c) $7x + y + 13 = 0;\ 41x - y = 93.$

55. $y = -4x + 3;\ y = \frac{1}{4}x - \frac{11}{2}.$

57. $(\frac{5}{2}, \frac{53}{4}).$

59. (a) $x + 4$;   (b) $\frac{1}{2}x + 4 + \dfrac{24}{x}$;   (c) $\frac{1}{2} - \dfrac{24}{x^2}.$

61. $\dfrac{-20000}{(50 + x)^2}$

63. $R' = \dfrac{dR}{dQ} = P \cdot \dfrac{d}{dQ}(Q) + Q\dfrac{d}{dQ}(P)$
$$= P \cdot 1 + Q \cdot P'$$
$$= P + Q \cdot P'.$$

64. $R = (4 - Q^2)Q,\ R' = 4 - 3Q^2,\ P' = -2Q.$

65. $\dfrac{50}{(x - 1)^2}.$ **67.** $2.25.$

72. (c) $n[f(x)]^{n-1}f'(x).$

Exercise 3.2

1. $10(x^2 + x)^9(2x + 1).$

3. $\left[3\left(\dfrac{1}{x^2 - 4}\right)^2 + 2\right]\dfrac{-2x}{(x^2 - 4)^2}.$

5. $20(5x + 2)^3.$

7. $3(3t^3 - 4t + 1)^2(9t^2 - 4).$

9. $(v^2 + 3)(-5v^2 + 4v - 3).$

11. $\dfrac{-6x}{(3x^2 - 7)^2}.$ **13.** $\dfrac{-2x}{3\sqrt[3]{(3 - x^2)^2}}.$

15. $12x^{4/3}(3 + 4x)^2 + \frac{4}{3}x^{1/3}(3 + 4x)^3 = \frac{4}{3}x^{1/3}(3 + 4x)^2(13x + 3).$

17. $3 - \dfrac{\sqrt{3x}}{2x^2}.$

19. $-\dfrac{1}{\sqrt{7 - 2x}} + \dfrac{1}{\sqrt{(7 - 2x)^3}}.$

21. $\dfrac{10(x-4)}{(2x-3)^3}$.

22. $\dfrac{2\sqrt{x}+1}{4\sqrt{x}\sqrt{x+\sqrt{x}}}$.

23. (a) $6(3x-5)$; (b) $y = 6x - 11$;
(c) yes, at $x = \frac{5}{3}$.

24. (a) $3(x+3)(x-1)$; (b) $f'(x) = 0$ at $x = -3$ and $x = 1$; $y = 0$,
$y = -32$; (c) $x > 1$ or $x < -3$; (d) $-3 < x < 1$.

25. (a) $\dfrac{x_1}{\sqrt{(x_1)^2+1}}$; (b) $m = 0$, $y = 1$ is a horizontal tangent.

27. (a) 0; (b) $x = 1$.

29. $\dfrac{10}{\sqrt{2Q+5}} - \dfrac{10}{Q^2}$, 1.9.

31. 12, 18.

33. 0, $\dfrac{2}{\sqrt{5}}$, $\dfrac{4}{\sqrt{17}}$.

34. $50\pi\dfrac{\text{in}^3}{\text{min}}$.

Exercise 3.3–3.4

1. $12x^2 - 6$.

3. $2a$.

5. $2(2 - 3Q)$.

7. $-\dfrac{1}{(\sqrt{2x})^3} = -\dfrac{1}{2x\sqrt{2x}}$.

9. $\frac{-2}{45}(1-x)^{-5/3}$.

11. (a) $3x^2 - 4x - 4 = (x-2)(3x+2)$;
(b) 2, $-\frac{2}{3}$, horizontal line;
(c) $x > 2$ or $x < -\frac{2}{3}$; (d) $-\frac{2}{3} < x < 2$;
(e) $6x - 4$; (f) $\frac{2}{3}$;
(g) $x > \frac{2}{3}$; (h) $x < \frac{2}{3}$.

12. (a) $v(t_0) = 3t_0^2 - 5$, $v(3) = 22$, $a(t_0) = 6t_0$, $a(3) = 18$;
(c) $v(t_0) = \dfrac{t_0}{\sqrt{t_0^2+9}}$, $v(4) = \frac{4}{5}$, $a(t_0) = \dfrac{9}{(t_0^2+9)(\sqrt{t_0^2+9})}$, $a(4) = \frac{9}{125}$.

13. (b) $v = 2(t - 4)$; (c) $a = 2$.

14. (b) $D_x^n x^n = n!$

15. (a) Implicit; (b) explicit;
(c) implicit; (d) implicit.

17. $\dfrac{3x}{1-y}$.

19. $\dfrac{x-y}{x-3y}$.

21. $-\dfrac{(2xy^2+3y)}{(2x^2y+3x-4)}$.

23. $-\sqrt{\dfrac{y}{x}}$.

25. $-\dfrac{y^2}{x^2}$.

26. $3x + 4y = 25,\ 3x - 4y = 25$.

27. $-\frac{2}{3}$.

Exercise 3.5–3.6

1. $f(x) > 0$ for $x > 7$ or $x < -1$; $f(x) < 0$ for $-1 < x < 7$.

3. $f(x) > 0$ for $2 < x < 3$; $f(x) < 0$ for $x > 3$ or $x < 2$.

5. $f(x) > 0$ for $x > -2$; $f(x) < 0$ for $x < -2$.

7. $f'(x) < 0$ for all x.

9. $f'(x) > 0$ for all x except $x = 3$.

11. $f'(x) > 0$ for $x > 4$ or $x < \frac{2}{3}$; $f'(x) < 0$ for $\frac{2}{3} < x < 4$.

13. $f''(x) > 0$ for all x.

15. $f''(x) > 0$ for $x > 3$; $f''(x) < 0$ for $x < 3$.

17. $f''(x) > 0$ for all x except $x = -1$.

19. (a) $f'(x) = 1 - \dfrac{1}{x^2}$;

    (b) $x = 1,\ x = -1,$ horizontal tangent lines;

    (c) $f'(x) > 0$ for $x > 1$ or $x < -1$, $f'(x) < 0$ for $-1 < x < 0$ or $0 < x < 1$;

    (d) $f''(x) = \dfrac{2}{x^3}$;

    (e) $f''(x) > 0$ for $x > 0$, $f''(x) < 0$ for $x < 0$.

21. $\frac{1}{2}$.

23. $\sqrt{7}$.

25. 1.

Review Exercise

1. 0.

3. $2x + 4$.

5. $2ax + b$.

7. $30x^5 - 6x^2 + 1$.

9. $54t^2(6 + t^3)^2$.

11. $-3x^2 - 14x + 2$.

13. $12t$.

15. $(2x + 3)^3(10x + 3)$.

17. $-\dfrac{1}{t^2}$.

19. $-\dfrac{26}{x^3}$.

21. $\dfrac{-x}{5\sqrt{9 - x^2}}$.

22. $\dfrac{-30x}{\sqrt[3]{(3x^2)^4}} = \dfrac{-10}{x\sqrt[3]{3x^2}}$.

23. $\dfrac{x - 1}{2x\sqrt{x}}$.

25. $\dfrac{-6x^2}{(\sqrt{x^2 + 3})^3}$.

27. $\dfrac{1}{\sqrt{2x}} - \dfrac{\sqrt{2x}}{2x^2}$.

29. $\dfrac{7x - 20}{x^3}$.

31. $\dfrac{2(3x^4 + 7)}{x^2}$.

33. 0.

35. $\dfrac{1 - x}{2\sqrt{x}(x + 1)^2}$.

37. $\dfrac{-4x}{(x^2 - 1)^2}$.

39. $\dfrac{4(x - 1)}{(x + 1)^3}$.

41. $\dfrac{-34(7 + x)}{(2x - 3)^3}$.

43. $\dfrac{14x^2 + 112x + 89}{(x + 4)^2}$.

45. $-\dfrac{x}{y}$.

47. $\dfrac{-(y + 7)}{x}$.

49. $\dfrac{-2xy}{x^2 + 16}$.

51. $3x + 2y = 12$.

53. $6(3x - 2)$; $f(x) = (3x - 2)^2$.

55. (a) $\dfrac{2}{\sqrt{4x + 13}}$;

(b) $2x - 5y + 19 = 0$;

(c) no;

(d) $\dfrac{-4}{(\sqrt{4x + 3})^3}$;

(e) no.

56. (a) $3x^2 + 6x - 9$;

(b) $1, -3$, horizontal line;

(c) $x > 1$ or $x < -3$;

(d) $-3 < x < 1$;

(e) $6(x + 1)$;

(f) $x = -1$;

(g) $x > -1$;

(h) $x < -1$.

57. 1507, 303.

59. $\dfrac{3x^2 + 30x}{(x + 5)^2}$.

61. (a) $v = -32t_1$;

(b) $-32, -64, -96$.

63. $3x + y = 12$.

64. (a) 2;

(b) 3.

Chapter 4

Exercise 4.2

1. For $x_1 < x < x_2$, $x_2 < x < x_3$, $x > x_5$, $\uparrow$; for $x < x_1$, $x_3 < x < x_4$, $x_4 < x < x_5$, $\downarrow$.

3. For all x, $\uparrow$.

5. For $x > 3$, $\uparrow$; for $x < 3$, $\downarrow$; horizontal tangent at $x = 3$.

7. For $x > 5$, $\uparrow$; for $x < 5$, $\downarrow$; horizontal tangent at $x = 5$.

9. For $x > \frac{1}{2}$, $\uparrow$; for $x < \frac{1}{2}$, $\downarrow$; horizontal tangent at $x = \frac{1}{2}$.

11. For $x > 2$ or $x < \frac{1}{3}$, $\uparrow$; for $\frac{1}{3} < x < 2$, $\downarrow$; horizontal tangents at $x = \frac{1}{3}, 2$.

13. For $x > 1$ or $x < -1$, $\uparrow$; for $-1 < x < 1$, $\downarrow$; horizontal tangents at $x = \pm 1$.

15. For $x > -2$ except $x = 1$, $\uparrow$; for $x < -2$, $\downarrow$; horizontal tangents at $x = -2, 1$.

17. For all x except $x = 2$, $\downarrow$.

18. (13.)   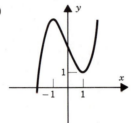   (17.)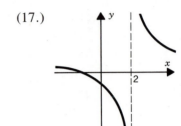

19. (a) When $m > 0$, $\uparrow$;   (b) when $m < 0$, $\downarrow$;
    (c) the function is a horizontal line.

20. (a) When $x > \dfrac{-b}{2a}$, $\uparrow$; (b) when $x < \dfrac{-b}{2a}$, $\downarrow$; (c) there is a horizontal tangent.

21. Decreasing; decreases; increases.

23. For all $t > 0$, $\uparrow$.

25. (a) When $t > 36$, $s > 0$; when $0 < t < 36$, $s < 0$;
    (b) when $t > 16$, $v > 0$; when $0 < t < 16$, $v < 0$.

27. (a) 2000;   (b) $2000 + \dfrac{5200}{x}$;

    (c) $\dfrac{-5200}{x^2}$;   (e) decreasing.

29. (a) Increasing; (b) decreasing; (c) decreasing; (d) decreasing.

Exercise 4.3–4.4

1. No relative extrema; for all x, $\uparrow$.

3. Relative maximum at $(3, 16)$; for $x < 3$, $\uparrow$; for $x > 3$, $\downarrow$.

5. Relative maximum at $(5, 4)$; for $x < 5$, $\uparrow$; for $x > 5$, $\downarrow$.

7. Relative maximum at $(0, 8)$; for $x < 0$, $\uparrow$; for $x > 0$, $\downarrow$.

9. Relative minimum at $(2, -6)$; relative maximum at $\left(\dfrac{1}{3}, \dfrac{-37}{27}\right)$; for $x < \dfrac{1}{3}$ or $x > 2$, $\uparrow$; for $\dfrac{1}{3} < x < 2$, $\downarrow$.

11. Relative minimum at $(1, 1)$; relative maximum at $(-1, 5)$; for $x < -1$ or $x > 1$, $\uparrow$ for $-1 < x < 1$, $\downarrow$.

13. Relative minima at $(-1, -\frac{1}{2})$ and $(1, -\frac{1}{2})$; relative maximum at $(0, 0)$; for $-1 < x < 0$ or $x > 1$, $\uparrow$; for $x < -1$ or $0 < x < 1$, $\downarrow$.

15. Relative minimum at $(-2, -3)$; for $x > -2$, except $x = 1$, $\uparrow$; for $x < -2$, $\downarrow$.

17. No relative extrema; for all x, except $x = 2$, $\downarrow$.

19. Relative minimum at $(0, 3)$; for $x > 0$, $\uparrow$; for $x < 0$, $\downarrow$.

21. (5.) 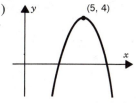   (13.)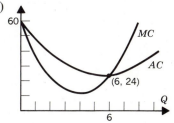

22. (a) $AC = Q^2 - 12Q + 60$;   (b) $(AC)' = 2Q - 12$, $Q = 6$;
     (c) $MC = 3Q^2 - 24Q + 60$.   (d)

23. (a) $AC = Q^2 - 10Q + 40$;   (b) $(AC)' = 2Q - 10$, $Q = 5$;
     (c) $MC = 3Q^2 - 20Q + 40$.

25. For $A < 50$, $\uparrow$; for $A > 50$, $\downarrow$; maximum performance at $A = 50$.

27. Absolute maximum at $(1, -12)$; absolute minimum at $(3, -16)$.

29. Absolute maximum at $(-1, 5)$; absolute minimum at $(-3, -15)$.

31. Absolute maximum at $(8, 7)$; absolute minimum at $(0, 3)$.

33. (a) Relative maximum at $x = x_3$; relative minimum at $x = x_1$;
     (b) relative minimum at $x = x_3$.

Exercise 4.5–4.7

1. $f''(x) = 2$; no point of inflection; for all x, $\cup$.

3. $f''(x) = 6(x - 1)$; $(1, 2)$; $x > 1$, $\cup$; $x < 1$, $\cap$.

5. $f''(x) = 12x$; $(0, -2)$; $x > 0$; $\cup$; $x < 0$, $\cap$.

7. $f''(x) = (x - 3)(x - 1)$; $(3, 3)$, $(1, \frac{5}{3})$; $x < 1$ or $x > 3$, $\cup$; $1 < x < 3$, $\cap$.

9. Relative minimum at $(3, -16)$.

11. Relative minimum at $(-4, 0)$.

13. No relative extrema.

15. Relative minimum at $(2, 5)$; relative maximum at $(1, 6)$.

17. Relative minimum at $(0, \frac{9}{4})$.

19. $f'(x) = 2(x - 4)$; $f''(x) = 2$; relative minimum at $(4, -4)$; $x > 4$, ↑; $x < 4$, ↓; for all x, ∪.

21. $f'(x) = 2(x - 3)$; $f''(x) = 2$; relative minimum at $(3, 0)$; $x > 3$, ↑; $x < 3$, ↓; for all x, ∪.

23. $f'(x) = -2x$; $f''(x) = -2$; relative maximum at $(0, 9)$; $x < 0$, ↑; $x > 0$, ↓; for all x, ∩.

25. $f'(x) = 6x^2 - 18x + 12$; $f''(x) = 12x - 18$; relative minimum at $(2, 5)$; relative maximum at $(1, 6)$; $x < 1$ or $x > 2$, ↑; $1 < x < 2$, ↓; point of inflection at $(\frac{3}{2}, \frac{11}{2})$; $x > \frac{3}{2}$, ∪; $x < \frac{3}{2}$, ∩.

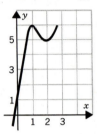

27. $f'(x) = 3x^2 - 12$, $f''(x) = 6x$; relative minimum at $(2, -15)$; relative maximum at $(-2, 17)$; $x < -2$ or $x > 2$, ↑; $-2 < x < 2$, ↓; point of inflection at $(0, 1)$; $x > 0$, ∪; $x < 0$, ∩.

29. $f'(x) = x^2 - 2x$; $f''(x) = 2x - 2$; relative minimum at $(2, \frac{5}{3})$; relative maximum at $(0, 3)$; $x < 0$ or $x > 2$, ↑; $0 < x < 2$, ↓; point of inflection at $(1, \frac{7}{3})$; $x > 1$, ∪; $x < 1$, ∩.

31. $f'(x) = 3x^2 - 18x + 15$; $f''(x) = 6x - 18$; relative minimum at $(5, -25)$ relative maximum at $(1, 7)$; $x < 1$ or $x > 5$, ↑; $1 < x < 5$, ↓; point of inflection at $(3, -9)$; $x > 3$, ∪; $x < 3$, ∩.

33. $f'(x) = x^3 - 3x + 2$; $f''(x) = 3x^2 - 3$; relative minimum at $(-2, -3)$; $x > -2$, except for $x = 1$, ↑; $x < -2$, ↓; points of inflection at $(1, \frac{15}{4})$ and $(-1, -\frac{1}{4})$; $x < -1$ or $x > 1$, ∪; $-1 < x < 1$, ∩.

35. $f'(x) = 4x^3 - 6x^2$; $f''(x) = 12x^2 - 12x$; relative minimum at $(\frac{3}{2}, -\frac{27}{16})$; $x > \frac{3}{2}$, ↓; $x < \frac{3}{2}$, except for $x = 0$, ↓; points of inflection at $(0, 0)$ and $(1, -1)$; $x < 0$ or $x > 1$, ∪; $0 < x < 1$, ∩.

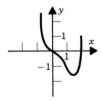

37. $f'(x) = 4x^3 - 4x$; $f''(x) = 12x^2 - 4$; relative minima at $(1, 0)$ and $(-1, 0)$; relative maximum at $(0, 1)$; $-1 < x < 0$ or $x > 1$, ↑; $x < -1$ or $0 < x < 1$, ↓; points of inflection at $(\frac{\sqrt{3}}{3}, \frac{4}{9})$ and $(-\frac{\sqrt{3}}{3}, \frac{4}{9})$; $x < -\frac{\sqrt{3}}{3}$ or $x > \frac{\sqrt{3}}{3}$, ∪; $-\frac{\sqrt{3}}{3} < x < \frac{\sqrt{3}}{3}$, ∩.

38. $f'(x) = 4(x^3 + 3x^2 - 4) = 4(x - 1)(x + 2)^2; f''(x) = 12x(x + 2);$ relative minimum at $(1, -9)$; $x < 1$, except for $x = -2$, $\downarrow$; $x > 1$, $\uparrow$; points of inflection at $(-2, 18)$ and $(0, 2)$; $x < -2$ or $x > 0$, $\cup$; $-2 < x < 0$, $\cap$.

39. $f'(x) = 15x^4 - 60x^2$; $f''(x) = 60x^3 - 120x$; relative minimum at $(2, -64)$; relative maximum at $(-2, 64)$; $x < -2$ or $x > 2$, $\uparrow$; $-2 < x < 2$, except $x = 0$, $\downarrow$; points of inflection at $(0, 0)$, $(-\sqrt{2}, 28\sqrt{2})$ and $(\sqrt{2}, -28\sqrt{2})$; $-\sqrt{2} < x < 0$ or $x > \sqrt{2}$, $\cup$; $0 < x < \sqrt{2}$ or $x < -\sqrt{2}$, $\cap$.

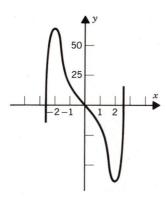

40. $f'(x) = \frac{2}{3}x^{-(1/3)}$; $f''(x) = -\frac{2}{9}x^{-(4/3)}$; relative minimum at $(0, 3)$; $x > 0$, $\uparrow$; $x < 0$, $\downarrow$; $x > 0$ or $x < 0$, $\cap$.

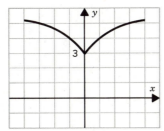

41. $f'(x) = \frac{1}{3}(1 + x)^{-2/3}$; $f''(x) = -\frac{2}{9}(1 + x)^{-5/3}$; for all x, except $x = -1$, $\uparrow$; point of inflection at $(-1, 0)$; $x < -1$, $\cup$; $x > -1$, $\cap$.

42. $f'(x) = 1 - \dfrac{1}{x^2}$; $f''(x) = \dfrac{2}{x^3}$; relative minimum at $(1, 2)$; relative maximum at $(-1, -2)$; $x > 1$ or $x < -1$, $\uparrow$; $-1 < x < 0$ or $0 < x < 1$, $\downarrow$; $x > 0$, $\cup$; $x < 0$, $\cap$.

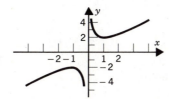

43. $f'(x) = 1 + \dfrac{1}{x^2}$; $f''(x) = -\dfrac{2}{x^3}$; for all x, except $x = 0$, $\uparrow$; $x < 0$, $\cup$; $x > 0$, $\cap$.

44. $f'(x) = \dfrac{3x + 2}{2\sqrt{x + 1}}$; $f''(x) = \dfrac{3x + 4}{4(x + 1)^{3/2}}$; relative minimum at $\left(-\dfrac{2}{3}, -\dfrac{2\sqrt{3}}{9}\right)$; $x > -\dfrac{2}{3}$, $\uparrow$; $-1 < x < -\dfrac{2}{3}$, $\downarrow$; $x > -1$, $\cup$.

45. $f'(x) = \dfrac{(1 - x)}{\sqrt{2x - x^2}}$; $f''(x) = \dfrac{-1}{(2x - x^2)^{3/2}}$; relative maximum at $(1, 1)$; $0 < x < 1$, $\uparrow$; $1 < x < 2$, $\downarrow$; $0 < x < 2$, $\cap$.

47.

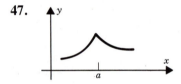

Exercise 4.8

1. 20 years.

3. 2 hr.

4. (a) \$4000; (b) \$1730.

5. 26; $6\frac{1}{4}$ days.

7. (a) $C'(x) = .2x - 15\sqrt{x}$; (b) 5625 items;
(c) 0; yes.

9. (a) $2 - 2x$; (b) $x > 2$; (c) 120.

11. 3 days.

13. (a) $(-32t_0 + 48)$ ft/sec; (b) $\frac{3}{2}$ sec;
(c) 64 ft; (d) -32 ft/sec^2; (e) $\frac{7}{2}$ sec.

15. (c) 150 boxes; (d) \$0.90 per box; (e) \$45.

16. (a) $R(x) = \$(150 - \frac{3}{2}x)x$;
(b) $P(x) = x(150 - \frac{3}{2}x) - \left(500 + 14x + \frac{1}{2}x^2\right)$;
(c) 34 chairs.

17. (a) $P(x) = 250x - \left(\frac{1}{4}x^2 + 56x + 3060\right)$;
(b) 388 microscopes.

19. (a) 155 sets; (b) \$6400.

21. 8, 8.

23. (a) length $= x$, width $= 120 - x$; (b) 60 ft $\times$ 60 ft

25. 14 ft $\times$ 14 ft.

27. $2a^2$ square units.

29. 15 ft $\times$ $7\frac{1}{2}$ ft.   **30.** (c) 4 ft $\times$ 4 ft $\times$ 2 ft.

31. $27.

32. Surface area $= 2\pi r^2 + \dfrac{48}{r}$; $r = \sqrt[3]{\dfrac{12}{\pi}}$.

33. $r = \dfrac{2}{\sqrt[3]{\pi}}$, minimal cost $72\sqrt[3]{\pi}$ cents.

35. Piece to form circle is $\dfrac{40\pi}{4 + \pi}$, piece to form square is $\dfrac{160}{4 + \pi}$.

37. $\sqrt{\dfrac{S}{6\pi}}$ where S is the surface area.

38. $\dfrac{24}{6 - \sqrt{3}}$ ft $\times$ $\dfrac{12(3 - \sqrt{3})}{6 - \sqrt{3}}$ ft.

39. Cylindrical container: $r = \left(\dfrac{V}{2\pi}\right)^{1/3}$; $h = \dfrac{V}{\pi}\left(\dfrac{2\pi}{V}\right)^{2/3}$; rectangular container: $l = w = V^{1/3}$; $h = V^{1/3}$; surface area, SA, cylindrical container $= 5.536$ ft^2; SA rectangular container $= 6.000$ ft^2; cylindrical container.

40. (b) 4000; (c) 10; (d) 26 days.

41. $\sqrt{\dfrac{2DS}{I}}$.   **42.** 880.

43. Run the power lines under water to a point on the bank $\frac{5}{8}$ miles from the factory.

45. 140 cases.

46. $3\frac{1}{2}\%$

48. 27 in^3/sec.   **49.** $\dfrac{1}{9\pi}$ in/min.

Exercise 4.9

1. $(2x - 4)\,dx$.   **3.** $(-9x^2 + 4x + 9)\,dx$.

5. $\dfrac{2x^2 - 1}{(2x^3 - 3x)^{2/3}}\,dx$.   **7.** $\dfrac{-3x^2 + 10x + 6}{(x^2 + 2)^2}\,dx$.

9. 4.9.   **11.** 3.979.

13. .0099.

15. $\Delta y = 1.25$; $dy = 1$; $E = \frac{1}{4}$; $\dfrac{E}{\Delta x} = \frac{1}{2}$.

17. $dy = \Delta y = -\frac{2}{3}$; no error; linear function.

19. $880. **21.** (a) .12 cu cm; (b) .24 sq cm.

23. 48 cu in.

Review Exercise

3. (a) $f'(x) > 0$ when $x_2 < x < x_4$, $x > x_5$; $f'(x) < 0$ when $x < x_2$, $x_4 < x < x_5$; $f'(x) = 0$ when $x = x_2$, x_4.
  (b) $f'(x) > 0$ when $x_1 < x < x_2$; $f'(x) < 0$ when $x_2 < x < x_3$, $x_5 < x < x_7$; $f'(x) = 0$ when $x = x_2$, $x_3 < x < x_4$, $x_4 < x < x_5$.

6. $f'(x) = 12x - 11$; $f''(x) = 12$; relative minimum at $(\frac{11}{12}, \frac{-289}{24})$; $x > \frac{11}{12}$, $\uparrow$; $x < \frac{11}{12}$, $\downarrow$; for all x, $\cup$.

8. $f'(x) = 3(x^2 - 3)$; $f''(x) = 6x$; relative maximum at $(-\sqrt{3}, 6\sqrt{3})$; relative minimum at $(\sqrt{3}, -6\sqrt{3})$; $x < -\sqrt{3}$ or $x > \sqrt{3}$, $\uparrow$; $-\sqrt{3} < x < \sqrt{3}$, $\downarrow$; point of inflection at $(0, 0)$; $x > 0$, $\cup$; $x < 0$, $\cap$.

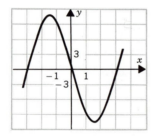

9. $f'(x) = 3x(x - 2)$; $f''(x) = 6(x - 1)$; relative maximum at $(0, 3)$; relative minimum at $(2, -1)$; $x < 0$ or $x > 2$, $\uparrow$; $0 < x < 2$, $\downarrow$; point of inflection at $(1, 1)$; $x > 1$, $\cup$; $x < 1$, $\cap$.

11. $f'(x) = (3x + 4)(x - 2)$; $f''(x) = 2(3x - 1)$; relative maximum at $(-\frac{4}{3}, \frac{230}{27})$; relative minimum at $(2, -10)$; $x < -\frac{4}{3}$ or $x > 2$, $\uparrow$; $-\frac{4}{3} < x < 2$, $\downarrow$; point of inflection at $(\frac{1}{3}, \frac{-20}{27})$; $x > \frac{1}{3}$, $\cup$; $x < \frac{1}{3}$, $\cap$.

13. $f'(x) = 9 + 6x - 3x^2 = 3(1 + x)(3 - x)$; $f''(x) = 6(1 - x)$; relative maximum at $(3, 28)$; relative minimum at $(-1, -4)$; $-1 < x < 3$, $\uparrow$; $x < -1$ or $x > 3$, $\downarrow$; point of inflection at $(1, 12)$; $x < 1$, $\cup$; $x > 1$, $\cap$.

14. $f'(x) = 4x(x - 2)(x + 2)$; $f''(x) = 4(3x^2 - 4)$; relative maximum at $(0, 16)$; relative minima at $(-2, 0)$ and $(2, 0)$; $-2 < x < 0$ or $x > 2$, $\uparrow$; $x < -2$ or $0 < x < 2$, $\downarrow$; points of inflection at $\left(\dfrac{2}{\sqrt{3}}, \dfrac{64}{9}\right)$ and $\left(-\dfrac{2}{\sqrt{3}}, \dfrac{64}{9}\right)$; $x < \dfrac{-2}{\sqrt{3}}$ or $x > \dfrac{2}{\sqrt{3}}$, $\cup$; $\dfrac{-2}{\sqrt{3}} < x < \dfrac{2}{\sqrt{3}}$, $\cap$.

16. $f'(x) = 4x(x + 3)^2; f''(x) = 12(x + 1)(x + 3)$; relative minimum at $(0, 0)$; $x > 0$, ↑; $x < 0$, except $x = -3$, ↓; points of inflection at $(-3, 27)$ and $(-1, 11)$; $x < -3$ or $x > -1$, ∪; $-3 < x < -1$, ∩.

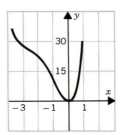

17. $f'(x) = \dfrac{1}{2\sqrt{x + 2}}; f''(x) = \dfrac{-1}{4\sqrt{(x + 2)^3}}$; no relative extrema; $x > -2$, ↑; $x > -2$, ∩.

19. $f'(x) = \dfrac{-6}{x^3}; f''(x) = \dfrac{18}{x^4}$; no relative extrema; $x < 0$, ↑; $x > 0$, ↓; $x < 0$ or $x > 0$, ∪.

21. $f'(x) = \dfrac{1}{3\sqrt[3]{x^2}}; f''(x) = \dfrac{-2}{9\sqrt[3]{x^5}}$; no relative extrema; $x < 0$ or $x > 0$, ↑; point of inflection at $(0, 0)$; $x < 0$, ∪; $x > 0$, ∩.

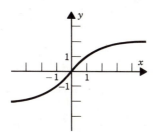

22. (a) No; (b) yes.

23. No.

24. True.

25. False.

26. True.

27. False.

28. True.

29. False.

30. True.

31. False.

32. True.

33. True.

35. (a) Maximum rate = 134, minimum rate = 70; (b) 86.

37. 10 days.

39. 25 trees.

41. (a) 1 hr;   (b) 1 hr;
(c) at $t = 1$ hr.

42. (a) 8 ft/sec; -8 ft/sec; (b) 2 sec; (c) 16 ft; (d) 4 sec; (e) -16 ft/sec.

43. 10.

45. (a) \$319,880; (b) 5000 sets;
(c) \$499,880.

47. $R(x) = x[95 - \frac{2}{5}(x - 75)]$; 156 passengers

51. $\frac{1}{4}P \times \frac{1}{4}P$.

53. 30 ft $\times \dfrac{15}{2}$ ft $\times$ 6 ft.

54. 20 $\sqrt{3}$ ft $\times$ $10\sqrt{3}$ ft.

55. $\sqrt[3]{\dfrac{50}{\pi}}$. **56.** $S = \dfrac{2V}{r} + 2\pi r^2$; $r = \sqrt[3]{\dfrac{V}{2\pi}}$.

57. $\frac{33}{20}$ units per month, $\frac{63}{40}$ units per month.

58. $\dfrac{1}{4\pi}$ in/min.

59. 25 mi/hr.

60. (a) $\frac{82}{27} = 3\frac{1}{27}$; (b) $\frac{80}{27} = 2\frac{26}{27}$.

61. $-.8$. **62.** $S = 4\pi r^2$; 8π.

63. 75 bacteria.

Chapter 5

Exercise 5.2

1. $5x^2 + c$. **3.** $\frac{3}{2}x^2 - 4x + c$.

5. $5x - x^2 + c$. **7.** $x^3 - \frac{1}{2}x^2 + 2x + c$.

9. $3x^3 + 3x^2 + x + c$. **11.** $\frac{1}{3}\left(\dfrac{x^4}{4} - 2x^3 + 3x^2 - 2x\right) + c$.

13. $-\dfrac{2}{x^2} - \dfrac{5}{x} + 20x + c$. **15.** $\dfrac{-2}{3x} + \dfrac{3}{4x^2} + c$.

17. $\frac{1}{2}x^2 + x + \dfrac{1}{x} + c$. **19.** $\frac{4}{3}x^{3/2} + c = \frac{4}{3}x\sqrt{x} + c$.

21. $\frac{2}{5}x^{5/2} + \frac{3}{5}x^{5/3} + c = \frac{2}{5}\sqrt{x^5} + \frac{3}{5}\sqrt[3]{x^5} + c$. **23.** False.

25. True. **27.** False.

29. False. **31.** $C(x) = .2x^3 + x^2 + 3x + 420$.

33. $P(x) = 20x - .2x^3 - 34.2$.

35. (a) $\overline{C}(x) = \dfrac{x}{4} + \dfrac{17}{x} + \dfrac{207}{4}$; (b) $C(x) = \dfrac{1}{4}x^2 + \dfrac{207}{4}x + 17$.

37. $y = \sqrt{x}$. **39.** $s(t) = \frac{1}{3}t^3 - t^2 + 4t$.

41. (a) $v(t) = -fgt + v_0$;   (b) $t = \dfrac{v_0}{fg}$;

     (c) $s(t) = \dfrac{-fgt^2}{2} + v_0 t$;   (d) $\dfrac{v_0{}^2}{2fg}$.

43. $\int x^{-1}\, dx$ would yield $\dfrac{x^0}{0}$, and division by zero is impossible.

Exercise 5.3

1. $1 + \frac{1}{2} + \frac{1}{3} + \frac{1}{4}$.   **3.** $1 + 3 + 6$.

5. $f(x_1) + f(x_2) + f(x_3) + f(x_4) + f(x_5)$.

7. $f(x_1)\,\Delta x + f(x_2)\,\Delta x + \cdots + f(x_n)\,\Delta x$.

9. $\displaystyle\sum_{n=1}^{5} 2n$.   **11.** $\displaystyle\sum_{i=1}^{4} x_i$.

13. $\displaystyle\sum_{i=1}^{4} f(x_i)$.   **15.** $\displaystyle\sum_{i=1}^{n} x^{2i-1}$.

17. $\displaystyle\sum_{n=1}^{3} \dfrac{1}{n(n+1)}$.   **19.** $\displaystyle\sum_{i=1}^{n} f(x_i)\,\Delta x$.

21. (a) $\displaystyle\sum_{k=1}^{5} 13 = 13 + 13 + 13 + 13 + 13 = (13)(5)$;

     (d) $\displaystyle\sum_{i=1}^{n} f(x_1)\,\Delta x = f(x_1)\,\Delta x + f(x_2)\,\Delta x + \cdots + f(x_n)\,\Delta x =$

$$\Delta x[f(x_1) + f(x_2) + \cdots + f(x_n)] = \Delta x \sum_{i=1}^{n} f(x_i).$$

23. (a) $\frac{15}{8}$; (b) $\frac{3}{2}$; (c) $\frac{3}{2}$.   **24.** (a) $\frac{91}{8}$; (b) 9.

25. (a) $\frac{9}{2}$; (b) 4; (c) 4.   **28.** $\frac{5}{2}$.

30. 1.   **33.** 5.

34. 5.

Exercise 5.4–5.5

1. True.   **3.** True.   **5.** True.   **7.** False.   **9.** True.

11. False.   **13.** 9.   **15.** 12.   **17.** 16.   **19.** $\frac{2}{3}$.

21. $\frac{22}{3}$.   **23.** $\frac{4}{3}$.   **25.** $\frac{16}{15}$.   **27.** $\frac{27}{8}$.   **29.** 9.

31. $5 + 8\sqrt{2}$.   **33.** 15.   **35.** 30,000 people.

39. 336,000 pounds.

40. (a) $\frac{1}{5}$, most equally distributed; (b) $\frac{1}{3}$; (c) $\frac{3}{5}$

41. 372

43. (a) 302 bacteria; (b) 1201 bacteria.

44. (b) x^2; (c) x^2.   **45.** $\sqrt[3]{x}$.   **47.** $\dfrac{1}{x^3}$.   **49.** 4.

51. True.   **53.** True.   **55.** False.

Exercise 5.6

1. $\frac{4}{3}$.   **3.** $\frac{32}{3}$.   **5.** $\frac{125}{6}$.   **7.** $\frac{9}{2}$.   **9.** 9.   **11.** $\frac{16}{3}$.

13. $\frac{27}{4}$.   **15.** $\frac{40}{3}$.

17. $\displaystyle\int_{-1}^{2} [(x + 2) - x^2]\, dx = \frac{9}{2}$.

19. $\displaystyle\int_{-2}^{4} [(x + 2) - (x^2 - x - 6)]\, dx = 36$.

21. $\displaystyle\int_{-3}^{2} [(-x^2 + 9) - (x^2 + 2x - 3)]\, dx = \frac{125}{3}$.

23. $\frac{51}{4}$.   **25.** 8.

26. $\displaystyle\int_{0}^{4} (8\sqrt{x} - x^2)\, dx$ or $\displaystyle\int_{0}^{16} \left(\sqrt{y} - \frac{y^2}{64}\right) dy = \frac{64}{3}$.

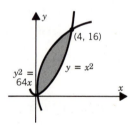

27. $\frac{32}{3}$.

28. $\displaystyle\int_{-2}^{4} [\tfrac{1}{2}(y + 4) - (\tfrac{1}{4}y^2)]\, dy = 9$.

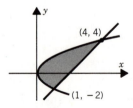

29. $\frac{9}{2}$.   **31.** $\frac{21}{2}$.

33. $C = \$6.67;\ P = \$6$.   **35.** $\$0.67$.

37. 3680.

38. (a) $R(x) = 225x - \frac{5}{2}x^2$; (b) $D(x) = 225 - \frac{5}{2}x$; (c) \$85.

39. (a) $\frac{5}{16}$; (b) $\frac{67}{256}$. **41.** (a) 1; (b) $\frac{3}{4}$; (c) $\frac{3}{5}$.

Exercise 5.7

1. $\frac{1}{5}(x^3 - 5)^5 + c$. **3.** $\frac{52}{3}$.

5. $\frac{1}{12}$. **7.** $\dfrac{-1}{4(2x + 3)^2} + c$.

8. $\dfrac{1}{2(n + 1)}(2x + 3)^{n+1} + c$. **9.** $\dfrac{1}{a(n + 1)}(ax + b)^{n+1} + c$.

11. $\frac{9}{4}$. **13.** $\frac{1}{2}\sqrt{x^4 - 4x + 4} + c$.

15. $-\sqrt{a^2 - x^2} + c$. **17.** $\frac{1}{6}(2x^3 + 5)^5 + c$.

19. $\dfrac{(x^3 + 6x)^5}{15} + c$.

21. $\frac{1}{10}(2x + 3)^{5/2} - \frac{1}{2}(2x + 3)^{3/2} + c = \frac{1}{5}(2x + 3)^{3/2}(x - 1) + c$.

23. $\frac{5}{49}[\frac{1}{7}(7x + 5)^7 - \frac{5}{6}(7x + 5)^6] + c$.

25. $\frac{149}{15}$. **27.** $\frac{64}{3}$. **29.** $\frac{8}{3}$. **31.** $\frac{1}{2}$.

33. 60,000 tons; will meet environmental standards.

34. $C = \$32$; $P = \$3$.

35. $C(x) = \frac{2}{15}(x + 1)^{3/2}(3x - 2) + \frac{19900}{3}$.

37. $\frac{1}{6}\sqrt{(2x - 3)^3} + c$. **39.** $\dfrac{-\sqrt{25 - x^2}}{25x} + c$.

41. $3\left[\dfrac{-1}{2(9 - x^2)} + \dfrac{9}{4(9 - x^2)^2}\right] + c$.

43. $\frac{2}{15}(15x^2 + 12x + 8)\sqrt{(x - 1)^3} + c$.

45. $\dfrac{-2}{x}\sqrt{2x^2 + 7x} + c$.

Exercise 5.8–5.9

Note. D stands for divergent integral.

1. $\frac{1}{2}$. **3.** D. **5.** $\frac{3}{2}$. **7.** D. **9.** D. **11.** D.

15. 1. **17.** $k = 1$. **19.** (a) $c = 3$; (b) $\frac{3}{2}$; (c) 3.

Review Exercises

5. (a) $\frac{5}{4}$; (b) 1; (c) 1. **6.** (a) $\frac{217}{27}$; (b) $\frac{20}{3}$.

7. $\dfrac{x^4}{2} + \dfrac{7x^2}{2} - 9x + c$. **9.** $\frac{49}{2}$.

11. 39. **13.** $4\sqrt{x} + \dfrac{1}{2x} + c.$

14. $\frac{3}{4}\sqrt[3]{x^4} + \frac{3}{2}\sqrt[3]{x^2} - 3x + c.$ **15.** 6540. **17.** 15.

19. 6. **21.** 24.

23. $4[\frac{1}{5}(16 - x^2)^{5/2} - \frac{16}{3}(16 - x^2)^{3/2}] + c.$

25. $\frac{1}{2}(\sqrt{x} - 1)^4 + c.$ **27.** $\frac{2}{5}(x - 1)^{5/2} + \frac{4}{3}(x - 1)^{3/2} + c.$

29. $\frac{9}{2}$. **31.** $\frac{9}{2}$. **33.** $\frac{32}{3}$. **35.** $\frac{17}{3}$.

37. $C = \$9; P = \$18.$ **39.** $C = \$10.67; P = \$4.00.$

41. (b) (2, 16); (c) $C = \$5.33; P = \$25.60.$

42. (a) \$2.00; (b) \$1.00; (c) 200, \$4.00; (d) $P = \$5.33.$

43. \$13,343; \$8,877. **45.** 33 fish.

47. $y = x^3 + x^2 - x + 1.$ **49.** \$61,891.67.

51. $Q(t) = .15t^2 + 60t + c.$

53. $s = f(t) = 2t^3 + t^2 - 30t + 32.$

55. $s = f(t) = \frac{1}{6}\sqrt{(t^4 - 6t^2)^3} + 10; \frac{27}{2}\sqrt{3} + 10.$

57. 352 feet. **59.** True. **61.** False. **63.** True.

65. False. **67.** True. **69.** True. **71.** True.

Chapter 6

Exercise 6.1

1. 125. **3.** 27. **5.** 10^9 **7.** 16. **9.** $e^6.$

11. 1.1461. **13.** 0.5441. **15.** 1.6902. **17.** $-0.3010.$

19. $-1.7960.$ **21.** (a) 40; (b) 60; (c) 120.

23. 3. **25.** 1. **27.** $\frac{1}{9}$. **29.** $\frac{1}{2}$. **31.** $\frac{1}{3}$.

Exercise 6.2

1. $\dfrac{6x + 2}{3x^2 + 2x - 5}.$ **3.** $\dfrac{2}{x}.$ **5.** $\dfrac{6x}{x^2 - 7}.$ **7.** $\dfrac{10}{x}.$

9. $\dfrac{-2}{x}.$ **11.** $\dfrac{-x}{4 - x^2}.$ **13.** $\dfrac{1}{x\sqrt{\ln(3x^2)}}.$ **15.** $\dfrac{2(1 - \ln x^2)}{x^3}.$

16. $\dfrac{2 \ln x(1 - \ln x)}{x^3}.$ **17.** $\dfrac{2(x - 4)}{x(x - 2)}.$ **18.** $\dfrac{1}{x \ln x}.$

19. $\dfrac{2x(x + 1) \ln(x + 1) - (x^2 - 1) \ln(x^2 - 1)}{(x + 1)(x^2 - 1) (\ln(x + 1))^2}.$

21. $\dfrac{-y(2y + 1)}{2x(2y \ln(2x) + \ln \sqrt{x})}.$

23.
$$y = x^n$$
$$\ln y = \ln x^n = n \ln x$$
$$\frac{1}{y} y' = n\frac{1}{x}$$
$$y' = n\frac{y}{x} = n\frac{x^n}{x} = nx^{n-1}.$$

25. $\dfrac{x}{x^2 + 144}.$

27. $\dfrac{1 + 3x^2}{3(17 + x + x^3)}.$

29. $\dfrac{5(16x^3 + 9x^2 + 14)}{(2x + 3)(7 + 5x^3)}.$

31. $\dfrac{3x^2 + x - 4}{(2x + 1)(x^2 - 4)}.$

33. $\dfrac{2(x - 4)}{x(x - 2)}.$

34. $\dfrac{14x}{(x^2 - 7)(x^2 + 7)}.$

35. $\dfrac{1250}{x}.$

36. $\$e^4 \approx \$54.60; \$16,379.$

37. $\dfrac{2x + 40}{x^2 + 40x} + 200.$

39. (a) 11.7 years; (b) 11.7 years.

41. 4.25 years.

43. $y = 2x - 1.$

45. $y = 3x - 2.$

47. $y' = \dfrac{2x}{x^2 + 1}$; $y'' = \dfrac{2(1 - x^2)}{(x^2 + 1)^2}$; relative minimum at $(0,0)$; $x > 0$, ↑; $x < 0$, ↓; points of inflection at $(1, \ln 2)$ and $(-1, \ln 2)$; $-1 < x < 1$, ∪, $x > 1$ or $x < -1$, ∩.

49. $y' = \dfrac{1}{x}$; $y'' = -\dfrac{1}{x^2}$; $x > 0$, ↑; $x < 0$, ↓; for all x, except $x = 0$, ∩.

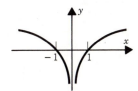

50. $y' = 1 + \ln x$; $y'' = \dfrac{1}{x}$; relative minimum at $\left(\dfrac{1}{e}, -\dfrac{1}{e}\right)$; $x > \dfrac{1}{e}$, ↑; $0 < x < \dfrac{1}{e}$, ↓; $x > 0$, ∪.

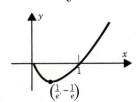

Exercise 6.3

1. $2e^{2x}$.　　**3.** $2e^{2x+3}$.　　**5.** $e^x\left(\dfrac{2}{x} + \ln x^2\right)$.　　**7.** $\dfrac{2x - x^2}{e^x}$.

9. $9(e^{3x-1})^3$.　　**11.** $\dfrac{1}{2\sqrt{x+4}}e^{\sqrt{x+4}}$.　　**13.** 1.

14. $\dfrac{1}{x}e^{\ln x} = \dfrac{1}{x}x = 1$.　　**15.** $\dfrac{x+1}{x}$.

17. $e^{x\sqrt{x+1}}\left(\sqrt{x+1} + \dfrac{x}{2\sqrt{x+1}}\right) = e^{x\sqrt{x+1}}\left(\dfrac{3x+2}{2\sqrt{x+1}}\right)$.

19. $\dfrac{y(xe^x - xe^y - 1)}{x(xye^y + 1)}$.

21. $y' = e^x$; $y'' = e^x$; intercept at $(0,1)$; for all x, $\uparrow$; for all x, $\cup$.

23. $y' = 2xe^{x^2}$; $y'' = 2e^{x^2}(2x^2 + 1)$; relative minimum at $(0,1)$; $x > 0$, $\uparrow$; $x < 0$, $\downarrow$; for all x, $\cup$.

25. $y' = (1 + x)e^x$, $y'' = (2 + x)e^x$; relative minimum at $\left(-1, -\dfrac{1}{e}\right)$; $x > -1$, $\uparrow$; $x < -1$, $\downarrow$; point of inflection at $\left(-2, -\dfrac{2}{e^2}\right)$; $x > -2$, $\cup$; $x < -2$, $\cap$.

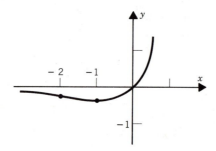

26. (a) $x < 0$, $\uparrow$; $x > 0$, $\downarrow$; (b) $\left(0, \dfrac{1}{\sqrt{2\pi}}\right)$;

(c) $x < \dfrac{-\sqrt{2}}{2}$ or $x > \dfrac{\sqrt{2}}{2}$, $\cup$; $\dfrac{-\sqrt{2}}{2} < x < \dfrac{\sqrt{2}}{2}$, $\cap$;

(d)

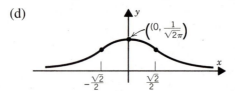

27. (a)

t	0	1	2	3	4	5
$K(t)$	5	1.8394	.6767	.2489	.0916	.0337

(c) 3 hours; (d) $2.996 \approx 3$ hours.

28. (a) 18%, 63%, 86%; (e) 34 days.

31. $100(e^{.5} - 1) \approx 65$; $100(e - e^{.5}) \approx 107$; $50e^{.25} \approx 64$; $50e^{.75} \approx 106$.

33. First National, interest of $790.85.

34. 1% decline.

35. $e - 1$.   **37.** $\ln(e - 1)$.   **41.** $2xa^{x^2} \ln a$.   **43.** $(3e)^x \ln(3e)$.

45. $a^x(x \ln a + 1)$.   **47.** $\dfrac{3^{2x}(2x \ln 3 - 1)}{x^2}$.

49. $y' = (\frac{1}{2})^x \ln \frac{1}{2}$; $y'' = (\frac{1}{2})^x (\ln \frac{1}{2})^2$; for all x, $\downarrow$; for all x, $\cup$.

51. $y' = 1 + 2^x \ln 2$; $y'' = 2^x (\ln 2)^2$; for all x, $\uparrow$; for all x, $\cup$.

Exercise 6.4–6.5

1. $\frac{1}{2}e^{2x} + c$.   **3.** $2(e - 1)$.   **5.** $-10 \ln |x| + c$.

7. $2 \ln 7$.   **9.** $\frac{1}{2} \ln 2$.

11. $\frac{1}{3} \ln |x^3 - 3x + 7| + c$.   **13.** $\dfrac{-1}{2(1 + e^x)^2} + c$.

15. $\frac{1}{2}(\ln 2)^2$.   **17.** $\frac{1}{2}x^2 + 4x - 3 \ln |x| + c$.

18. $\ln |\ln x| + c$.   **19.** $\frac{1}{3}(1 + \ln x)^3 + c$.

21. $\frac{1}{2}x^2 \ln x - \frac{1}{4}x^2 + c$.

23. $x \ln x - x + c$.   **25.** $\ln 16 - \frac{3}{2}$.

27. $-2e + \dfrac{16}{e}$.   **29.** $\ln 2$.

31. $\frac{1}{2}(e^4 - 1)$.   **33.** $-300(e^2 - e)$.

35. $1000 \ln 10$.   **37.** $R(x) = xe^{-x}$.

38. (b) $18 - 6 \ln 4$; (c) $6 \ln 4 - 6$.

39. 50.   **41.** $440 + 40 \ln 21$.

43. Relative maximum at $(-1, -2)$; relative minimum at $(1, 2)$;
$x < -1$ or $x > 1$, $\uparrow$, $-1 < x < 0$ or $0 < x < 1$, $\downarrow$; $x > 0$, $\cup$;
$x < 0$, $\cap$; area $= \frac{15}{8} + \ln 4$.

44. (a) $-\frac{2}{15}(4 - x)^{3/2}(8 + 3x) + c$; (b) $\frac{1}{15}(x^2 + 7)^{3/2}(3x^2 - 14) + c$.

45. $5x^6 \left(\dfrac{\ln x}{6} - \dfrac{1}{36} \right) + c$.   **47.** $e^x(x^2 - 2x + 2) + c$.

49. $\frac{3}{4}e^{2u}(2u - 1) + c$.   **51.** $\ln |\ln 2x| + c$.

53. (a) Diverges; (b) diverges.

55. (a) $\displaystyle\int_0^\infty e^{-x}\,dx = 1$; (b) $\ln 2$.

56. (a) $k = 1$; (b) $1 - \dfrac{2}{e}$; (c) $\dfrac{2e - 3}{e^2}$.

Exercise 6.6–6.8

1. $y = x^2 + 2x + c$.   **3.** $\ln x = \ln y + c$.

5. $\frac{1}{2}x^2 - \ln x = -\dfrac{1}{y} + c$.   **7.** $\ln y = e^x + c$.

9. $e^y = \frac{2}{3}\sqrt{x^3 + 3} + c$.   **11.** $M = 434e^{-.09t}$; 176.45.

12. $N(t) = 200e^{(1/10)t} + 100$; 503.

13. (a) $y = x^3 + x + c$; (b) $y = x^3 + x + 2$; (c) 4.

15. (a) $y = 2e^{2x} + c$; (b) $y = 2e^{2x} + 3$.

17. 5 days.   **19.** $v = -32t + 64$; $s = -16t^2 + 64t + 75$; 139 feet.

21. .69 hours   **23.** 6.6 hours.

24. 13.86 years; 21.97 years.   **25.** 7%.

27. 27,183.   **29.** 4.58%.

30. National Bank, $1055.76; Community Bank, $1053.90; National Bank.

31. (a) $k = .086$; (b) 11.65g.   **32.** 226 curies.

33. 14 days.   **35.** 7.7 months.

37. $N = (5t + 15)^2$.   **39.** $N = 200e^{-.02t}$.

41. 49.5 years.

Review Exercises

1. 2.   **3.** 4.   **5.** 1.   **6.** 8.   **7.** 15.5 years.

9. $686.39.   **11.** $\dfrac{35x^4 - 3}{7x^5 - 3x}$.   **13.** $(35x^4 - 3)e^{7x^5 - 3x}$.

15. $\dfrac{1}{2x}$.   **16.** $\dfrac{1}{2x\sqrt{\ln x}}$.   **17.** $\dfrac{e^{\sqrt{x}}}{2\sqrt{x}}$.   **18.** $\frac{1}{2}\sqrt{e^x}$.

19. $\dfrac{1}{x} + 2e^{2x}$.   **21.** $e^x\left(\dfrac{1}{x} + \ln x\right)$.

23. $e^x + exe^{-1}$.   **25.** $\dfrac{e^x + e^{-x} - 2}{(1 + e^{-x})^2}$.

26. $\dfrac{2(1 - x^2 \ln x^2)}{xe^{x^2}}$.   **27.** $-\dfrac{y}{x}$.   **29.** $\ln |x| + e^x + c$.

31. $\ln \frac{4}{3}$.   **33.** $\frac{2}{3}(1 + \ln x)^{3/2} + c$.

35. $\frac{1}{3}(e^5 - e^2)$.   **37.** $4\sqrt{1 + e^x} + c$.

39. $xe^x - e^x + c.$　　**41.** $y = 2x - 1.$　　**43.** $\frac{1}{2} \ln 3.$

46. $-\dfrac{kT}{m} e^{-mv^2/2kT} + c.$

47. (a) 100; (b) 1.39 hours.　　**48.** (a) $e^y = \dfrac{x^2}{2} + 2 \ln |x| + c.$

49. (a) $y = \frac{1}{2} \ln(x^2 + 1) + c$; (b) $y = \frac{1}{2} \ln(x^2 + 1).$

51. 200 cells.　　　　　　　**53.** $2699.72; $699.72.

55. 6.9 years.　　　　　　　**57.** (a) $x > 4$;　(b) $x > 3$　or　$x < -3$;

(c) all x; (d) all x.

59. True.　　　　　　　　　**61.** False.

63. True.　　　　　　　　　**64.** True.

65. True.　　　　　　　　　**67.** True.

Chapter 7

Exercise 7.2–7.4

3. $f(0,1) = 1; f(1,2) = 2e + 1 + \ln 2.$

5. $f(-1,4) = -2; f(8,9) = 6; f(64,64) = 32.$

7. (a) 9; (c) $2\sqrt{21}.$　　**11.** 25.

13.

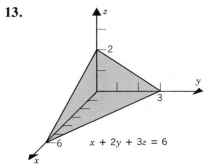

16.

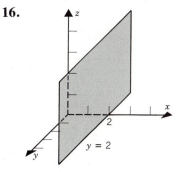

22.

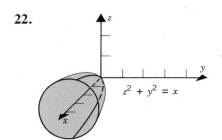

23.

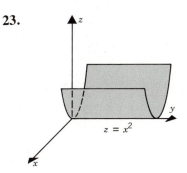

Exercise 7.5–7.6

1. $f_x(x,y) = 3x^2 - 6xy$
   $f_y(x,y) = -3x^2 + 2$
   $f_{xx}(x,y) = 6x - 6y$
   $f_{yy}(x,y) = 0$
   $f_{xy}(x,y) = -6x$
   $f_{yx}(x,y) = -6x.$

3. $f_x(x,y) = ye^{xy} + \dfrac{1}{x}$

   $f_y(x,y) = xe^{xy} + \dfrac{1}{y}$

   $f_{xx}(x,y) = y^2 e^{xy} - \dfrac{1}{x^2}$

   $f_{yy}(x,y) = x^2 e^{xy} - \dfrac{1}{y^2}$

   $f_{xy}(x,y) = e^{xy}(xy + 1)$
   $f_{yx}(x,y) = e^{xy}(xy + 1).$

5. $f_x(x,y) = e^{x^2-y^2}(2x^2 + 1)$
   $f_y(x,y) = -2xye^{x^2-y^2}$
   $f_{xx}(x,y) = 2xe^{x^2-y^2}(2x^2 + 3)$
   $f_{yy}(x,y) = 2xe^{x^2-y^2}(2y^2 - 1)$
   $f_{xy}(x,y) = -2ye^{x^2-y^2}(2x^2 + 1)$
   $f_{yx}(x,y) = -2ye^{x^2-y^2}(2x^2 + 1).$

7. $f_x(x,y) = 12x^2(x^3 + y^2)^3$
   $f_y(x,y) = 8y(x^3 + y^2)^3$
   $f_{xx}(x,y) = 108x^4(x^3 + y^2)^2$
   $\qquad\qquad + 24x(x^3 + y^2)^3$
   $f_{yy}(x,y) = 48y^2(x^3 + y^2)^2$
   $\qquad\qquad + 8(x^3 + y^2)^3$
   $f_{xy}(x,y) = 72x^2y(x^3 + y^2)^2$
   $f_{yx}(x,y) = 72x^2y(x^3 + y^2)^2.$

9. $f_{xy}(x,y) = f_{yx}(x,y) = -\dfrac{1}{y^2} + 4xye^{x^2}.$

11. $f_{xy}(x,y) = f_{yx}(x,y) = \dfrac{2}{y} + \dfrac{2x}{(x^2 + 2)^2}.$

13. $\dfrac{\partial^2 z}{\partial x^2} = \dfrac{4y^2 - 4x^2}{(x^2 + y^2)^2}; \dfrac{\partial^2 z}{\partial y^2} = \dfrac{4x^2 - 4y^2}{(x^2 + y^2)^2}.$

15. (a) $\dfrac{\partial Q}{\partial K} = \dfrac{20}{3}\dfrac{L^{1/3}}{K^{1/3}}; \dfrac{\partial Q}{\partial L} = \dfrac{10}{3}\dfrac{K^{2/3}}{L^{2/3}}.$

17. $\dfrac{\partial R}{\partial x} = 90 - 3x^2; \dfrac{\partial R}{\partial y} = 100 - 8y.$

19. $\dfrac{\partial P}{\partial x} = 2ax + by; \dfrac{\partial P}{\partial y} = bx + 2cy.$

21. (a) $f_x(x,y) = 4x + 3y;$ (b) $f_y(x,y) = 3x + y;$
    (c) $f_x(x,12) = 4x + 36;$ (d) $f_y(20,y) = 60 + y;$
    (e) $f_x(20,12) = 116;$ (f) $f_y(20,12) = 72.$

23. $\dfrac{\partial V}{\partial r} = 2\pi rh;$ $\dfrac{\partial V}{\partial h} = \pi r^2;$ $\dfrac{\partial^2 V}{\partial h\,\partial r} = 2\pi r;$ $\dfrac{\partial^2 V}{\partial r\,\partial h} = 2\pi r;$ $\dfrac{\partial^2 V}{\partial r^2} = 2\pi h;$

   $\dfrac{\partial^2 V}{\partial h^2} = 0.$

25. $\dfrac{\partial H}{\partial U} = 1;\ \dfrac{\partial H}{\partial V} = P.$

26. (a) Exact, $\dfrac{\partial M}{\partial y} = \dfrac{\partial N}{\partial x} = 2y^2$; (b) not exact, $\dfrac{\partial M}{\partial s} = 6ts,\ \dfrac{\partial N}{\partial t} = 4ts.$

Exercise 7.7

1. Relative minimum at $(\frac{3}{2},0)$ where $f(\frac{3}{2},0) = -\frac{9}{4}$.

3. Relative maximum at $(2,2)$ where $f(2,2) = 8$.

5. Relative maximum at $(0,0)$ where $f(0,0) = 0$.

6. Relative maximum at $(-5,-2)$ where $f(-5,-2) = \frac{513}{2}$; relative minimum at $(4,3)$ where $f(4,3) = -233$.

7. $x = 20$ at \$90 each; $y = 30$ at \$155 each; $P = \$3650$.

9. $x = \frac{8}{3};\ y = \frac{5}{3}.$ **11.** $x = 9;\ y = 76;\ P(9,76) = \$5369.$

13. For every dollar spent on newspaper advertising, $\frac{2}{3}$ dollars should be spent on radio advertising, and 2 dollars should be spent on television advertising.

15. $x = 3,\ y = 4.$

Exercise 7.8

1. Maximum value at $(22,24)$ where $f(22,24) = 1220$.

2. Minimum value at $(\frac{2}{3},\frac{1}{3})$ where $f(\frac{2}{3},\frac{1}{3}) = 0$.

3. Maximum value at $(4,4)$ where $f(4,4) = 32$; maximum value at $(-4,-4)$ where $f(-4,-4) = 32$; minimum value at $(4,-4)$ where $f(4,-4) = -32$; minimum value at $(-4,4)$ where $f(-4,4) = -32$.

5. $K = 40;\ L = 75.$

7. Model I machines, 3; Model II machines, 9.

9. $\frac{2}{3}\sqrt{6}$ in. $\times \frac{2}{3}\sqrt{6}$ in. $\times \frac{1}{3}\sqrt{6}$ in.

11. Length and width of floor, approximately 132.6 feet; height approximately 56.8 feet, cost of approximately \$316,400.

13. $x = \frac{1}{2},\ y = \frac{1}{2},\ z = -\frac{1}{2}.$

Exercise 7.9

1. 6.   **3.** 16.   **5.** a^3.   **7.** 22.   **9.** $\frac{28}{3}$.   **11.** 4.

13. 28.   **15.** $\frac{2}{3}a^3$.   **17.** $\frac{2}{3}$.   **18.** $4 - \ln 5$.   **19.** 8 cu units.

21. 36 cu units.   **23.** $\frac{32}{3}$ cu units.   **25.** $\frac{16}{3}$ sq units.   **27.** 14.

29. 4.   **31.** $\frac{8}{3}$.   **33.** $\frac{1}{2}(e^2 - 2e + 1)$.

Review Exercise

3. $f(3,1) = 12 - \ln 3; f(4,2) = 18 - \ln 8.$

9. $f_x(x,y) = 2xy - y^2 + 10$
$f_y(x,y) = x^2 - 2xy$
$f_{xx}(x,y) = 2y$
$f_{yy}(x,y) = -2x$
$f_{xy}(x,y) = 2x - 2y$
$f_{yx}(x,y) = 2x - 2y.$

11. $f_x(x,y) = e^y - \dfrac{1}{x}$

$f_y(x,y) = xe^y - \dfrac{1}{y}$

$f_{xx}(x,y) = \dfrac{1}{x^2}$

$f_{yy}(x,y) = xe^y + \dfrac{1}{y^2}$

$f_{xy}(x,y) = e^y$

$f_{yx}(x,y) = e^y.$

13. $\dfrac{e^{xy}(xy + y^2 - 1)}{(x + y)^2}$

16. (b) $f_{xy}(x,y) = f_{yx}(x,y) = \dfrac{x - y}{(x + y)^3}.$

17. $\dfrac{\partial Q}{\partial x} = 2xy, \ \dfrac{\partial Q}{\partial y} = x^2 + 1.$

19. Relative minimum at $(2,2)$ where $f(2,2) = -8.$

21. Relative maximum at $(20,20)$ where $f(20,20) = 1200.$

23. $x = 615, y = 385.$

25. $\sqrt[3]{V} \times \sqrt[3]{V} \times \sqrt[3]{V}.$

26. 8.

27. 16.

29. $\frac{1}{4}(e - 1).$

31. $\frac{1}{2}(e^4 - e - 3).$

33. 8 cubic units

34. 4 cubic units.

35. 162.

Exercise A.I.1

1. $4x(y + 2z)$

2. $2(x^2 - 5x + 8).$

3. $-3x(2x - y).$

4. $(x + 3)(x - 3).$

5. $(4x + w)(4x - w).$

6. $3(5 + 3x)(5 - 3x).$

7. $(x^2 + y^2)(x + y)(x - y).$

8. $(x + 1)^2.$

9. $(x + 2)^2.$

10. $3(2x - 1)^2.$

11. $(x - 3)(x + 1).$

12. $(x + 3)(x - 1).$

13. $x(x + 4)(x - 4).$

14. $(3x + 2)(2x - 5).$

15. $(x + y)(3z - 1).$

16. $2x^2(x + 1)(x + 3).$

17. $2(4x^2 + 9y^2).$

18. $(c - d)(a + c).$

19. $y^2(x + 6)(x - 6).$

20. $(2x + 3y)(x + 2y).$

21. $(x - 4)(x - 2).$

22. $x(3 + 4y - 7x).$

23. $15(x + 2)(x - 2)$.

24. $(w^2 + 2)(w + 1)(w - 1)$.

25. $(t - 2)^2$.

26. $x^3(x + 6)(x - 1)$.

27. $(x + 1)(x - 1)(x + 3)(x - 3)$.

28. $z^2(x + 5z)(x - 5z)$.

29. $3(x^2 + 2x + 4)$.

30. $(x + 1)(x + 2)$.

31. $xy(x + y)(x - y)$.

32. $(u - 4)^2$.

33. $x^2y^2(y - x)$.

34. $2(2x - 1)(x - 2)$.

35. $(x + 4)(x + 3)$.

36. $(5x + 6y)^2$.

37. $(3t + 2v)^2$.

38. $(5a + 9b)(a + b)$.

39. $(2x + 3y)(x + 2y)$.

40. $3(2x + 3)(x - 1)$.

Exercise A.I.2.

1. $\dfrac{xy^2}{8}$.

2. $\dfrac{7xy - 12x + 4y}{6x^2y^2}$.

3. $\dfrac{4(x + 2)}{x}$.

4. $\dfrac{3}{2y}$.

5. $\dfrac{2x + 3}{x - 2}$.

6. $\dfrac{2x^2 + x + 6}{(x - 2)(x + 2)}$.

7. $\dfrac{3x + 6y - 10}{2(x - 2y)(x + 2y)}$.

8. $\dfrac{4}{x - 2}$.

9. $\dfrac{2x - 5y}{3x}$.

10. $\dfrac{4x}{x + 1}$.

11. $\dfrac{7x}{4(x - 5)}$.

12. $\dfrac{9x - 1}{6x^2}$.

13. $\dfrac{3}{x(x - 2)}$.

14. $3(a - 2)$.

15. $\dfrac{(b - a)(2a - 3b)}{a + b}$.

16. $\dfrac{y}{x}$.

17. $\dfrac{x + 2}{2}$.

18. $2(x - 4)$.

19. $\dfrac{1}{xy(x - y)}$.

20. $\dfrac{x - 3}{3y^2}$.

Exercise A.I.3

1. $-\frac{7}{3}$.

2. -4.

3. $-\frac{13}{3}$.

4. $\frac{11}{2}$.

5. $\frac{7}{5}$.

6. $6, -4$.

7. $4, -\frac{1}{2}$.

8. $5, -2$.

9. $0, -3$.

10. $3, -3$.

11. 3.

12. -3.

13. $1, 9$.

14. $1, -9$.

15. $\frac{4}{3}, -\frac{4}{3}$.

16. $-\frac{2}{3}$.

17. $\dfrac{1 \pm \sqrt{5}}{2}$.

18. $\frac{4}{3}, -\frac{1}{2}$.

19. $2 \pm \sqrt{2}$.

20. $\dfrac{-1 \pm \sqrt{7}}{3}$.

21. $0, \frac{3}{2}, -\frac{3}{2}$.

22. $5 \pm \sqrt{5}$.

23. $\dfrac{5 \pm \sqrt{3}}{2}$.

24. $\dfrac{-3 \pm \sqrt{21}}{6}$.

25. $-3 \pm \sqrt{17}$.

26. $\dfrac{-2 \pm \sqrt{5}}{6}$.

27. $\frac{9}{2}, -1.$ **28.** $-1 \pm \sqrt{2}.$

29. $1, -\frac{1}{3}.$ **30.** $-2 \pm \sqrt{3}.$

Exercise A.I.4

1. $8a^6.$ **2.** $\frac{1}{2}.$ **3.** $x^3y^2.$ **4.** $32.$ **5.** $5^4 = 625.$

6. $16.$ **7.** $2^8 = 256.$ **8.** $\frac{1}{4}.$ **9.** $2^{15}.$ **10.** $25.$

11. $27.$ **12.** $30x^2.$ **13.** $\dfrac{b^2}{a^2}.$ **14.** $2^6 = 64.$

15. $2^5 = 32.$ **16.** $12.$ **17.** $\frac{1}{8}.$ **18.** $\dfrac{4x^2}{y^2}\sqrt[3]{x}.$ **19.** $3.$

20. $7.$ **21.** $x^6.$ **22.** $x^{-6} = \dfrac{1}{x^6}.$ **23.** $8.$ **24.** $y + x.$

25. $1.$ **26.** $\frac{4}{9}.$ **27.** $2^{5n}.$ **28.** $7.$ **29.** $2.$ **30.** $\dfrac{y}{x}.$

31. $5.$ **32.** $7.$ **33.** $3.$ **34.** $0.$ **35.** $3.$

36. $\dfrac{3x + 1}{\sqrt{3x}}.$ **37.** $0.$ **38.** $1.$

39. (a) True; (b) False; (c) False; (d) False; (e) True; (f) False.

40. (a) True; (b) False; (c) True; (d) False; (e) True; (f) False; (g) True; (h) True.

Exercise A.II

1. $0 < |x - 1| < \frac{1}{2}.$

3. (a) $|(3x - 4) - 2| < 3 \leftrightarrow |3x - 6| < 3 \leftrightarrow 3|x - 2| < 3 \leftrightarrow |x - 2| < 1;$

(b) $|(3x - 4) - 2| < \epsilon \leftrightarrow |3x - 6| < \epsilon \leftrightarrow 3|x - 2| < \epsilon \leftrightarrow |x - 2| < \dfrac{\epsilon}{3};$

(c) $\delta = \dfrac{\epsilon}{3}.$

5. $\dfrac{\epsilon}{4}.$ **7.** $7;$ let $\delta \leqslant \epsilon.$ **9.** $2;$ let $\delta \leqslant 3\epsilon.$ **11.** $0;$ let $\delta \leqslant 2\epsilon.$

Exercise A.IV.1

1. (a) $\frac{3}{2}\pi;$ (c) $\frac{1}{6}\pi.$

2. (a) $90°;$ (c) $60°.$

3. (a) $\frac{1}{2}, \dfrac{\sqrt{3}}{2}, \dfrac{\sqrt{3}}{3};$ (c) $\dfrac{\sqrt{3}}{2}, \frac{1}{2}, \sqrt{3}.$

5. $\frac{1}{6}\pi.$ **6.** $\frac{1}{4}\pi.$

Exercise A.IV.2

1. (a) $\dfrac{d}{dx}(\csc x) = \dfrac{d}{dx}\left(\dfrac{1}{\sin x}\right) = \dfrac{-\cos x}{\sin^2 x} = -\csc x \cot x;$

    (c) $\dfrac{d}{dx}(\sec x) = \dfrac{d}{dx}\left(\dfrac{1}{\cos x}\right) = \dfrac{\sin x}{\cos^2 x} = \sec x \tan x;$

    (e) $\dfrac{d}{dx}(\cot x) = \dfrac{d}{dx}\left(\dfrac{\cos x}{\sin x}\right) = \dfrac{-\sin^2 x - \cos^2 x}{\sin^2 x} = -\csc^2 x.$

3. (a) $2 \cos 2x$; (c) $3 \cos 3x \cos 2x - 2 \sin 3x \sin 2x;$

    (e) $\dfrac{-(1 + \cos^2 x)}{\sin^3 x}$; (g) $e^x(\cot x - \csc^2 x);$

    (i) $3 \tan^2 x \sec^2 x.$

4. (a) $5 \cos 5x$; (c) $x \sec^2(\tfrac{1}{2}x^2)$; (e) $\dfrac{-2 \csc^2 2x}{\cot 2x};$

    (g) $12 \sin^2 x \cos x$; (i) $3x^2 \sec x^3 \tan x^3.$

5. $P'(t) = \dfrac{\pi}{36} \cos\left(\dfrac{\pi}{180}t\right)$; $P''(t) = -\dfrac{\pi^2}{6480} \sin\left(\dfrac{\pi}{180}t\right)$; relative maximum at $t = 90$; relative minimum at $t = 270$; $0 \leqslant t < 90$ or $270 < t \leqslant 360$, $\uparrow$; $90 < t < 270$, $\downarrow$; $0 \leqslant t < 180$, $\cap$; $180 < t \leqslant 360$, $\cup$.

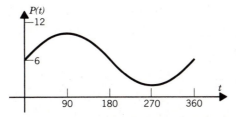

7. (a) $y' = \cos x$; $y'' = -\sin x$; relative maxima at $(\tfrac{1}{2}\pi + k \cdot 2\pi, 1)$, relative minima at $(-\tfrac{1}{2}\pi + k \cdot 2\pi, -1)$; k, any integer;
    $-\tfrac{1}{2}\pi < x < \tfrac{1}{2}\pi$, $\tfrac{3}{2}\pi < x < \tfrac{5}{2}\pi \ldots$, $\uparrow$;
    $-\tfrac{3}{2}\pi < x < -\tfrac{1}{2}\pi$, $\tfrac{1}{2}\pi < x < \tfrac{3}{2}\pi$, $\ldots$, $\downarrow$;
    points of inflection at $(0 + k\pi, 0)$;
    $-\pi < x < 0$, $\pi < x < 2\pi$, $\ldots$, $\cup$;
    $0 < x < \pi$, $\ldots$, $\cap$.

    (c) $y' = \cos x - \sin x$; $y'' = -\cos x - \sin x$

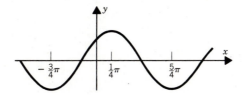

Exercise A.IV.3

3. $-\frac{1}{3} \ln |\cos 3x| + c.$

5. $\frac{1}{2} \sin x^2 + c.$

7. $\sec(x + \pi) + c.$

9. $-\dfrac{1}{\sin x} + c.$

11. $-\dfrac{1}{\sin x} + c.$

13. $-x \cos x + \sin x + c.$

15. $\frac{1}{2}(x - \frac{1}{2} \sin 2x) + c.$

17. $\frac{1}{2}\pi + 1.$

19. $\frac{1}{3}.$

20. $\frac{8}{3}.$

21. $\pi.$

23. $\displaystyle\int_{(1/4)\pi}^{(5/4)\pi} (\sin x - \cos x)\, dx = 2\sqrt{2}.$

25. \$144,000.

Exercise A.V

1. (a) $\{\frac{2}{1}, \frac{3}{4}, \frac{4}{9}, \frac{5}{16}\}$, converges;
 (c) $\{-\frac{3}{7}, -\frac{3}{26}, -\frac{1}{21}, -\frac{3}{124}\}$, converges;
 (d) $\{\frac{1}{2}, \frac{1}{4}, \frac{1}{8}, \frac{1}{16}\}$, converges.

2. (a) $\dfrac{n}{n + 1}$, converges; (c) $\dfrac{1}{2 - \dfrac{2n - 1}{n}}$, diverges.

3. 15 feet. **5.** \$38,974.34. **6.** \$12,155.06; \$55,256.

7. (a) $\frac{3}{2}.$ **8.** (a) $\frac{2}{3}$; (c) $\frac{14}{99}.$ **9.** Jones $\frac{2}{3}$, Smith $\frac{1}{3}.$

11. \$15,864.03.

Exercise A.VI

1. $\sqrt{2} + \dfrac{\sqrt{2}}{4}(x - 2) - \dfrac{\sqrt{2}}{32}(x - 2)^2 + \dfrac{\sqrt{2}}{128}(x - 2)^3 - \dfrac{5\sqrt{2}}{2048}(x - 2)^4.$

3. $x - \dfrac{x^2}{2} + \dfrac{x^3}{3} - \dfrac{x^4}{4}.$ **5.** $e + \frac{1}{2}e(x - 2) + \frac{1}{4}e\dfrac{(x - 2)^2}{2!} + \frac{1}{8}e\dfrac{(x - 2)^3}{3!}.$

INDEX

Differentiation Forms

$$\frac{d}{dx}(x^n) = n \cdot x^{n-1}$$

$$\frac{d}{dx}[g(x) \pm h(x)] = g'(x) \pm h'(x)$$

$$\frac{d}{dx}[g(x) \cdot h(x)] = g(x) \cdot h'(x) + h(x) \cdot g'(x)$$

$$\frac{d}{dx}[c \cdot h(x)] = c \cdot h'(x)$$

$$\frac{d}{dx}\left[\frac{g(x)}{h(x)}\right] = \frac{h(x) \cdot g'(x) - g(x) \cdot h'(x)}{[h(x)]^2}$$

$$\frac{d}{dx}[f(x)]^n = n \cdot [f(x)]^{n-1}f'(x)$$

$$\frac{d}{dx}(\ln u) = \frac{1}{u} \cdot \frac{du}{dx}$$

$$\frac{d}{dx}(e^u) = e^u \cdot \frac{du}{dx}$$

Integration Forms

$$\int_a^b f(x)\,dx = F(b) - F(a)$$

$$\int u^n\,du = \frac{u^{n+1}}{n+1} + c;\ n \neq -1$$

$$\int \frac{1}{u}\,du = \ln|u| + c$$

$$\int e^u\,du = e^u + c$$

$$\int u\,dv = uv - \int v\,du$$